Acknowledgments

Dement's Sleep & Dreams

Second Edition

RAFAEL PELAYO, M.D

WILLIAM C. DEMENT, M.D., Ph. D

with Krystle A. Singh, LVN

Important Note

This is an educational textbook. It is not intended to take the place of medical advice from your healthcare provider. Readers are advised to consult a qualified professional for any sleep concerns. Good sources for additional information are the American Academy of Sleep Medicine and the National Sleep Foundation.

AN INDEPENDENT PAPERBACK

Designed by Rafael Pelayo
Cover copyright © 2017 by Rafael Pelayo
Cover Photo by Rafael Pelayo
Copyright © 2017 by Rafael Pelayo and William Dement

All rights reserved. No part of this book may be reproduced or transmitted in any form or by any means, electronic or mechanical, including photocopying, recording, or by any information storage and retrieval system, without the written permission of the authors, except where permitted by law.

Printed in the United States of America
January 2017 version 2

Acknowledgments

This book had its genesis in the need created by my involvement with undergraduate teaching starting in the Fall of 1970 when I was a resident fellow and taught a course on sleep for the students in my dormitory. The course began to attract other students and because of this I decided to offer a formal presentation in Winter Quarter 1971 and I have offered a course on sleep annually up until Winter Quarter 2003. Due to high student demand, we resurrected the course in Winter Quarter 2006. Around 1980, the various handouts developed into a quasi-textbook and by the middle of the '90s, the first Stanford Sleep Book was born.

The creation of Sleep and Dreams is one of my many proud accomplishments at Stanford. I find comfort knowing that Dr. Rafael Pelayo is at my side and he is a worthy successor for the course going in the future. I am also grateful for the support and leadership of our sleep center director Emmanuel Mignot. With regard to acknowledgments, I am most indebted to all the students, more than 25,000, who over all these years inspired me, gave feedback, and in general made teaching Sleep and Dreams an annual adventure. All of the Teaching Assistants in the early years of teaching Sleep and Dreams, as well as Doctors Mary Carskadon and Merrill Mitler played crucial roles.

A special thanks to ResMed Corp. who continues to contribute and support Sleep and Dreams every year. In the past decade, my colleague, Krystle Singh, has not only played a major role in the revision of this book with her inspiration, support and dedication of an incredible amount of work and time, but she has also along with Anna Castorena, largely sustained me.

Finally, I would like to acknowledge with gratitude, the support and love of my family, and in particular the strength and encouragement from my late wife Pat.

William C. Dement
Stanford University, 2017

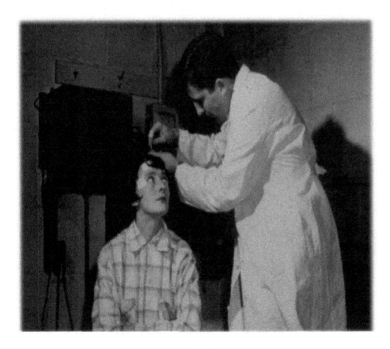

This photo was taken in Chicago (1955). I miss her, but I see her in my dreams. WCD

Teaching *Sleep and Dreams* with the irreplaceable Bill Dement has been one of the highlights of my professional life. I know I am extremely lucky he has become my friend.

I am often asked how I sleep. The truth is I sleep great. I fall asleep easily, sleep through the night, and wake up refreshed. One reason is that I am one of the few sleep doctors that gets to sleep next to another sleep doctor. No, it is not Dr. Dement. I am fortunate to sleep with Dr. Kin Yuen. She and our children are the reason I look forward to waking up. I am a better person for having them in my life. Thank you!

I am grateful to my patients. They teach me what is not in the books.

Krystle Singh is the stick that stirs the drink. She knows our quirks and how to get us to do what has to be done. She made even the tedious portions of book editing fun.

I must thank our department chair, Laura Roberts, Associate chair, Alan Louie, division chief, Emmanuel Mignot, administrator Stephanie Lettieri and clinic director, Clete Kushida for their support in providing me with the time to devote to the Stanford Sleep and Dreams course. I am privileged to work with the strongest group of sleep clinicians in the entire world starting with the master, Christian Guilleminault and including Drs. Fiona Barwick, Robson Capasso, Joseph Cheung, Michelle Cao, Emmanuel During, Makoto Kawai, Scott Kutscher, Stanley Liu, Mitch Miglis, Anstella Robinson, Logan Schneider, and Shannon Sullivan! Thank you to our small army of front desk personnel, medical assistants, administrators, sleep medicine fellows, and sleep technologists that run the Stanford Sleep Clinic.

Our teaching assistants edited this edition of the book and are indispensable to Sleep and Dreams. THANK YOU!

I met Bill Dement through sleep medicine, but what really brought us closer together was our mutual love of music. Bill was an accomplished jazz musician before he was a doctor. Sleep and dreams have always inspired musicians. At our annual sleep conference, our mutual friend Ron Richard organizes a Blue's Night that is often kicked off by Bill singing and sometimes playing standup bass. Lacking any musical talent, Ron still sometimes tolerates my being the MC for the evening. From that friendship, I was grateful to meet Jeff and Moe' Chimenti and the music has never stopped. There would be no book without all these friends.

Rafael Pelayo, MD

Table of Contents

Acknowledgments	3
Table of Contents	5
Preface	6
CHAPTER 1 Why Learn About Sleep?	11
CHAPTER 2 What is this thing called Sleep?	21
CHAPTER 3 Sleep Quality & Sleep Debt	54
CHAPTER 4 Alert Wakefulness from Inside Ourselves: The Biological Clock	84
CHAPTER 5 The Opponent Process Model	97
CHAPTER 6 Drowsiness RED Alert!!	114
CHAPTER 7 Exercise, Nutrition & Sleep: The Triumvirate of Health	134
CHAPTER 8 Sleep Deprivation: Consequences & Counter Strategies	147
CHAPTER 9 Dreaming and the problem of Consciousness and Useful Fiction	170
CHAPTER 10 The Psychophysiology of Dreaming	190
CHAPTER 11 The Content of Dreams	206
CHAPTER 12 Dream Anarchy Rules!	235
CHAPTER 13 Lucid Dreaming: a special case	260
CHAPTER 14 Sleep Disorders & Walla Walla	269
CHAPTER 15 Insomnia	287
CHAPTER 16 Restless Legs & Sleep related movement disorders	315
CHAPTER 17 Obstructive Sleep Apnea	321
CHAPTER 18 Narcolepsy	347
CHAPTER 19 Parasomnias and Sleep Related Movement Disorders	361
CHAPTER 20 Pediatric Sleep Disorders	377
CHAPTER 21 Circadian Rhythm Sleep Disorders	400
CHAPTER 22 Sleep Education & Public Policy	410
Coda:	443
Glossary	444
Index	465

Doctor: "How do you know you have a sleep problem?"

10-year-old Veve Pigg: "When I am sleeping and wake up, I want to keep sleeping."

Preface

"America is a vast reservoir of ignorance about sleep, sleep deprivation and sleep disorders." This statement was made by, the late great United States Senator Mark Hatfield on January 6, 1993, in the U.S. Capitol at a press conference organized by the American Academy of Sleep Medicine to publicize the findings and recommendations of the National Commission on Sleep Disorders Research. Aside from its stunning truth, this utterance was a defining moment of validation for me because it demonstrated that national policy makers and legislators could recognize the gigantic human and dollar cost of this ignorance.

Dr. Dement leading the National Commission on Sleep Disorders Research

One of several reasons I have become involved in educational initiatives on sleep that are somewhat off the beaten path for a university professor is to save lives. I do not wish to be overly dramatic, but this is true. During the several years, (1991-1994) I served as Chairman of the Congressionally mandated National Commission on Sleep Disorders Research, the Commission's work exposed me to areas of concern that under ordinary circumstances would not have come within my purview. At public hearings, Commissioners would listen intently to the testimony of witnesses telling their stories about the many ways in which sleep deprivation and sleep disorders destroyed their lives or the lives of loved ones. When I looked into the eyes of grieving parents while they were telling us about a beloved son or daughter who fell asleep driving home from college and died in a head-on collision. Also, Sudden Infant Death mothers describing the terrible grief and pain of losing their babies, or the victims of narcolepsy describing years of failure in school due to unappreciated pathological sleepiness, my perspective about my work was changed forever. Even to this day, my eyes still tear. I could truly understand the impact of sleep deprivation and sleep disorders on individual lives.

Through access to surveys and other types of large population studies, Commissioners could simply multiply these stories a million times and, by so doing, finally grasp the full magnitude of sleep-related problems in our society. Finally, the National Commission concluded that the root cause of pervasive sleep deprivation and unidentified sleep problems is the continued low level of public and professional awareness due to failure to give sleep topics adequate attention in the educational system.

From all this and more, I simply came to believe that no one can, in good conscience, continue to sit on the sidelines of the national awareness process. I am now on the playing field, and one result is this book on sleep, specially prepared first for students.

Although there is no required tradition for the teaching of material related to sleep — its mechanisms and functions, and its impact on waking life — are part of the educational process at any level. Surprisingly, this includes a lack of adequate instruction about these issues and the pathologies of sleep even in medical schools. Given this state of affairs, there is no consensus whatsoever about what high school and college students should be taught and what the public should know, and no precedent regarding which of the scholarly disciplines is the best candidate to assume responsibility for the education of the public about sleep. In terms of the traditional academic organization of colleges and universities, the obvious homes are psychology departments and biology departments. For high schools, psychology, health education, biology, and drivers' education classes are the most logical bases. The rest of the public must absorb knowledge from those of us who understand the importance of sleep.

The amount of sleep at night and whether or not it is healthy are fundamental and crucial determinants of daytime performance and mental activity. The full range of the dimension of waking sleepiness/alertness extends from very impaired, essentially zero function, to peak performance.

Since sleep and its impact on waking alertness and function has been previously totally disregarded in the educational process, huge numbers of individuals are relegated to spending their daytime existence in the portion of the alertness dimension furthest removed from peak performance. Moreover, even in today's technologically and medically advanced society, millions of people with sleep disorders are suffering and dying because the symptoms of their preventable and treatable illnesses are not being recognized by doctors or any other health professionals. Most important of all, they remain unrecognized even by their victims.

The American society remains almost totally ignorant about sleep, sleep deprivation, and sleep disorders, and, as of today, I am aware of no institution that has made sleep education a top priority. To confront this lack of systematic sleep education, this book is an attempt to provide the essentials of sleep knowledge in a concise yet comprehensive form for the public and college students everywhere.

Fortunately, things are starting to change. We are now seeing more teaching of sleep materials at other universities and high schools. At Stanford, the course remains healthy and vibrant with the collaboration of my longtime colleague Rafael Pelayo who shares my enthusiasm for teaching the course. "Welcome to the mysterious, exciting and sometimes worrisome world of sleep." You will see many useful applications in your daily life as a student, and perhaps even exciting new horizons for your future. ~Dr. William C. Dement

Preface

Personal introduction from Dr. Dement

To the Reader, Welcome to my world...

Far and away the major purpose of this book is to impart much needed sleep knowledge and to elevate sleep to a higher level of importance in your life. Also, at whatever extent possible, to persuade you to "put sleep first." You know many things that are vital to your life and health. Some are simple and totally ingrained. You do not run into a busy street without looking. You avoid hornets, black widow spiders, and uninsulated, high voltage cables. You want to know the contents of a bottle before you drink. In addition, you almost certainly possess lots of knowledge about nutrition, geography, human anatomy, physical fitness, and sports.

In stark contrast to such commonplace knowledge possessed by everyone is their near total lack of awareness about the important facts of sleep. During Winter, Spring, and Fall Quarters of 2009, I talked to students in sixteen different Stanford University residences. In the "all freshman" residences I did not encounter a single student who possessed even one iota of the knowledge about sleep that I now consider to be absolutely vital and essential for every person on earth. This lack of awareness exhibited by Stanford students is surely representative of society. The materials in this book are not difficult to understand. We are merely dispensing knowledge that everyone should possess, but, as far as I can tell, nearly everyone does not.

The goal is to impart this essential knowledge to as many people as possible. At the end of this book we hope you will be thoroughly familiar with the essentials, and we certainly hope you know yourselves better than you did when you were introduced to this book and the associated course. The most important thing in regard to the latter is that you will understand much more clearly the role your nightly sleep plays in determining the way you feel during the day and the way you perform your daily tasks. We hope you apply what you learn as often as possible for as long as you live. I have taught "Sleep and Dreams" at Stanford University for more than 40 years, since the fall of 1970 when I was a resident fellow in Cedro House of the Wilbur Hall complex. This was a turbulent time. I decided to give a special seminar on sleep for all my freshman as a way of keeping them at home while waves of violent confrontation were engulfing the campus including tear gassing students. Although I knew much less about sleep and sleep disorders than I do today, this first course was quite successful. Indeed, it was so successful that huge numbers of students from other residences wanted to crash the party. In order to avoid the possibility of yet another wave of campus turmoil right in our own living room, so to speak, I agreed to repeat the course in Winter Quarter of 1971 for any Stanford student who wished to register.

Amazingly, nearly 600 students signed up! As a professor in the Medical School, I was unfamiliar with undergraduate teaching. Accordingly, I had not thought about where I would teach this new course. I was very dismayed to discover that all the large lecture halls were taken and I would have to chop about 300 students from the roster. Back then, such an act could have precipitated a riot, a sit-in or other form of protest. Happily, Dean of the Chapel, Davie Napier, came to my rescue and allowed me to teach Sleep and Dreams in the Stanford University Memorial Church. What an experience! I sometimes speculate that my evangelistic approach to teaching about sleep is because I somehow became infected by the hallowed atmosphere of this beautiful "classroom." At any rate, presenting this course for undergraduates was so much fun and so rewarding that I have offered it each year, with one or two exceptions ever since. Over the years, a series of traditions have emerged. One I am particularity proud of is the outreach projects where Stanford student volunteers share their knowledge with other students. One of the more unorthodox traditions is letting students fall asleep in class and waking them up with water guns as a way of demonstrating retrograde amnesia. We look forward to new traditions emerging in the future as the course continues to grow.

"Dreamweaver I believe you can get me through the night..."-Gary Wright

CHAPTER 1 Why Learn About Sleep?

"I decided I can make a better living as a mediocre physician, than as a mediocre musician" -William C. Dement, 1948

Case History #1

I was speaking to a small group of students in an after-dinner residence presentation. At the end of the talk, I asked, "Do any of you feel that you might have a real sleep problem?" One student raised her hand. "I can't sleep. I can't ever sleep," she said. "I go to bed and lie awake for hours. It's awful." The student, Lena S., looked very tired and somewhat depressed. "Do you finally fall asleep?" I asked. "Yes." "About what time?" "Usually between seven and eight in the morning, sometimes later." "What happens then?" "Well, I usually get up in the afternoon." "When do you go to class?" "I don't." There was a kind of shocked silence. I looked quizzically at the dormitory's resident fellow who had invited me. He also looked somewhat shocked. Lena had become very dismayed and even frightened because she was unable to solve her problem. Her inability to attend morning classes had become a grave concern. She was also very fatigued, felt completely defeated, and was considering leaving school. Her problem was an extraordinarily severe delayed sleep phase syndrome.

Case History #2

On another occasion, in a different residence, two students lingered after the formal question-and-answer period for a private meeting. "Our room is next to someone who snores so loud it disturbs half the floor," one said. "His snoring makes our walls rattle. We can't sleep at all." I suggested they arrange for me to meet with the snoring student. At the meeting, I learned he was tired all day long and to my surprise, he had been told he had abnormally high blood pressure. This occurred when he donated blood and a routine blood pressure measurement was made. There was no question in my mind that this student had obstructive sleep apnea syndrome as well as one of its most dangerous complications, cardiovascular disease, which was present at the unusually early age of twenty.

Case History #3

This interaction took place at Cornell University in Ithaca, New York, where for many years I gave an annual lecture to approximately 1,700 students in Introductory Psychology. I always lingered an hour or two after the lecture answering questions.

One female undergraduate, a sophomore, said she thought she might have narcolepsy. "I had a sleep test in my home town and they told me I did not have narcolepsy." She had been told the sleep test did not show the results necessary to confirm the diagnosis of narcolepsy. I asked, "Do you have cataplexy?" "I don't know what that is," she said. Careful questioning elicited recall of an incident which made narcolepsy almost certain. Several months earlier, she was at a party where someone told a joke she thought was very funny. As she started to laugh, she realized she could not move. She remembered being greatly alarmed by her paralysis, but then it vanished. As far as she could recall, this was the only such episode. "You experienced a very typical cataplectic attack at the party," I told her. "I'm afraid your sleepiness is caused by narcolepsy. Please stay in touch, because without treatment, you will probably have great difficulty with your studies. With proper treatment, however, you will be okay."

Case History #4

After I finished a lecture to a group of practicing physicians about the dangers of sleep deprivation, one doctor lingered until he was the only one who remained. Dr. Jonathon T. was a middle-aged gastroenterologist. "I have a story for you," he said. "When I was a freshman in college, I was driving back to school with my three best friends after a long weekend in the mountains. I fell asleep and the car went off the road and crashed into a ravine. I was thrown clear and my friends were all killed. I bear the remorse and the sorrow every day of my life." He began to sob softly.

The Education Gap

In recent decades, sleep scientists have learned an enormous amount about sleep and its important role in our everyday lives. While research continues, the challenge today is to deliver this knowledge to the entire general public. Planning our lives and our schedules inevitably involves decisions about when to go to bed, when to get up, and finding an optimal balance of time spent asleep and awake.

CHAPTER 1 Why Learn About Sleep?

As noted, it is almost certain that anyone reading this book will have had little education about sleep in high school, college, or any other formal manner of education, particularly about sleep deprivation and how it affects waking behavior. Yet, as you the reader will learn, young people are generally very sleep deprived. This has a very negative impact on their alertness, creating a serious risk of falling asleep at the wheel just as they are becoming new drivers. All of this pertains to college students and the general public as well. Before reading this book, many people may have one or more notions about sleep that might be called "old wives' tales," but it will be an extremely rare person who already knows what everyone should know about sleep. Much of what sleep experts feel should be commonplace knowledge in our society is simple and straightforward and could easily be incorporated into college, high school, or even middle school curricula.

Each year, students in our course on sleep at Stanford University are surveyed at the beginning of the quarter to determine what, if anything, they learned in high school about sleep as well as several other topics for comparison. The questions focus mainly on nutrition, physical fitness, and sleep, three basic areas we think of as the "triumvirate of health." Well over 90 percent of Stanford students in "Sleep and Dreams" had high school teaching about nutrition and the importance of a proper diet. They also learned about physical fitness and how to achieve it. But the vast majority had received little teaching whatsoever about sleep. The third member of the crucial triumvirate of health has been greatly ignored.

As the accumulated knowledge about sleep increases in the scientific community, a continued absence of prioritized teaching about the subject to our students is unacceptable. Fortunately, more medical schools are teaching about life-threatening sleep disorders as a matter of public policy. Serious sleep disorders are among America's most common illnesses. If unrecognized and untreated, serious sleep disorders can kill people. Moreover, most serious sleep disorders can be easily diagnosed and effectively treated.

At the present time, teaching about sleep is only starting to permeate the mainstream educational system. Millions of people around the world are suffering and dying each year even though effective treatments and cures are readily available. Some of the human beings who are suffering and dying without knowing why may even be your parents or other members of your family. Every year we teach this class and students recognize sleep disorders among themselves or in their family members. As we wait for our educational system to change and our public policymakers to wake up, we hope this textbook and our course, "Sleep and Dreams," will provide an alternative method to spread the word, and may, in fact, accomplish an enormous amount of good.

Perhaps the most appropriate reason to be well informed about sleep is that sleep is such a fundamental part of human existence and behavior. Fortunately, we now know an enormous amount about our sleep. The purpose of *Dement's Sleep & Dreams* is to initiate the assimilation of this knowledge and to guide readers in understanding how the daily cycle of sleep and wakefulness experienced by all human beings actually works. This textbook will also offer guidelines as to how readers may best manage their sleep in the context of their often demanding and stressful lives.

This book will present the essentials of sleep and dreams as clearly as possible so that its readers will feel comfortable that they understand, and eventually can help others understand the fundamental aspects of sleep that affect us all.

A common problem with teaching about sleep is that most people believe things about sleep that are actually false. An incredibly commonplace yet erroneous assumption involving sleep is that eating a large lunch can directly cause us to become drowsy and can actually cause us to fall asleep. We believe this not because research has proven it, but merely because people frequently become drowsy after lunch. A similar conclusion might be drawn by a rooster who thinks his crowing causes the sun to rise every morning. He has not, of course, done the experiment of not crowing to see if the sun does not rise.

In today's world, sleep is regarded by many people as very mysterious. However, this attitude exists simply because the facts about sleep are not widely included in our education systems. Sleep can be demystified by presenting clear and simple explanations of how sleep and wakefulness are regulated. We will relate these explanations as closely as possible to everyday life because both ordinary life circumstances and even specific sleep disorders can disturb people's sleep or cause them to be excessively sleepy and tired throughout the day. Such problems can, in turn, interfere with people's ability to be attentive at work and remain alert while driving.

It is quite possible that this book will be the first systematic exposure to sleep knowledge for most readers. We have therefore avoided the temptation to be overly technical. Advanced knowledge about brain mechanisms, specialized tests, very rare sleep disorders, and much more is available for those who wish to delve further on their own.

Why You Should Know About Sleep

Sleep is Important for Every Human Being: There are No Exceptions

Everyone sleeps, but most people take their daily sleep completely for granted. A few wonder how and why it happens every single day of their lives. A very few wonder each night if they will wake up the next morning safe and sound. If our sleep is "good," most of us expect to feel rested and energetic in the daytime and to be able to cope effectively with the challenges of the waking world. If our sleep is "bad," we are not surprised to experience diminished energy and alertness the next day. If we do not sleep at all, we may feel miserable the next day, and we may find that the simplest task seems overwhelming. Despite all of this, most people have no clear certainty about why their sleep sometimes feels good, feels bad, or sometimes feels very bad. Sleep must be studied. Not only does sleep consume a third of human existence, but unhealthy sleep can also severely impair the other two-thirds. Although sleep is a natural process, more than half of all human adults claim they have difficulty sleeping. Unfortunately, very few physicians and other health professionals, let alone the average person, have a clear idea of the many specific causes of these complaints and there are, as of yet, few physicians who specialize in the diagnosis and treatment of sleep disorders. Thankfully the tide is turning.

The Cycle of Sleep and Wakefulness Organizes Our Lives

Do we live to sleep or sleep to live? Sleep and wakefulness are complementary phases in the daily cycle of existence. In order to foster a harmonious existence, we need to know how best to manage these fundamental behavioral states.

The temporal anchors of getting up in the morning and going to bed at night determine what we do and when we do it. We must continually monitor our progress through the day and judge how effectively we carry out our scheduled activities and tasks. We must always allow for an end of the day and plan for the new beginning.

Another issue, also taken mostly for granted, is that we must have a place to sleep. Houses, apartment buildings, hotels, motels, encampments, villages, towns, and cities all evolved primarily because human beings need a safe and sheltered place to sleep. Safety is necessary because a primary characteristic of the sleeping organism is vulnerability. It is a deplorable fact that in the inner portions of our large cities, we have begun to see large numbers of individuals who are homeless and who do not have a comfortable, safe place to sleep. The sociological implications of sleep in various cultures and situations of poverty remain largely unexplored. Locating a safe place to sleep is not unique to the human species. A roving troupe of baboons will always find a tree before night overtakes them. Safe from nocturnal predators in the high branches, they can momentarily surrender their waking vigilance. Burrows, nests, and caves all exist as havens in which to spend the sleeping hours. Some larger animals, such as lions and tigers, do not need to be concerned about safety. Consequently, they tend to sleep anywhere they wish and at any time of the day or night. Most species are either nocturnal (awake and active at night), diurnal (awake and active in the daytime) or crepuscular (active at dawn and dusk). Humans and most primates are diurnal because they depend almost entirely upon vision to guide their activities. Moreover, humans are more vulnerable in the night. Therefore, the daily cycle of light and dark as the earth turns on its axis has profoundly organized human behavior in an evolutionary sense. Waking up marks the beginning of the day as we face the world. The waking state serves the overall biological purpose of coping with our environment and managing ourselves within its constraints for the purpose of survival: feeding ourselves, procreating, planning the future, and carrying out all of our social interactions.

As the waking day wanes, an inexorable drive toward sleep develops. Just as we become hungrier and thirstier as time without food or drink increases, we become sleepier and think more and more about our beds as the amount of time we have been awake increases. This show cases the process of homeostasis that we will describe later. As the light of day fades and darkness arrives, we begin to prepare for sleep, performing the unique human and individual rituals that precede actually getting into bed.

We Now Live in a "24/7" World

In today's world, the evolutionary adjustment of the human race to the daily occurrence of light and dark no longer applies. As a result of modern science and industry, we now live in a world that runs 24 hours a day, seven days a week. Many of our institutions - hospitals, police forces, fire departments, and factories - operate 24 hours a day. Modern technology has enabled everyone to be online around the clock. We are in constant communication with the other side of the earth. These advances have created the need for major departures from the normal schedule of human sleep and wakefulness. "Normal," in this case, refers to sleeping at night and being awake in the day. Currently, more than 20 percent of the American work force consists of shift workers, so called because they work hours other than the conventional 9 a.m. to 5 p.m. This component of the work force includes factory workers, doctors, nurses, street cleaners, pilots, and the police, among many others. The impact of this social change upon the normal rhythms of sleep and wakefulness has far-reaching consequences. Not only must shift workers try to sleep at the "wrong" times, but many persons in these jobs must work very long hours. Shift workers are sleep deprived and more vulnerable to errors and accidents as a result of their impaired alertness. Finally, some of the worst industrial catastrophes of all time have involved errors committed by sleepy shift workers.

There are entities that attempt to address the gigantic need for awareness about sleep. One of these is, The National Sleep Foundation, which conducts a poll each year exploring connection with sleep. This poll explores the association between Americans' use of communication technologies and sleep habits. Many Americans report dissatisfaction with their sleep during the week. The 2011 poll found that 43% of Americans between the ages of 13 and 64 say they rarely or never get a good night's sleep on weeknights. More than half (60%) say that they experience a sleep problem every night or almost every night (e.g., snoring, waking in the night, waking up too early, or feeling unrefreshed when they get up in the morning.) Communications technology use before sleep is pervasive. Americans report very active technology use in the hour before trying to sleep. Almost everyone surveyed, 95%, uses some type of electronics like a television, computer, video game or cell phone at least a few nights a week within the hour before bed.

Not Getting Enough Sleep

Most students and many other people as well have bragged about pulling one or more "all-nighters." This is a dubious honor often self-awarded by students that are well acquainted with the misery of severe sleep deprivation. Before the advent of the light bulb, people probably did not voluntarily choose to stay awake at night because there was nothing to do, and if there was something to do, people could not see to do it.

In today's 24-hour society, we can do all sorts of things at night - watch television, go to parties, work, shop - and therefore we may sleep somewhat less than our ancestors. Certainly, we seem to sleep less. However, we do not try to forgo sleeping entirely because we cannot. Sooner or later sleep will overcome us, or we will be too miserable to keep going. On the other hand, giving up a small amount of sleep each day may not produce a sufficiently uncomfortable response to prevent us from continuing a schedule of chronic sleep loss. Successive surveys carried out since the 1800s show that people in industrialized nations are sleeping less than they did a century ago. One study, carried out in Japan, in which citizens were surveyed each year, showed an overall reduction in the daily amount of sleep of one-and-a half hours since 1920. There are detrimental life and health consequences of habitually not getting enough sleep, though they may take place over weeks, months, and years, and are therefore not obvious.

Sleep and Stress

We all assume that the more slowly paced existence of the distant past was relatively free from the extraordinary pressure that exists in our modern industrialized society. Today we live in the "pressure cooker" of extended hours of work, constant examinations in the educational process with the ever-present threat of failure, and the incessant effort to complete too many assigned tasks. Stress and pressure have become an unavoidable part of modern existence and without a doubt, are the leading causes of insomnia. The exact mechanism by which stress disturbs sleep remains only partially understood. However, there is plentiful evidence, both objective and subjective, proving that sleep is very sensitive to stress and anxiety. It is highly unlikely that stress and anxiety can be removed entirely from our lives. Therefore, it is important to know how to reduce their detrimental effects on sleep.

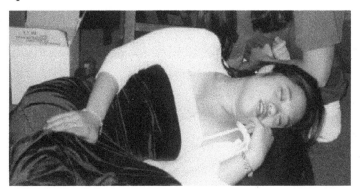

The Need for Sleep Awareness Is Urgent Even Today

The gap created by the absence of adequate sleep education is causing problems in all components of our society. When a young driver falls asleep while driving and is killed or disabled for life, it is rare that anyone fully realizes that such tragic events are preventable. When a 50-year-old executive dies of a heart attack, very few realize that an undiagnosed sleep disorder may have been the culprit. The educational activities that could reduce and eventually eliminate such tragedies must be increased. We hope that the reader will help to spread this knowledge.

We contend that the amount of sleep debt that has been accumulated by an individual is a hidden factor with great influence on behavior and mental processes in otherwise normal human beings. If this were widely known and accepted, we would be able to account for otherwise puzzling variables in performance. It is quite possible that even IQ scores have erroneously been assigned to individuals that were tested when they were unknowingly, significantly impaired by habitual sleep loss.

Everyone should understand those aspects of sleep that affect their daily lives. Individuals must be more sensitive to their own personal domain of sleepiness/drowsiness and alertness. We now know that the process that regulates our sleep operates continuously around the clock, night, and day.

In spite of our subjective feelings, we must understand that the only change in our continuously active sleep-inducing process is intensity. We are always vulnerable to falling asleep. The only variable in question is the strength of this vulnerability at any particular moment. Every human being should know explicitly how much sleep he or she needs every night to feel wide awake and alert the entire day, and every human being should have the capacity and the knowledge to schedule himself or herself in an optimal fashion. A student who might someday manage a complex institution, such as a large hospital that operates on a 24-hour basis, would be well advised to become knowledgeable about the interplay of shift work, sleep deprivation, and errors in the workplace.

"I feel like this is a dream, and I apologize for how I dressed some of you." - Ray Romano

CHAPTER 2 What is this thing called Sleep?

"If sleep doesn't serve a vital function, it is the biggest mistake evolution ever made."
-Allen Rechtschaffen

The Essential Characteristics of Sleep

Sleep is a naturally occurring restorative process. Sleep and wakefulness occur as complementary phases in the daily cycle of existence. This daily cycle is the fundamental adjustment of human life to the earth's rotation. This chapter presents a description of the sleep phase of the daily cycle, including its nature, and some of its changing manifestations over the course of a typical night.

Sleep is often defined and described by comparing and contrasting it with wakefulness. Its relative quiescence has led many people to look upon sleep as merely a periodic cessation or interruption of the waking state. There is an inclination to assume that there is a great reduction in brain activity as well as bodily movement. This view greatly oversimplifies the complex and fascinating world of sleep. Scientific studies of the sleeping brain have revealed exciting new information about how the human organism functions during sleep, and also how sleep processes affect wakefulness. For example, we now know that during certain periods of sleep, some areas of the brain are more active than when we are awake. Perhaps we will someday find it more useful to think of "wakefulness" as merely a periodic cessation or interruption of the sleeping state!

In ideal circumstances, we only fall asleep when we want to. We must make a voluntary and deliberate reduction of our activity and movement. In other words, people ordinarily do not fall asleep while they are walking around. For this reason, it was erroneously believed for many years that sleep occurred passively as the result of a reduction in bodily activity and external stimulation. When careful observations of brain activity were carried out by measuring the discharge rates of individual neurons during sleep, it was discovered that the overall discharge rates of the majority of neurons do not change substantially when we fall asleep. This means that our brain is essentially as active while we sleep as it is while we are awake, and therefore, falling asleep is not the passive result of a marked reduction in brain activity.

From an outside observer's vantage point, sleep is characterized by, other than breathing, an absence of movement. Of course, most people do not watch someone sleeping for long periods of time. If they did and meticulously kept track of all movements made during an entire night, most people would be quite surprised at the large number.

Normal sleep is defined by two essential characteristics. First, perceptual disengagement and second, reversibility. The first essential defining characteristic of sleep is a complete **perceptual disengagement** from the environment. A sleeping person does not make meaningful responses to stimuli. Although a brief stimulus may awaken a sleeper, the specific nature of the stimulus is never recognized or remembered. Perceptual disengagement means that as compared to the waking state in which we are continuously conscious of a veritable infinity of sights, sounds, and smells in the physical world around us, during sleep we have essentially zero awareness of the physical world around us. This must result in a restorative state.

This disengagement from the environment is an active process in which sensory input is deflected or modified in a way that results in a complete perceptual shutdown. Even if our eyelids are taped open, we stop seeing when we fall asleep. In other words, at the moment of sleep we instantaneously become totally blind. However, this does not mean we become totally unconscious. Though we are disengaged from the physical world during sleep, at the same time we are engaged with an inner world—we experience a peculiar sleep consciousness, and at certain times we dream.

Calling an awake and attentive person in a very low voice just above the threshold of hearing should elicit a response. Calling to a sleeping person at the same low sound level will elicit no response, or very rarely the sleeper might awaken. When the sleeper's name is called more loudly, or very loudly, he or she will always awaken. The latter result brings us to the second essential, defining characteristic of sleep which is reversibility. When sleep is very deep, the awakening stimulus must be much more intense and might even require shaking or poking the sleeper. The latter is rarely necessary under ordinary circumstances. People almost always awaken quickly in response to reasonable signals such as an alarm clock, the calling of their name, knocking on the door, and so forth. It is this reversibility that clearly differentiates normal sleep from many sleep-like states such as coma, anesthesia, and hibernation.

In summary, the two essential properties that define the sleeping state are (1) perceptual disengagement from the environment and (2) reversibility.

Two Kinds of Sleep

Our journey through the night is very complex. During any lengthy period of sleep, we exist in two fundamental and markedly different states of being which alternate rhythmically throughout the night. One state is called **REM** sleep, which is characterized by dreaming along with rapid eye movements, active brain wave patterns, complete loss of muscle tone, irregular breathing, and irregular heart rate. The other fundamental state, **non-REM** sleep, which is longer, lacks rapid eye movements, exhibits unique brain wave patterns, and maintains more regular patterns of breathing, heart rate, and other functions.

Early Ideas About Sleep

In previous centuries, all states characterized by immobility and unresponsiveness to environmental stimuli tended to be viewed as various manifestations of sleep. These included coma, hibernation, hypnosis, anesthesia, drunken stupor, and, perhaps, even death. As the composer, Joseph Haydn wrote, "Sleep is a short, short death, and death is a long, long sleep." Though past beliefs about sleep were often mystical or religious, they were sometimes derived from factual observations. For example, one of the first and most prevalent theories of sleep from the 19th century based on animal research was the hypnotoxin theory. This theory proposed that a toxic fatigue product builds up during wakefulness and is then removed during sleep. The misery experienced by people who were prevented from sleeping for long periods of time seemed to resemble a toxic state and this was taken as confirmation of the theory. Although now taken for granted, it should be clear that not until brain waves could be monitored continuously in a reliable manner could sleep be meticulously observed and measured without disturbing the sleeper. The discovery of spontaneous electrical activity in the brain of animals was made by the Scottish biologist Richard Caton in 1875. This activity was only later referred to as brain waves. It was not until the late 1920s and early 1930s that German psychiatrist Hans Berger recorded spontaneous electrical activity from the scalp of human subjects and correctly concluded the activity was generated by their brain. Berger was also the first to report that the brain wave patterns in humans who fell asleep were different from brain wave patterns in the same individuals when they were awake.

Figure 2-1: This photograph appeared in the Chicago Tribune in 1956. Here, Dr. Dement applies electrodes to his wife Pat to perform some of the earliest all night continuous EEG recordings.

The first conceptualization of sleep and wakefulness that was supported by scientific experimentation was the notion that sleep occurred passively as a result of a reduction in brain activity. According to this view, continuous sensory bombardment of the brain was required to maintain the awake state. In the 1930s, Frederick Bremer, a Belgian physiologist, transected the brain stem of experimental animals at the level of the midbrain, a procedure which interrupted nearly all sensory input to the cerebral cortex. There was an immediate appearance of brain wave patterns resembling sleep in the cerebral cortex. Bremer erroneously concluded that this result confirmed the "passive theory" of sleep— that without sensory input, the brain cannot maintain itself in a state of wakefulness.

The development of the ink-writing polygraph which allowed brainwave activity (now called electroencephalograms or EEGs) to be recorded on moving paper was the culminating step in this crucial technological progress. Although a few investigators described interesting brain wave pattern changes during sleep, the technique of **electroencephalography** was utilized mainly as a clinical test to identify epileptic seizure patterns or evidence of brain damage. It was not until the discovery in 1952 of rapid eye movements during sleep that the ink-writing polygraph was also widely applied to the study of sleep.

The Brain Is Never Turned Off

One of the reasons that we have a sleep problem in our society is the common misunderstanding that going to sleep is turning off the brain. With the discovery of rapid eye movements during sleep in 1952, centuries-old beliefs about the fundamental nature of sleep were altered forever. Sleep could no longer be viewed as a quiet and passive state. The discovery of REM sleep and associated dreaming is particularly important to one of the authors. Dr. Dement was a second-year medical student when he participated in this research and co-authored the lead publication. This event changed the course of his life to the benefit of all of us.

In order to comprehensively describe the occurrence of rapid eye movements during sleep, it was necessary to monitor eye movements continuously in sleeping subjects. The ink-writing oscillograph used to record brain waves was ideally suited to this monitoring challenge. Both the electroencephalogram (EEG) and the **electro-oculogram** (EOG) could be continuously recorded on moving paper throughout the night without disturbing the sleeper.

Continuous, all-night recordings of sleeping subjects revealed that rapid eye movement periods lasted a number of minutes and alternated with periods of sleep in which there were no rapid eye movements. In sleeping cats, rapid eye movements were found to coincide with brain wave patterns (EEG) identical to patterns associated with a very active waking state. By monitoring the specific discharge of nerve cells, researchers further discovered that the brain is often more active during REM sleep than during wakefulness. Following these observations, it was no longer possible to think of sleep as a state of inactivity and an absent or low level of consciousness. To the extent that anyone conceives of sleep as the brain "turned off," this misunderstanding should be corrected. It is more accurate to say that the brain is never turned off. It is always active as long as we are alive and healthy.

Sleep Is Part of a Circadian Rhythm

To re-emphasize the fundamental principle stated in the beginning of this chapter, sleep and wakefulness are complementary phases in the daily cycle of existence. In addition, the daily cycles of sleep and wakefulness are the outstanding manifestation of what are called biological or circadian rhythms. **Circadian** is a term that means "close to 24 hours." It was coined to account for the fact that in isolation from all cues about the time of day, the human daily cycle of sleep and wakefulness and other biological rhythms assume a periodicity that is close to, but not exactly, 24 hours. In humans, the period length is usually slightly more than 24 hours with small individual differences. This circadian oscillation is under the control of a brain mechanism called the "biological clock" (see Chapter 5). It is now known that this mechanism is housed in the **suprachiasmatic nuclei**, two small bilateral clusters of neurons near the midline at the base of the brain. Both the mechanism and the neuroanatomical locus are encompassed by the term "biological clock." (Figure 2-2)

Since the location of the biological clock is now known, it can be selectively lesioned in experimental animals. When this is done, circadian rhythms entirely disappear. Though it mediates the circadian timing of sleep and wakefulness, the biological clock is not necessary for organisms to fall asleep or to awaken from sleep. Hundreds of observations have clearly demonstrated that animals without a biological clock fall asleep and wake up in a completely normal manner. However, there is no predictable circadian modulation of any bodily function, and when the organism is isolated from the environment, periods of wakefulness alternating with periods of sleep are evenly distributed throughout the 24-hour day. There is no daily long period of wakefulness and no daily long period of sleep. In summary, the major role of the biological clock is to amplify and reinforce sleep and wakefulness as alternating phases in the daily 24-hour cycle of existence.

Sleep Research Methods

Considerations in Studying Normal and Abnormal Sleep

As with all scientific disciplines, the study of sleep has its own special methodology as well as special challenges. A particular challenge is that the researcher must not disturb the sleep of the subject. A frequent exception is when the nature of the disturbance and the response to it are the focus of the study. The care required to avoid any disturbance of the sleeping subject varies depending on the circumstances. For example, when one wishes to describe the sleep of patients with insomnia, great precautions must be taken because such patients are often inordinately sensitive to noise or other environmental disturbance. On the other hand, adolescents can sleep through great noise.

In order to measure an individual's sleep, a sleep test had to be developed. Sleep tests usually require recording over extended periods of time. A typical period of observation is an entire night of sleep. Other recordings may include continuous observation over 24 hours. In addition to the all-night sleep study, the researcher might also evaluate the impact of the night of sleep on waking performance during the following daytime period. For observations geared toward evaluating the effectiveness of a particular sleeping pill, a study will more likely last several or many days and nights.

Studies of sleep and wakefulness which require large numbers of consecutive 24-hour periods are usually carried out in experimental animals. Attempts have been made to observe sleep and wakefulness continuously for the major portion of the life span of an individual organism.

Figure 2-2: The biological clock. There is a circadian time keeping mechanism in the brain of every human being and every animal on the face of the earth. In humans, this mechanism plays a major role (see Chapter Five) in the regulation of the daily cycle of sleep and wakefulness. It should be clear to all readers that your biological clock does not in any way resemble what is depicted in this figure ☺

Longitudinal observations of a normal human being's sleep over many years have rarely been attempted. Even shorter durations of a few years are almost non-existent. Although we know that sleep changes substantially across the lifespan, most observations are cross-sectional. The studies involve observations on groups of individuals at different age levels.

Figure 2-3: "Here come the trodes." Several young campers at the Stanford University Summer Sleep Camp with their electrodes attached even though they are not being recorded. Because it was impractical to attach and remove electrodes many times each day, they were left in place as long as possible, including when campers sallied forth to the Student Union Building and various other locations on campus. The sleep campers soon became widely known as the "trodes".

One important true longitudinal study of healthy volunteers extended beyond several years. Dr. Mary Carskadon, now a professor at Brown University, studied a group of 24 children in a comfortable, home-like sleep laboratory at Stanford University for ten consecutive years (1976-1985). The study began when the children were 10-12 years old and ended when they became adults (figures 2.3).

Figure 2-4: Dr. Mary Carskadon studied adolescent maturation in a home-like setting called the "summer sleep camp." The subjects were monitored with Polysomnography (PSG) and wore the electrodes around the clock.

Each summer, during the second decade of their lives, the young subjects returned to the sleep laboratory and went through exactly the same research protocol. Time in bed and total amount of nightly sleep were the same in each successive year, but the growing youngsters became progressively sleepier in the daytime. The original conclusion from these monumental observations was that the need for sleep does not decrease from the beginning to the end of the second decade of life. It may even increase!

This unprecedented historical work has been commemorated by a plaque at the site of the original research. The research was carried out at a Stanford residential dorm now named Jerry.

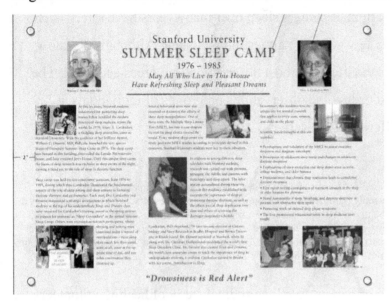

Observation vs. Intervention

Many studies of sleep in humans and animals are conducted with little or no intervention by the researcher. The individual goes to sleep and various aspects of sleep are measured as unobtrusively as possible. Much information about normal sleep must be gathered by purely descriptive observation.

Alternatively, researchers will use experimental manipulation in order to see how subject's response is modified by the states of sleep. One of the most common interventions involves awakening a sleeper during a period of REM sleep and asking whether or not he or she was dreaming. Originally, this was done to show that dream recall was much more likely to occur from REM period awakenings than from awakenings during non-REM sleep. Once this relationship had been established, REM period awakenings were usually carried out to provide a richer source of dream content for those interested in the meaning and other aspects of dream content. Another common intervention is to administer a medication/substance at bedtime and observe the subsequent changes in sleep. Typical examples of this are to test whether or not a particular sleeping pill is effective, or to understand if a medication given for some other illness such as asthma or epilepsy will produce a serious disturbance of sleep as an adverse side effect. Many people think that it would be very difficult to sleep in a laboratory situation. However, the reality is quite the opposite. Even with electrodes attached to their scalps and around their eyes and in spite of the fact that they are in a strange place, normal volunteers typically fall asleep quite easily. In a clinical situation, the likelihood that sleep will occur depends largely on the patient's condition. Insomniacs are occasionally not able to fall asleep or ironically will sleep better in the laboratory than at home.

In a dramatic experiment, several decades ago, University of Edinburgh Professor Ian Oswald assessed the ability of subjects to fall asleep under extreme levels of stimulation. The subjects were placed in a chair and barraged with bright, flashing lights and loud, cacophonous noise. The chair rocked and jerked, a fan periodically blew cold air in their faces, and noxious smells were released. As if all this were not enough, Professor Oswald made the subjects' legs jerk by administering electric shocks to a peripheral nerve. In spite of this unprecedented amount of stimulation, subjects eventually fell asleep. Such is the power of our sleep drive.

Physiological Measures of Sleep

Modern technology allows many physiological variables to be measured during sleep. Some variables, however, are more difficult and expensive to measure than others — intra-arterial blood pressure, for example. Consequently, these variables are measured only when strongly implicated in a health problem and the information is considered to be important. They are usually not studied in healthy volunteers.

When we first began human physiology research, people were usually examined with the sole objective of making a simple comparison with the waking state. A common procedure was to take a measurement in the waking state, allow the subject to fall asleep for a few minutes, and repeat the measurement. We underestimated the complexity of sleep. We now know that there are many dramatic changes within the night of sleep. Eye movement activity, brain waves, muscle potentials and levels of oxygen in the blood are examples of variables whose changes can be continuously monitored all night long by means of noninvasive methods.

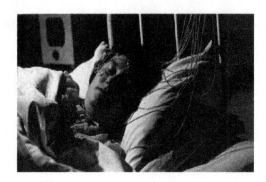

Polysomnography

In order to describe human sleep in a way that is more useful, sleep has been divided into precisely defined states and stages. At Stanford, the term "**polysomnography**" was coined in 1974 to describe continuous all-night recording of multiple variables. This term is now accepted around the world. A term which was initially controversial because it combined both Latin and Greek etymology.

Polysomnography was designed to be flexible in the parameter measured as the field advanced. While many physiological variables can be continuously recorded from a sleeping subject, the activity of three systems is always included: the brain, the eyes, and muscle tone. Originally, these three systems were monitored simultaneously by a paper and ink-based device called the polygraph in a manner that does not disturb the sleeping subject.

Electrodes attached to the scalp and face of the subject convey the bioelectric signals to the polygraph. These signals were recorded on moving chart paper by pens that automatically move up or down in response to changes in electrical potentials emanating from the sleeper. A single night of recording could have resulted in as much as a mile of recording paper. Stacks of paper needed to be warehoused as we accumulated more subjects. Fortunately, these recordings are now made using digital media.

Though a myriad of physiological systems can be, and have been, monitored using the polygraph the following three measures are required to identify and quantify the states and stages of sleep: electroencephalogram, electrooculogram, and electromyogram

Electroencephalogram (EEG) *Brainwaves = EEG*

The EEG is a measurement of surface brainwaves, which are a manifestation of the bioelectric activity generated by the brain. For brain recordings to be optimally useful and comparable from laboratory to laboratory, standard recording methods have been developed. For human sleep recordings, the EEG is monitored from electrodes attached to the scalp at standard locations. Continuous recording of the EEG is the most important method in the study of sleep.

Electrooculogram (EOG) *EYE MOVEMENTS = EOG*

The standard method of recording eye movements during sleep is called electrooculography, or EOG. The EOG takes advantage of the fact that electrical activity can be measured when the eyes move under closed eyelids. A stable voltage difference exists across the eyeball. The front (cornea) of the eye is electrically positive relative to the back (retina) of the eye. We should not be surprised that electrical activity is present in the retina. As the eyes move left, right, up, or down, the position of the electrically positive cornea and the electrically negative retina will change their positions relative to the electrodes. This results in a change in the voltage of one electrode with respect to the other. This voltage change will be registered on the polygraph or computer, and the change indicates that the eyeballs have moved to a new position.

Electromyogram (EMG) *MUSCLE ACTIVITY (EMG)*

All muscle activity is accompanied by biochemical reactions which are readily detectable. In human sleep recordings, the EMG is usually monitored from electrodes placed on the skin over the chin muscles. A continuous minimal contraction of muscle fibers (tone) generates the electrical activity that is recorded during sleep. This activity is generally of a high frequency, and the relative amount of activity is judged by the amplitude of the signal. When the muscle fibers contract more vigorously, the amplitude of the EMG activity increases; when the muscle relaxes, fewer fibers are firing and the overall amplitude of the EMG activity decreases. Since a powerful and active suppression of muscle activity and associated EMG potentials is associated with REM sleep, this physiological variable is routinely measured in sleep studies.

Figure 2-5: Electro-oculography during sleep. The eyeball is a battery: the cornea is the positive pole, and the retina is the negative pole. If electrodes are attached above and below the eyes, as well as on the sides, the exact pattern and direction of eye movements can be recorded. This figure illustrates the electrode placement typically used when the purpose is merely to detect whether or not rapid eye movements are present. Their detection is crucial in identifying the occurrence of REM sleep.

Even when an individual is very relaxed as in nonREM sleep, a number of muscle fibers continue to contract providing a readily detectable, continuous, low level EMG activity. However, in REM sleep, and *only* in REM sleep, EMG potentials disappear completely as the dreaming brain actively suppresses all motor activity (see figure 2.6). Thus, though REM sleep and waking EEG and EOG patterns may be similar, REM sleep can be readily differentiated from relaxed wakefulness by the decrease of EMG activity during the former in contrast to the relatively high levels seen in the latter.

Other Polysomnographic Variables

Although it is possible to measure any number of physiological processes during sleep, typically three additional variables are routinely measured. These are breathing, heart rhythm, and leg muscle movement. The recording of these three additional parameters has crucial importance in diagnosing and treating patients in sleep disorders centers where the total array is called clinical polysomnography. Historically, breathing was usually measured by a nasal device that senses the change in temperature as warm air was exhaled and cooler air was inhaled. A more modern technique to detect breathing utilizes nasal pressure changes via cannula. This more sensitive technique allows detection of important subtle breathing events. In addition, sensors are often attached to the chest which record its expansion and contraction as patients breathe in and breathe out. Heart rate and rhythm are measured by continuously recording the EKG. Leg movements are measured similar to the previously described methods measuring muscle movement around the jaw-- that is, by placing electrodes over the leg muscles.

The Nature of Sleep

There Are Two Entirely Different States of Sleep

Sleep consists of two, entirely different behavioral states. These states are called **REM (rapid eye movement)** sleep and **non-REM** sleep. Both states appear to exist in all mammals. Common to both states are certain features that define sleep in general. These include a lack of perceptual awareness of the external world, a non-waking consciousness, an absence of continuous fine and gross body movement, a relatively relaxed voluntary musculature, and a decrease in sensitivity to external stimulation.

Precise definitions have been developed that make it possible for sleep researchers to determine whether any defined epoch (a period of time usually defined as 10 to 30 seconds) in the life of a human being is REM sleep, non-REM sleep, or wakefulness. These definitions and polysomnogram scoring rules, together with numerous examples of raw recording data, are set forth in "A Manual of Standardized Terminology, Techniques and Scoring System for Sleep Stages of Human Subjects," which was first developed and published in 1968. This manual was widely adopted around the world. Remarkably, no revisions were made until 2007. The new version has been adapted for computerized scoring. One of the major changes in the new manual was the changes of stage 3 and 4 sleep. They are now consolidated into one state called slow wave sleep.

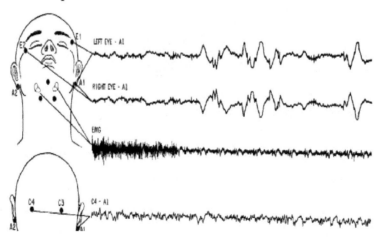

Figure 2-6: This diagram demonstrates the transition into REM sleep with rapid eye movements and decreased muscle tone.

The Fundamental Nature of REM Sleep

The simplest definition of REM sleep is the natural cyclic occurrence of a highly active brain in a paralyzed body. The paralyzed condition does not include every single muscle group. The diaphragm continues to function in a relatively normal fashion, although there is some evidence that its contractions may be slightly weaker in REM sleep. The throat (airway) muscles continue to function though at a somewhat more relaxed level, and, of course, the extraocular muscles (the muscles that rotate the eyes in various directions) appear to be completely unaffected by the REM sleep paralysis. The activity of smooth muscles, e.g., such as the gut, also appears to ontinue in a normal fashion throughout all states of sleep. Many smooth muscle groups in the body are relatively inaccessible and have not been well studied during sleep.

One of the most characteristic features of REM sleep — from which its name is derived — is the occurrence of binocularly synchronous, rapid eye movements. These rapid eye movements are essentially indistinguishable from eye movements which occur during wakefulness. Many sleep researchers believe that the movements are the result of the brain's efforts to scan the events in the dream world, a theory known as the "**scanning hypothesis**." However, there are features of eye position and slow movement during REM sleep that are not entirely similar to wakefulness. Eye centering and fixation, extremely important in wakefulness, is not optimal in REM sleep. There are occasional slow, drifting movements that never occur in the waking state. The direct observation of rapid eye movements during sleep in babies or family members is quite easy because the corneal bulge can be seen altering the surface of the eyelid as it shifts from side to side and up and down. Early morning is the best time to see rapid eye movements in one's bed partner or roommate because REM periods get longer as the night wears on, occupying about 50 percent of the last third of the night.

Motor Activity During REM Sleep

All human sleep is characterized by a marked reduction in bodily movements in comparison with wakefulness. In non-REM sleep, this appears to be a mainly passive phenomenon whereas in REM sleep there are specific brain mechanisms which actively suppress muscle activity. Indeed, a characteristic and necessary feature of normal REM sleep is an actively maintained blockage (inhibition) of the final nerve pathways to the muscles. One result of this inhibition is the complete disappearance of tonic EMG potentials from the skeletal (chin) muscles during REM sleep. This absence of *tonic* EMG potentials provides another very important way of judging, with a high degree of certainty, whether REM sleep is present or absent. The suppression of motor activity during REM sleep is briefly interrupted every five minutes or so by movement of a limb or a minor shift of position. Tendon reflexes may be elicited in non-REM sleep but, are absent in non-REM sleep.

Muscular twitching occurs during REM sleep and is easily seen in non-primate mammals. Muscle twitches are termed *phasic* because they are very brief. (The word tonic is used when activity is continuous throughout the state). In dogs and cats, bursts of muscle twitching involving the face, the eyelids, the whiskers, and the paws, are highly characteristic of REM sleep and very easy to see. Generally, muscle twitches occur in association with bursts of rapid eye movement.

REM Sleep Paralysis:
Too Little, Too Much, Just Enough

Students are encouraged to consider the phenomenon of **REM sleep paralysis** at some length. Why must we be able to elaborate the same nerve impulse messages to the muscles in our dream activities that, if we were awake, would propel us to carry out exactly the same activities? The fact that these neural commands travel all the way down to the alpha motor nerve cells in the spinal cord means that they must be blocked at the same spinal cord level. This requires a special neural system that must always "turn on" during only REM sleep to prevent the alpha motor neurons from responding to the brain's commands. Inhibiting the alpha motor neurons creates a flaccid paralysis of all voluntary muscles. Only the eye muscles (which are commanded by several of the cranial nerves) and the diaphragm are exempted from this paralysis. For the most part, human beings are entirely unaware of the remarkable and entirely natural phenomenon of being nearly completely paralyzed for about two hours every night.

Occasionally we become aware that we are paralyzed as we emerge from REM sleep. Almost everyone has experienced a nightmare in which they are being pursued by some terrifying apparition and their legs feel heavy and very difficult to move. Another commonly experienced manifestation of the REM paralysis is occasionally waking up in the middle of the night and being unable to move for a few seconds. Later in this book, the two most dramatic pathological manifestations involving REM paralysis will be described in detail. In the first, the REM paralysis intrudes into the waking state as part of the syndrome of narcolepsy. Patients with narcolepsy have frequent attacks of REM paralysis which are collectively called cataplexy. In the second manifestation, REM behavior disorder, there is a partial or complete failure of the REM atonia system to function during REM periods, and the patient acts out the dream. The latter two disorders are not uncommon. However, as manifestations of an abnormality specifically involving a basic sleep process, they deserve careful study. A great deal has been learned about the neurological substrate of REM sleep paralysis.

Brain Activity During REM Sleep

The activity of individual nerve cells in a variety of regions of the brain is often spectacularly elevated during REM sleep even compared to the active waking state. Mental activity during REM sleep is at a very high level. This is understandable considering the phenomenon of dreaming in which the brain is not only perceiving stimuli but creating the stimuli it perceives as well! Let us clarify this point— when an individual dreams, the experience seems to be real. The dreamer really tastes, touches, hears, sees, converses, etc. within the world of the dream. This means that the brain is not only perceiving and responding to the environment (as it does during waking experience), but the brain is also creating that environment! It is not unreasonable that a higher level of brain activity would be required to create and perceive the complex multi-colored dream world than the level of brain activity required during wakefulness when the environment is already there and information about it is received passively.

Nocturnal penile tumescence (abbreviated as NPT)

All healthy males, including infants, experience nocturnal penile tumescence. NPT is a spontaneous erection usually three to five times during the night, typically during REM sleep. Since REM sleep dominates the last half of the night this explains why males wake up with erections.

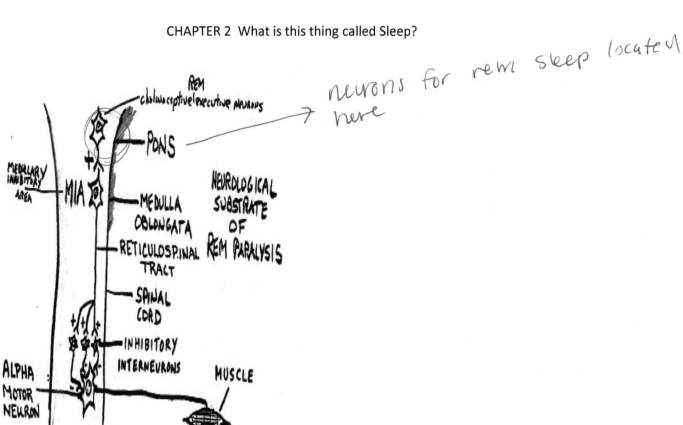

Figure 2-7: Neurological Substrate of REM Paralysis. Proceeding rostrally into the brain from the spinal cord, the first brain stem structure is the medulla oblongata, then the pons, and then the mesencephalon. The executive machinery (neurons) for REM sleep is located in the pons. One part of this system carries out the suppression of voluntary movement. The crucial cells in the pons are cholinoceptors (excited by acetylcholine). Their axons travel into the medulla oblongata synapsing in a neuronal network called the medullary inhibitory area. Nerve cells in this location send long axons down the spinal cord in the reticulospinal tract, branch extensively, and terminate on inhibitory interneurons at every level. During REM sleep, the activity of this system completely suppresses alpha motor neurons whose axons leave the spinal cord and travel to the voluntary muscles in the peripheral nerves. The neural pathways by which commands are delivered to the muscles in the waking state descend in a different part of the spinal cord. These pathways are trying to stimulate the motor neurons when we are dreaming, but cannot.

The Nature of Non-REM Sleep

Non-REM sleep can be identified and quantified with reasonable accuracy by brain wave patterns alone. NonREM patterns are clearly different from both wakefulness and REM sleep, with the possible exception of transitions from wakefulness to non-REM sleep where some brief ambiguity may occur.

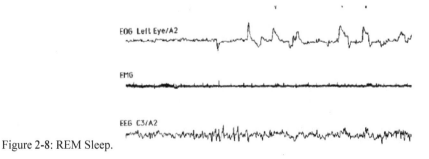

Figure 2-8: REM Sleep.

The Stages of Non-REM Sleep

Non-REM sleep was traditionally subdivided into four EEG stages but now is consolidated into three. They are all associated with an absence of rapid eye movements and the continuous presence of low-level tonic EMG activity from the neck and jaw muscles in humans. Slow, rolling eye movements may be seen in non-REM sleep, typically during non-REM Stage 1 and occasionally during Stage 2. Body movements can occur in non-REM sleep, though very infrequently in comparison with wakefulness. If the movements are sufficiently vigorous, they are usually associated with a brief awakening. Sample EEG tracings illustrating the three stages of non-REM sleep are shown in Figures 2-9 through 2-11.

N1(Stage 1) sleep is characterized by a relatively low voltage, mixed frequency EEG with a prominence of activity in the 4-8 cycles per second (cps) range. Stage 1 characterizes the transition from wakefulness to the other sleep stages (or typically follows body movements during sleep) and is usually short-lived, lasting about 1-7 minutes. Rapid eye movements are absent from non-REM Stage 1, but slow, pendular eye movements are a typical feature of this stage particularly as sleep is just beginning. EMG activity is usually a little higher than the remainder of sleep, but occasionally may be very low.

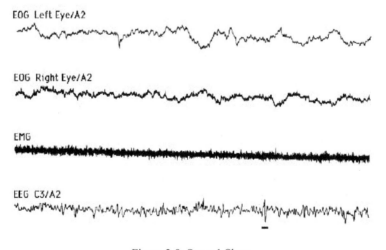

Figure 2-9: Stage 1 Sleep

N2(Stage 2) sleep is defined by the presence of sleep spindles and/or K-complexes in the EEG and the absence of high amplitude, slow EEG activity more characteristic of Stages 3 and 4. A sleep spindle is an EEG waveform in which a basic sinusoidal rhythm of 12-14 cps waxes and wanes in a highly characteristic fashion over 1-2 seconds. Sleep spindles occur at fairly regular intervals of 10 to 30 seconds during non-REM Stages N2 and N3. However, high amplitude slow waves in the latter stage N3 make them difficult to see. The K-complex is an EEG waveform having a well-delineated negative (upward deflection) sharp wave that is immediately followed by a slower positive (downward deflection) component. **K-complexes** can occur in response to a stimulus, but they frequently occur in the absence of detectable stimuli as well. Stage 2 usually accounts for a larger percentage of an all-night sleep recording in the human adult than any other stage, roughly half of the total sleep time. The term **K-complex** was coined by the American scientist Loomis. Why he chose this particular letter of the alphabet is unknown.

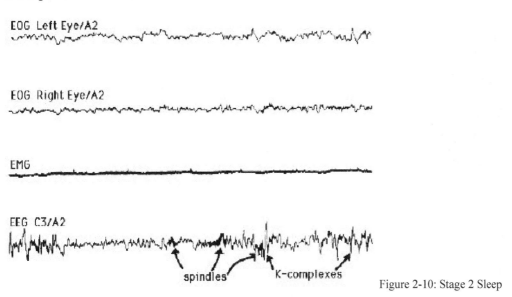

Figure 2-10: Stage 2 Sleep

Slow wave sleep (N3) is also referred to as **delta sleep** since the characteristic slow waves are in the delta frequency (less than 4 cycles per second, CPS) which is defined by a high amplitude EEG recording, slow waves are present 20 percent or more of the epoch. Identifiable K-complexes are not counted in the high amplitude slow activity. Sleep spindles may or may not be detectable in slow wave sleep. Tonic EMG activity is usually present, but its presence or absence is not relevant to the definition.

Delta waves have frequencies of 2 cps or slower and amplitudes greater than 75 microvolts peak to peak (the difference between the most negative and positive points of the wave). Sleep spindles, although characteristic, may not be present in slow wave sleep. During slow wave sleep the heart rate and the respiratory rate are at their slowest at this time of the night. Slow wave sleep occupies about 10-15% of the total sleep time but also may be undetectable in older individuals.

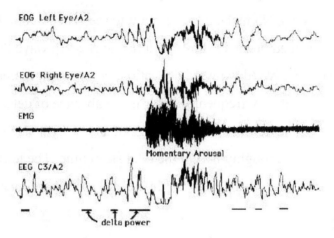

Figure 2-11: Stage 3 Sleep

REM versus Non-REM:

When rapid eye movements during sleep were first discovered, sleep was thought to be a single homogeneous state. Therefore, it was initially assumed that the eye movements indicated the occurrence of an inexplicable lightening or disturbance in this state. However, as more studies were done and more data on sleep was accumulated, the view that sleep was a single homogenous state had to be modified. First, periods of rapid eye movements occupied a predictable and large portion of a night of sleep. Second, there were extraordinary differences in their mental and physiological attributes as compared to periods of sleep without rapid eye movements. Accordingly, a new view emerged — the concept of REM sleep as an independent biological state, totally different from non-REM sleep.

Similarities of REM and Non-REM

Perception of the outer world is essentially absent in both non-REM sleep and REM sleep. Environmental stimuli can occasionally enter a dream but they are usually transformed and incorporated into the content of the dream. Real perception of the outer world does not occur. It may be very difficult to awaken a sleeper at 1 a.m. when he or she is in slow wave sleep, and just as difficult ten minutes later when he or she is in REM sleep. The ease of awakening the sleeper does not appear to be fundamentally different between the two states of sleep. Arousal threshold is much more a function of time of night.

Do both states serve equally to restore alertness? The data on this point are not conclusive. One study suggests REM sleep does not serve to restore alertness. People often ask, "What kind of sleep is best?" At the present time, we cannot answer this question. However, we do know that non-REM sleep is best for restoring alertness.

Differences of REM and Non-REM

Brain regions in REM sleep appear to be more active with respect to both non-REM sleep and quiet wakefulness. It is more difficult to generalize about comparisons with active wakefulness because brain activity is very difficult to observe when human subjects or animals are moving about. Recent studies of human brain metabolism during sleep utilizing modern imaging techniques suggest that certain specific brain areas become more active in REM sleep than in wakefulness and other specific areas become less active. A vigorous excitatory motor outflow from brain to spinal cord is elaborated in both REM sleep and wakefulness as opposed to non-REM sleep. However, during REM periods there is a powerful and generalized inhibition of spinal motor neurons and muscle activity and the body remains quiet. This is why typical deep tendon reflexes are usually absent in REM sleep. Rapid eye movements do not occur in non-REM sleep, or they are extremely rare. Activity in the autonomic nervous system is greatly increased in REM sleep. The variability and irregularity of a number of measures, e.g., heart rate and breathing, increases during REM sleep. Gross body movements can occasionally occur at any time during sleep, but REM sleep is a time when many smaller twitching movements occur. Although in non-REM sleep the respiratory and heart rates are at the slowest point of the 24-hour day, in REM sleep your respiratory and heart rates are at their peak.

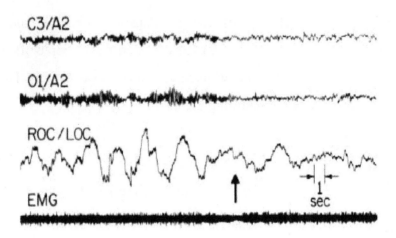

Figure 2-12: Change in brain wave patterns at the moment of sleep. When awake, human beings close their eyes, a sinusoidal brain wave pattern called alpha rhythm in the frequency range of 8-12 cps is enhanced. The moment of sleep onset is simultaneous with a disappearance or marked attenuation of the alpha rhythm. (This 8-12 cps rhythm is also somewhat attenuated when awake individuals open their eyes, but other indicators of the awake state are always very prominent.) Thus, the moment of sleep onset associated with an instantaneous perceptual shutdown is signaled by disappearance of alpha rhythm (vertical arrow). The alpha rhythm can usually be recorded over many areas of the scalp, but is most prominent in the posterior or occipital area.

If you have a sedentary lifestyle, then the biggest workout your heart gets is approximately the 2 hours when you are dreaming. As Stanford Professor Guilleminault says "your body goes to the Olympics every night." This simple fact reflects the importance of exploring why some people die in their sleep.

Journey Through the Night

The Onset of Sleep: EEG Changes

When the functional state of the brain shifts from awake to sleep, the most reliable and accurate indicator of this event is a characteristic change in brain wave patterns. Thus, at the moment we fall asleep, the activities of many hundreds of thousands, if not millions, of neural elements are abruptly altered. This moment is signaled by an easily monitored, relatively simple event, namely the disappearance of a characteristic waking brain wave pattern. This characteristic pattern is an 8-12 cps oscillation commonly known as "alpha rhythm" which is generated mainly from the occipital (posterior) region of the brain. A useful analogy to think about to understand the alpha rhythm is to think about a car's engine. The engine's revolutions per minute will fluctuate depending on the speed the car is moving.

Figure 2-13: Strobe light experiments. A subject lie on a gurney with a bright strobe light about 6 inches above his eyes. A microswitch is taped to his finger. Electrodes are in place to record brainwave patterns. In most subjects, the eyelids were taped open. The strobe light flash lasts 10 milliseconds and is extremely bright (50,000 lux). The task of the subjects is to press the microswitch when they see (are conscious of) the strobe light flash.

However, when a car is stopped the engine idles in a steady rhythm. The same way the alpha rhythm represents the idling of the visual cortex when the eyes are closed. Waking alpha rhythm is greatly enhanced by relaxation and eye closure. When the eyes are open during wakefulness, alpha rhythm is diminished but usually does not completely disappear. Since eyelid closure always precedes falling asleep in conventional situations, the disappearance of the enhanced alpha rhythm is a very good signal of the instant at which sleep replaces wakefulness. This characteristic **sleep onset** change is illustrated in Figure 2-12.

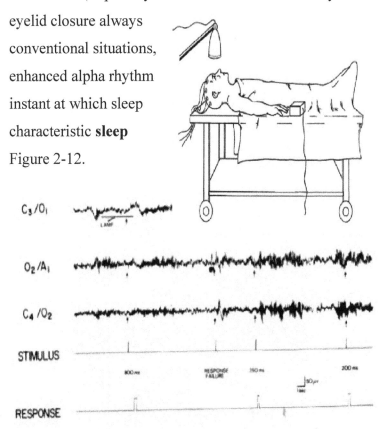

Figure 2-14: Sample recording from the strobe light experiment. The tracing labeled "stimulus" shows strobe light flashes. The "response" traces the microswitch presses made by the subject. No response follows the second flash. It can be seen that the response failure is associated very precisely with the disappearance of alpha rhythms. In hundreds of trials, whenever waking brain wave patterns were absent, subjects did not see the flashes. In other studies, clicks were presented; when waking brain wave patterns were absent, subjects did not hear the clicks. A very brief episode of sleep lasting two or three seconds to a minute or so is called a "microsleep." There is no standard definition that specifies the duration of microsleeps, but when sleep lasts five minutes or more, people begin to talk about napping. In summary, the transition from wake to sleep (the

moment of sleep) is almost instantaneous, and is precisely associated with an almost instantaneous shutdown of our awareness or consciousness of the outer world. We also refer to this as "perceptual disengagement" from the environment.

Slow Eye Movements

Another common feature of sleep onset recorded by the polygraph is the occurrence of slow, back and forth eye movements. If the eyelids are closed, these slow, rolling eye movements may sometimes begin a few seconds before the EEG change occurs and generally will continue for several minutes after the onset of sleep. In contrast to rapid eye movements of waking and REM sleep, these slow eye movements are not always well coordinated, i.e., not always binocularly synchronous. If you happen to see someone fall asleep with their eyelids partially open, these eye movements can be easily seen.

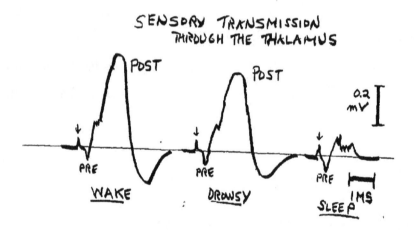

Figure 2-15: Postsynaptic potentials in three states. Sensory pathways are stimulated (in this case the optic nerve) and electrical activity is recorded by microelectrodes from the thalamus, specifically the visual relay nucleus, by means of tiny microelectrodes. The stimulus artifact is indicated by the small downward arrows. The pre-synaptic potential is small, but unambiguous. In the wake state, a large postsynaptic potential is seen as the thalamic relay neurons respond vigorously to the pre-synaptic excitation. This vigorous response continues even as behavioral evidence of drowsiness occurs. However, at the moment of sleep and in the sleeping state, the postsynaptic electrical activity almost entirely disappears while the pre-synaptic potentials are unchanged. These results suggest a major change in sensory processing occurs at the moment of sleep. The vertical bar indicates millivolts and the horizontal bar is one millisecond (Steriade et al.).

Perceptual Disengagement

The most salient feature of sleep onset is perceptual disengagement from the environment. With the onset of sleep, external stimuli are no longer perceived and there is no longer any conscious, meaningful interaction with the outer world. In other words, the occurrence of sleep initiates a total absence of awareness of the outside world. When a stimulus fails to elicit its accustomed perceptual response, the moment of sleep has occurred.

An experiment that dramatically demonstrates the change in consciousness at the moment of sleep is illustrated in Figures 2-13 and 2-14. Very bright strobe light flashes were directed into the eyes of adult volunteers whose eyelids were taped open as they were requested to press a microswitch when they saw a flash. Brainwaves were recorded continuously during the trials. When the EEG patterns of wakefulness were present, subjects invariably saw the strobe light flash and pressed the microswitch without exception. Whenever waking brainwave patterns gave way to NREM Stage 1, the subjects failed to press the microswitch and denied seeing the flash. This was consistently true even if the flash followed the brainwave change by less than one second. The instant waking brainwave patterns reappeared; subjects invariably saw the strobe light flashes again.

The absence of awareness does not mean that all the sensory organs have entirely ceased functioning. We know that the muscles in the middle ear are also functioning. For example, an auditory stimulus of sufficient loudness can always arouse the sleeper. A sleeping person can also be readily awakened by a wide range of stimuli, e.g. a touch, being called by name, the crying of a baby, an alarm clock, being shaken, repeated knocking, and so forth. Nonetheless, for all intents and purposes, the real world no longer exists in the mind of the sleeper.

In human beings, all sensory information (nerve impulses transmitted from the sensory apparatus) goes through a deep brain structure called the thalamus. In this major sensory structure, the axons from sensory nerves terminate upon relay neurons which then transmit the sensory information to appropriate locations in the cortex and cerebellum, most often in the cerebral cortex. Figure 2-15 illustrates dramatic change in the electrical response of these relay neurons associated with the moment of sleep. This is clear evidence that an independent brain process dramatically alters the processing of sensory information in association with the moment of sleep onset, but not before this moment.

There is a very rare genetic illness called Fatal Familial Insomnia in which victims stop sleeping and eventually die. Their final weeks are spent in a comatose state that resembles extremely severe sleep deprivation. Their polysomnogram is characterized by absence of sleep spindles. At autopsy, the only pathological findings are small lesions in the thalamus in areas outside the sensory relay nuclei. This is consistent with the notion that these areas are, in part, responsible for the alterations in sensory processing that are necessary for normal sleep onset to take place. It is hypothesized that when these areas are damaged, sleep onset cannot take place. Figure 2-16, a and b, are schematic drawings depicting a hypothetical neural mechanism that is consistent with the known facts.

In Figure 2-16a, the thalamic relay neurons are maximally excitable during wakefulness so that the flow of sensory information can be optimally processed in a manner that enables us to be fully conscious of our environment and the scenes and sounds it contains. In addition to the relay neurons, the thalamus contains large numbers of short axon inhibitory neurons. A brainstem network called the **reticular activating system (RAS)** amplifies the sensory processing and fosters the maintenance of wakefulness. It is hypothesized that the RAS exerts an excitatory effect on thalamic relay neurons and also on secondary inhibitory neurons which prevent other inhibitory neurons from suppressing the relay neurons. When sleep debt is low and the sleep homeostat is weakly active, sleep cannot occur.

Retrograde Amnesia

In addition to the abrupt cessation of perceiving the outer world, the onset of sleep has profound consequences for memory. The occurrence of sleep appears to close the gate between short-term memory and long-term memory. All information in short-term memory storage at the onset of sleep apparently fades away. Accordingly, although one is unequivocally aware of the environment (perceiving) before falling asleep, these perceptions are usually lost from memory because they are not transferred into more permanent memory storage.

The phenomenon of sleep-related **retrograde amnesia** (the forgetting of events that have occurred just before falling asleep) has a number of consequences that most people have experienced at one time or another. For example, it is very difficult to "capture" the moment of sleep in memory. The nearly instantaneous passage from an awake, aware individual to a sleeping, insentient being is virtually impossible to recall and examine retrospectively (figure 2-17.)

Other common experiences of this sleep-related memory loss occur during the course of a night's sleep: a telephone conversation in the middle of the night is completely forgotten in the morning; the spouse wakes the bed partner in the night with an exciting piece of news which cannot be recalled in the morning; an individual wakes up 30 minutes late for work and cannot remember having turned off the alarm though he or she remembers setting it the night before; a sleeper awakes in the morning with a hazy notion of a spectacular dream but cannot recall any of the details. Each of these experiences can be explained by the memory-processing properties of sleep.

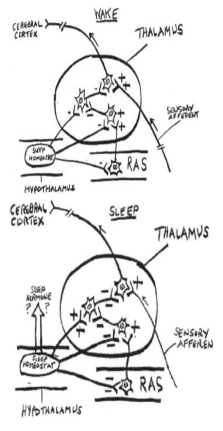

Figure 2-16a: All sensory neuronal pathways entering the brain terminate in a large subcortical structure called the thalamus where they synapse with the cell bodies of relay neurons whose axons travel on to the cerebral cortex.

Figure 2-16b: In Figure 2-16b, sleep debt is large and the hypothetical sleep homeostat is exciting the neurons which inhibit the relay neurons, and at the same time inhibiting the RAS and the interneurons that would oppose this action.

Hypnagogic Imagery

If individuals are awakened a few seconds to a minute or so after the onset of sleep, they can usually recall some mental activity. It usually consists of images — mostly vague, some vivid -- and disconnected thoughts. The images are referred to collectively as **hypnagogic imagery**. In the seconds preceding the moment of sleep abstract, thoughts progress in a kind of reverie and then, as sleep onset occurs, abstract thoughts are transformed into concrete images. Some people may attribute supernatural or mystical meanings to this phenomenon.

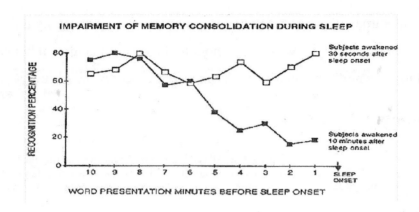

Figure 2-17: Retrograde amnesia. It is well known that people frequently do not remember something they did in the middle of the night. The classic example is the nurse phoning the intern in the middle of the night, receiving detailed instructions for a patient (which are often wrong), and the intern has no recall of this the next morning. This figure depicts one study to demonstrate the retrograde amnesia associated with falling asleep.

Sleep Starts

Often associated with sleep onset is a sensation of floating which is occasionally abruptly terminated by the sensation of falling with a startle, a jerk of the muscles, and a return to wakefulness. These **sleep starts**, also called "hypnic myoclonus" or "hypnic jerks," are normal and occur at all ages although they are said to be more frequent in adults than children. Sleep starts generally occur within five minutes of sleep onset and involve muscle contraction of the legs, arms, or, sometimes, the entire body. They typically do not disturb sleep during the remainder of the night.

The Basic Sleep Cycle

A night of sleep typically progresses through a more or less orderly succession of predictable changes called sleep cycles. In adult humans, sleep normally begins with the non-REM stages except in specific sleep disorders or when subjects are maintaining abnormal sleeping schedules. The first REM period always follows a period of non-REM sleep and the two sleep states continue to alternate throughout the night. A rhythmic alternation of the two fundamental basic states of sleep is

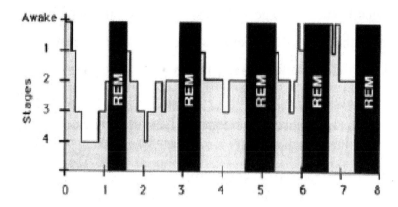

Figure 2-18: This graphic representation shows the relative amounts of each stage over time. Notice that as the night progresses, proportionately more time is spent in REM sleep.

seen in all mammals. In humans, this alternation is called the "basic sleep cycle." When very large numbers of such cycles are averaged for adults, the period length is found to be 90 minutes. It is slightly different in infants and the elderly, and varies widely among other species, averaging about 25 minutes in the cat, 12 minutes in the rat, and around 5 minutes in the mouse. A full night of human sleep will usually consist of 4-6 non-REM/REM sleep cycles. The composition of each cycle, in terms of the distribution and length of the various stages and states, varies in relation to many factors, including age, prior sleep history, ingestion of alcohol or other drugs, and most especially, time of night.

States and Stages Through the Night

The average individual has no conscious comprehension of the complexity of sleep and the numerous changes and events that take place in the course of a single night. In humans, the temporal sequences and the amounts of the different states and stages of sleep in the first half of the night do not precisely resemble those of the second half. In general, non-REM sleep predominates early in the night while the second half of the night contains about three fourths of the total nightly amount of REM sleep. Although there is substantial night-to-night and person-to-person variability, generally 75 percent of nightly sleep is non-REM sleep and 25 percent is REM sleep. (Figure 2-18)

The following description will exemplify the course of events during sleep in the "typical" adult human. Sleep is entered through non-REM Stage 1, which usually persists for only a few minutes. During sleep onset Stage 1, sleep is easily interrupted by very low intensity stimuli—

a door closing or a gentle nudge. Stage 2 sleep is signaled by the appearance of sleep spindles and/or K complexes in the EEG. The sleeper is more difficult to arouse in this stage; a stimulus that would produce wakefulness from Stage 1 will often evoke a K complex in the Stage 2 EEG, but not a complete awakening. During the course of the first five to fifteen minutes of Stage 2, high amplitude, slow frequency EEG activity gradually begins to appear. There are generally a few minutes that can be classified as slow wave sleep before high amplitude, slow waves are predominant in the EEG, defining slow wave sleep. In most adults, slow wave sleep of this first sleep cycle of the night and it is much harder to awaken the sleeper at this time than later in the night. Slow wave sleep typically continues for about 20 to 40 minutes, after which a series of body movements signals an "ascent" to lighter non-REM sleep stages. A general slowing of EEG before the slow wave sleep may occur, or there may be a direct transition to Stage 2 sleep. After 5 or 10 minutes, the first period of REM sleep appears. This first REM episode of the night is often short-lived, usually lasting 1 to 10 minutes.

In human beings, REM sleep can usually be recognized by its characteristic brainwave patterns. The **alpha rhythm** (8-12 cps), which characterizes quiet wakefulness in most human beings, is much less prominent. If present, it may have a slightly slower frequency. Unique patterns called **saw-tooth waves** are present while **sleep spindles** and highamplitude slow waves are absent. The saw-tooth waves help to differentiate REM sleep from Stage 1 non-REM sleep. In other non-primate mammalian species, the brain wave patterns of REM sleep cannot be distinguished from the patterns associated with active wakefulness. In cats and dogs, the brain wave patterns of active wakefulness and REM sleep are similar. In rats, the patterns associated with REM sleep resemble those seen with intense emotional arousal in the waking state. In humans, the end of a REM episode and the transition to non-REM sleep may be associated with some body movement, and a very brief arousal or the transition may occur with no movements at all and no arousal. Often, there is a change in body position such as a rolling over or a series of smaller adjustments.

Non-REM and REM sleep continue to alternate during the night. The duration of the first sleep cycle — from sleep onset to the end of the first REM episode — is typically 60 to 90 minutes. In the second sleep cycle, there is less slow wave sleep and more Stage 2 sleep. The REM portion of the second cycle increases. The duration of the second sleep cycle, measured from the end of the first REM sleep period to the end of the second, is generally longer than the first, averaging 100 to 110 minutes although cycles are usually 90 minutes in duration. In the third REM sleep period and beyond, slow wave sleep is usually entirely absent or present in very small amounts; the non-REM portion of these cycles is almost entirely Stage 2 sleep. REM episodes tend to become longer in later cycles, and the fourth or fifth REM episode typically lasts 30 to 45 minutes. Brief episodes of wakefulness tend to occur in later cycles, generally in association with transitions between Stage 2 and REM sleep. These brief arousals are usually not remembered in the morning.

In summary, a night of sleep in an adult human is characterized by a cyclic alternation of non-REM sleep and REM sleep. The average period of this cycle is typically 90 minutes, although the lengths of individual cycles show considerable variability over the night. Stages 3 and 4 sleep tend to predominate in non-REM during the first two cycles. Consequently, the first third of the night is usually considered the deepest sleep. REM sleep and non-REM Stage 2 predominate in the last third of the night. This why our most vivid dreams may occur just before we wake up.

In many cultures, the child is held almost continuously until three or four years of age. This Guatemalan child is asleep on his mother's back.

Other Aspects of Normal Sleep

Age Differences in Sleep Patterns

It is well known that sleep changes substantially over the human lifespan. In addition, certain sleep disorders are associated with specific age levels, and overall, sleep pathology tends to increase as humans get older. The relative distribution of sleep stages varies with age. Physiology during sleep is different from physiology during the awake state. At birth, REM sleep occupies as much of our lives as wakefulness. Children have greater amounts of slow wave sleep than older individuals. During slow wave sleep, early in the night, it may be virtually impossible to awaken young children, or they may take several minutes to achieve full awareness. In young children, sleep talking, sleepwalking, sleep terrors, and bedwetting are more likely to occur during this portion of the night. By age 60, slow wave sleep may be completely absent, particularly in men. In addition, sleep is generally interrupted by wakefulness much more frequently in older individuals. Although a great deal is known about age-related changes, much work remains in terms of collecting truly adequate samples of sleep/wake behavior at every age, from birth or even prematurity to old age. Critical periods of the life cycle, such as middle age or menopause, have yet to be studied adequately.

The Regulation of Vital Processes During Sleep

Given the daunting complexity of the human brain, we are not surprising that in wakefulness it can do many things simultaneously, both consciously and unconsciously. Similarly, many different things are going on in the sleeping brain. The brain is the executive organ and its activity determines the activity and responses of the body during sleep. There is now an encyclopedic compendium of knowledge about changes in the regulation of breathing, blood pressure, heart rate, temperature, and other vital processes during sleep. For example, if body temperature drops below a certain threshold in the waking state, brain mechanisms housed in the hypothalamus respond by increasing metabolism in order to generate more body heat. If an organism is in REM sleep and the body temperature drops below the same threshold or even lower, there is no response at all. In essences in REM we become coldblooded dreamers. This major change in the physiology is a direct consequence of the occurrence of REM sleep. Other regulatory processes also show dramatic REM sleep-related changes.

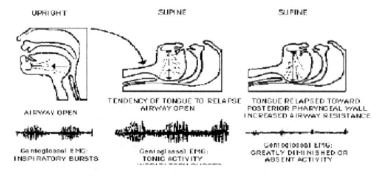

Figure 2-19: This schematic illustration shows the tendency for the tongue (genoglossas) to relapse in some sleepers. Snoring and even sleep apnea may result (see Chapter 17).

There are important differences in our breathing between awake and sleep states. There is general agreement that sleep leads to changes in the character and rate of respiratory movements but there are wide individual differences in the magnitude of these changes. Most people believe that the best indicator of the onset of sleep is a detectable change in the quality of breathing. Not uncommonly, there may be a sigh, a deepening and slowing and perhaps a soft snoring noise. In individuals who snore habitually, the noise of snoring can be an immediate and reliable signal that they have fallen asleep.

A very characteristic and important change associated with sleep is an increase in resistance to airflow when the throat muscles relax (see Figure 2-19).

Summary

Sleep researchers study sleep by means of polysomnography, a standard technique in which physiological variables are continuously recorded. The standard variables are the EEG, EOG, EMG, respiratory variables, heart rate, and leg movements.

Sleep consists of two fundamental and very different states, REM sleep and non-REM sleep. During the night, these two fundamental states alternate in a rhythmic fashion with an average cycle of 90 minutes. During REM sleep, the brain is very active. An active inhibitory process suppresses voluntary muscle activity, and consciousness is directed towards the dream world created. There is an active inhibitory process which suppresses volunteer activity. During nonREM sleep, the brain is somewhat less active, brain wave patterns show sleep spindles and slow waves, and muscle activity is not suppressed. The journey through the night begins at sleep onset which is associated with perceptual disengagement, retrograde amnesia, hypnagogic imagery, slow eye movements, and a cessation of alpha rhythm in the EEG. High amplitude slow wave EEG patterns dominate the first portion of the night and alternating Stage 2 with REM Sleep occur the latter portion.

People say, 'I'm going to sleep now,' as if it were nothing. But it's really a bizarre activity. 'For the next several hours, while the sun is gone, I'm going to become unconscious, temporarily losing command over everything I know and understand. When the sun returns, I will resume my life. **George Carlin**

CHAPTER 3 Sleep Quality & Sleep Debt

"For you to sleep well at night, the aesthetic, the quality, has to be carried all the way through." - Steve Jobs

Sleep somehow has a restorative role in facilitating an optimal level of waking function. We should wake up feeling refreshed and not tired. The components of waking function that are most affected by habitual sleep deprivation are:

1. Alertness and ability to maintain focus and attention 2. Cognitive performance including learning, memory, and creativity

3. Mood

4. Energy and motivation

5. Control, coordination, and impulsiveness

6. Pain

These areas of waking function may overlap with one another to some degree, but the overall effect of sleep deprivation is a progressive decline in all waking functions. This decrease is typically associated with subjective feelings of tiredness and fatigue. The underlying variable behind this decline in waking function is the strength of the tendency of the brain to go to sleep. Much of what has been learned over the years about how sleep affects waking performance has come from studies of sleep deprivation in which human volunteers were not allowed to sleep for some duration of time. For example, studies were done with subjects for 36 hours consecutive wakefulness. Other studies were done with subjects for 60 consecutive hours of wakefulness. With this degree of sleep deprivation, functioning in all the areas listed above noticeably deteriorates. Voluntary total sleep loss studies in human subjects have only rarely been carried out for longer periods of time due to practical and ethical issues.

As early as 1871 research was done on total sleep sleep deprivation in humans. Partial sleep deprivation research, which is more relevant to our society, was not carried out until later. In this textbook, the focus will be on partial sleep loss and its effect on waking functions. In contemporary society, there is not only a lack of adequate education about sleep, but also misconceptions that are entrenched and perpetuated about sleep.

Involuntary sleep deprivation is a sanctioned form of torture in some jurisdictions. In addition to having great difficulty remaining awake voluntarily, subjects fairly soon need to be kept awake by the investigators which in turn also becomes progressively more difficult.

Key Concepts:

Sleep Tendency
At any moment throughout our waking existence, there is a specific propensity or tendency to fall asleep. The strength of the ongoing sleep tendency at any particular moment is operationally defined by the speed of falling asleep in a standard situation.

Sleep Homeostasis
When the purpose of a physiological regulatory process is to maintain a constant level or flow, the process is deemed "homeostatic." One powerful component of sleep regulation is sleep homeostasis. In the simplest terms, this means that when sleep time is reduced, our tendency to fall asleep when awake is increased. When "extra" sleep is obtained, our tendency to fall asleep while awake is decreased. The purpose of this homeostatic process is to ensure that individuals will obtain the amount of sleep they need by continually adjusting their sleep tendency up or down in response to the prior amount of sleep.

Sleep Need
Each human being needs a certain amount of daily sleep to maintain optimal waking function and to avoid the consequences of chronic sleep loss. The daily sleep requirement of an individual person is operationally defined in terms of the formal properties of homeostatic sleep regulation, and the amount can be precisely determined by measurements carried out in a sleep laboratory. The detailed methods of such an individual assessment will be presented in a later section.
At this point, it is sufficient to state that the sleep need of any individual is the nightly amount which results in a level of daytime alertness which does not increase or decrease, but remains the same on the following days. If less than this nightly amount of sleep is obtained, then the tendency to fall asleep in the daytime will increase; if more than the needed nightly amount is obtained, then the tendency to fall asleep in the daytime will decrease. In other words, each

person needs a certain amount of sleep, which, if obtained on a daily basis, will maintain his or her sleep homeostatic equilibrium. If this is done, then no change in overall daytime alertness should occur.

Sleep Debt

Having defined individual sleep need as the nightly amount of sleep required to maintain the same daily level of alertness, **sleep debt** is therefore the accumulated amount of sleep less than the daily amount needed. For example, if an individual's daily sleep requirement is eight hours, sleeping six hours a night for one week will create a sleep debt of fourteen hours (7 nights x 2 hours less/night = 14 hours). Similarly, sleeping only 5 hours a night for a week will create a sleep debt of 21 hours.

The larger the sleep debt, the stronger the tendency to fall asleep in the daytime. The size of the sleep debt is the major determinant of the strength of the homeostatic drive to fall asleep. In turn, this homeostatic drive to sleep is one of the two major determinants of the strength of the tendency to fall asleep at any moment throughout the 24-hour day. The second major determinant is clock-dependent alerting (see below).

Sleep Drive

The concept of "drive" is usually applied to behaviors that are homeostatically regulated. It also underlies the psychological concept of motivation. Simply put, when humans are sleep deprived, they experience an increased motivation to sleep. They begin to think about taking a nap and may even look around for an appropriate place to sleep - for example, a comfortable chair in the library. If the drive to sleep is very strong, sleep may occur before a suitable place to nap has been located. The brain mechanism which regulates the strength of the drive to sleep is continuously active, 24 hours a day, 7 days a week, reducing the drive as we sleep and increasing the drive as we remain awake. The strength of the drive to sleep can vary widely depending upon the prior amount of sleep, in other words, the size of the sleep debt.

The Biological Clock

Often called the "body clock," the **biological clock** consists of the control and timing mechanisms by which the daily 24-hour oscillations in a host of bodily functions are maintained. The clock is located in a specific site of the brain called the suprachiasmatic nuclei (SCN).

Clock-Dependent Alerting

At certain times of the day, we feel more energetic and alert than we do at other times. This alerting influence comes from the biological clock, and is therefore called **clock-dependent alerting**. When this influence is strong, it is very difficult to fall asleep. On the other hand, when clock-dependent alerting is weak, it is easy to fall asleep. This apparently simple concept is often over looked. It is a common misconception that eating a heavy meal will often make you sleepy. If this was true, then why does not breakfast not make you sleepy? The reason that you're more or less alert is not related to the food intake but to this clock-dependent alerting phenomenon.

Opponent Process Model

The most important aspect of normal and abnormal sleep that readers must clearly understand is why people tend to fall asleep or stay awake at any particular time of the day or night. The answer is the interaction of the two independent determinants of alertness and behavior. The brain of every human being houses a process whose function is to induce and maintain sleep called sleep homeostasis. Every human being's brain also possesses a second process whose function is to induce and maintain alert wakefulness called clock-dependent alerting. The ongoing interaction of these two processes to produce the daily cycle of sleep and wakefulness is called the **opponent process model**.

Case History

W.J. is a commercial truck driver. He belongs to a subgroup of drivers that are often referred to as "overthe-road" or "long-haul" truckers which means they drive a single load often thousands of miles. They are generally away from home for much of the time. According to W.J., "Our life is on the road." W.J. was one of nine long-haul drivers who came to Stanford University in the summer of 2000 to receive fatigue management training and also included polysomnography and treatment if a sleep disorder was present. The most common and serious problem for long-haul commercial drivers is obstructive sleep apnea (see Chapter 18).

According to W.J., he was tired all the time. He had to take two or three naps a day while on the road, and when home for a weekend, he spent most of his days sleeping on the couch. Neither W.J. nor any of the other drivers had ever been taught anything about sleep, biological rhythms, sleep deprivation, or sleep disorders. When the group realized that the material being presented in the training sessions was absolutely crucial for safe driving and for the elimination of their constant fatigue, they became the best and most eager students we ever taught. All nine drivers were tested for obstructive sleep apnea and seven required treatment. Learning about sleep debt, being motivated to get extra sleep, and having their sleep disorders treated literally resurrected in the lives of the drivers. They went back on the road for a month and a half and then returned to Stanford for follow-up.

During their return, they expressed elation if not actual euphoria about the positive change in their lives. W.J., for example, no longer needed to nap on the road. He was never tired and slept well. When he was home he got things done, he did activities with his wife, and his family was very happy about the change in his energy level and participation in home life. He also said, "I know why I see so many trucks down in the ditch. Those guys are falling asleep." We had the opportunity to interview the group five years later including W.J. and they were still very happy, pleased, and alert. They said, "You saved our lives. This knowledge is a matter of life and death and should be available to everyone on the road.

Sleep and Daytime Function

The vast psychological and physiological spectrum that ranges from extremely wide awake and energetic to drowsiness, slow wave sleep and dreaming is one of the most important domains of human existence. The formal properties and behavioral mechanisms of this sleepiness and alertness dimension are now well established.

However, few make an effort to manage their schedules in a way that accounts for restorative sleep. In addition to being a serious threat to well-being and safety, impaired alertness is unhealthy. It may also be a key symptom of pathological sleep disorders. It is therefore particularly important that everyone is aware of the common sleep disorders, as well as the unhealthy behaviors that cause severe excessive daytime sleepiness. Sleep disorders will be covered in later chapters of this book.

It is common knowledge that if you stay awake all night, you will experience waves of drowsiness the following day. When drowsiness is severe, all but the most rudimentary waking functions are preempted by the struggle to stay awake. Performing our daily tasks safely and successfully requires some degree of alertness. Beyond this, there is an optimal level of alertness when we are completely wide-awake, energetic, and exhibiting peak performance and cognitive function all day long. Ideally, we would like to experience this optimal level every day. Unfortunately, inadequate sleep, or pathological sleep, or sometimes both, prevent the vast majority of us from consistently achieving optimal waking alertness. Many people, in fact, live their days in a kind of "twilight zone" of impaired alertness and, as a consequence, fail or only partially succeed in work, study, and human relationships.

Multiple Sleep Latency Test:
The Gold Standard for Quantifying Sleepiness

Introduction
The field of clinical sleep medicine was pioneered at Stanford University in the early 1970's. With it came a realization that we needed to measure the development between sleep at night and alertness in the daytime. Day-today consequences of this relationship are the most important issues in understanding the role of sleep in our lives. A stable, objective test was required to measure the strength of the daytime sleep tendency and to allow quantitative studies of the day versus night relationship. Prior attempts were unsuccessful and not widely accepted. It wasn't until our team at Stanford developed the Multiple Sleep Latency Test (MSLT) that there was an objective tool finally available. We noted that subjective sleepiness made before a sleep recording frequently predicted the sleep latency. In the spring of 1976 at Stanford we established sleep latency as an objective physiological measurement of sleepiness. This was done with a series of experiments that included two days of total sleep deprivation by Stanford students. From this work the protocol was established for the MSLT.

The MSLT went from a research tool performed on Stanford students to a clinical test accepted worldwide to objectively measure daytime sleepiness. The validity and reliability of the MSLT has become so well established that it is now a reimbursable medical procedure. By far the most widely accepted and utilized objective measure of sleepiness is the Multiple Sleep Latency Test (MSLT.)

The MSLT measures the average length of time it takes a person to fall asleep during the day, if that person is trying to fall asleep. As a desperately thirsty person will gulp down a glass of water in a few seconds, a person who is very sleep deprived will fall asleep very quickly, often in less than a minute.

We learned earlier in the prior chapter that it is possible to identify the moment of sleep onset with great precision. Sleep latency is the length of time it takes to fall asleep once a person intends to fall asleep. In a sleep lab setting it is typically measured once the subject is in bed from time lights are out to the moment of falling asleep. The MSLT measures daytime sleepiness by averaging the sleep latencies of a subject across a series of nap opportunities. These nap opportunities are usually spread out in 2 hour intervals during throughout the day. Different protocols for the MSLT have been used over time depending on the clinical question being asked.

The strength of the physiological tendency to fall asleep can be evaluated directly by measuring the length of time it takes a person to go from wakefulness to the moment of sleep. Sleep latencies are typically reported to the nearest half-minute. In many individuals, the onset of sleep can be pinpointed within a second or two. Figure 3-1 shows when the door of perception slams shut and the sleep world opens up. In this figure, the precise moment when the doors slams shut is demonstrated. In a few individuals, the transition from wake to sleep is not as sharp and clear. In these individuals longer stretches of time are required to be absolutely certain that sleep onset has occurred.

Usually people do not fall asleep if they are actively engaged in doing something that is interesting or exciting. The presence of noise and any disturbing factors will delay the onset of sleep. Obviously, someone is less likely to fall asleep in a roller coaster than in a comfortable bed. Therefore, a valid measurement of sleep latency must be made in a calm and quiet environment. Providing a monotonous environment will unmask the subject's sleepiness. During a **MSLT** the subject or patient is asked to lie quietly in a comfortable bed in a quiet darkened room. They wear comfortable loose clothing and meals are provided for them so they are neither hungry nor thirsty during nap opportunities. By providing a predictable environment the sleep latency can be more accurately recorded.

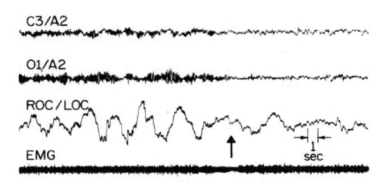

Figure 3-1: The transition from wakefulness to Stage I sleep. The most marked changed is visible on the two EEG channels (C3/A2 and O2/A1), where a very clear pattern of rhythmic alpha activity (8-12 cps) changes to a relatively low-voltage, mixedfrequency pattern at about the middle of the Figure. The level of EMG activity does not change markedly. Slow eye movements (ROC/LOC) are present throughout this episode, preceding the EEG change by at least 20 seconds. The change in EEG patterns to Stage I as illustrated here is the moment sleep occurs. (See arrow)

For all practical purposes, lying comfortably in bed without disturbing stimuli, measuring the speed of falling asleep becomes an accurate and reproducible measurement of the underlying physiological sleep tendency. In the research version of the MSLT, the sleep latency is measured every two hours. The fundamental concept is that the subject is trying to fall asleep and not willingly trying to stay awake. The lights are turned off and the subject is told to please close their eyes and fall asleep.

Brain waves are continuously recorded and scored in 30-second epochs, and the sleep latency is defined as the duration of the time interval from the lights out to the first 30-second epoch scored as sleep. Almost without exception, the first epoch is NREM Stage 1. When awake with eyes closed, most individuals' brain wave patterns show unambiguous alpha rhythm. The moment of sleep onset is associated with the abrupt disappearance of this alpha rhythm (see figure 3-1). The standard Multiple Sleep Latency Test requires five individual sleep latency measurements carried out at these two-hour intervals. The first measurement begins typically at 9 A.M. after which sleep latency measurements begin at 11 A.M., 1 P.M., 3 P.M. and 5 P.M. After lights out, the person is given 20 minutes to fall asleep. If the person does not fall asleep within those 20 minutes the nap opportunity is over, the person gets out of bed and waits for the next nap session.

When the MSLT is being used in research studies as opposed to clinical studies it is designed to measure the effect of sleep reduction on daytime sleep tendency and to illuminate the formal properties of sleep homeostasis. The ongoing test is terminated as soon as the brain wave changes indicate that sleep onset has occurred. In the clinical version of the MSLT additional time is added to measure the REM sleep which will be later discussed (see Narcolepsy chapter.) The intent is to allow no sleep to occur or, in reality, as little as possible. The subject is aroused and required to get out of bed and remain awake engaging in reading, watching TV, or another relatively quiet activity until the next test.

It is important to understand the distinction between the research and clinical protocol of the MSLT. Clearly if more than a few seconds of sleep were allowed, the accumulation of extra sleep over the course of the five individual tests could independently change the test results as the day progressed and invalidate the measurement. It must always be kept in mind that the indiviual nap sessions are terminated after the onset of sleep.

Reporting MSLT Results

An example of results from an MSLT are shown in Figure 3-2. Longer sleep latency values are plotted in the upward direction. which means that increasing sleep tendencies (shorter sleep latencies) are plotted in the downward direction. Decreased sleep latencies mean increased physiological sleep tendency and increased sleep latencies mean decreased physiological sleep tendency. The convention of plotting physiological sleep tendency is consistent with the widespread image that we go "down" into sleep, and we wake "up." The convention also conforms to the principle that increasing sleep tendency is associated with decreasing ability to perform in the waking state. The standard clock times of administering the sleep latency tests are also shown in Figures 3-2.

When physiological alertness/sleepiness needs to be described over several days, the five values of the individual nap sessions are averaged for the day. The daily averages are plotted for the MSLT profile. The MSLT profile is widely regarded as the gold standard indicator of the ongoing sleep tendency throughout the day.

Understanding the Results of the MSLT

Relating Test Scores to Real Life

In Figure 3-3, MSLT cutoff values are presented. The *"twilight zone"* designates the range of very short sleep latencies and indicates a very strong sleep tendency. This is always associated with impaired mental function and inability to maintain attentiveness. When patients with sleep apnea and narcolepsy sleep disorders characterized by persistent excessive daytime sleepiness are tested, the MSLT scores are typically within this range. Persons whose MSLT profiles are in the top "good alertness" range of the graph they are not expected to inappropriately fall asleep.

Extraneous Factors Must Not Be Allowed To Confound MSLT Results

The MSLT protocol is designed to facilitate sleep as much as possible. The dark, quiet environment minimizes external stimulation and subjects are told to lie quietly and try to go to sleep. Above all, they must not try to stay awake. In the MSLT situation, there should be no conflicting motivations. A potential source of a false negative would be somebody purposely trying to stay awake. If falling asleep is being resisted, or if people are engaging in vigorous muscular activity, their internal physiological sleep tendency cannot be accurately measured.

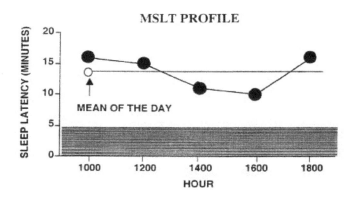

Figure 3-2: The results of an idealized Multiple Sleep Latency Test (MSLT). It is well established that the speed of falling asleep is directly related to the size of the sleep debt. However, in order for the speed of falling asleep (sleep latency) to be a valid and replicable scientific measurement, the events prior to the start of the individual test must be highly standardized and meticulously carried out.

Both physiological sleep tendency and subjective sleepiness can be affected by high motivation to remain awake. Unless the underlying sleep drive is very strong, an individual can voluntarily temporarily prevent falling asleep at almost any time simply by engaging in physical activity. However, it is important to understand that in all of these stimulating situations, the homeostatic drive to sleep will be completely unchanged. It remains the major factor in determining the strength of the tendency to fall asleep at any particular time and will express itself when stimulation is removed.

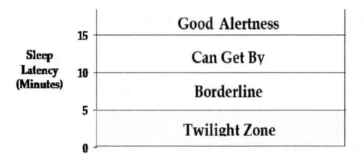

Figure 3-3: There is a general correspondence between the MSLT mean score and the overall level or degree of daytime alertness. If an individual falls asleep in less than five minutes on every test (sometimes less than a minute), this indicates a very strong sleep tendency and a very strong sleep drive. The label "twilight zone" is meaningful in this respect because memory and clarity of thinking is usually substantially impaired in individuals whose MSLT score is less than 5.

Effect of Prior Sleep Loss on Daytime Sleep Tendency

Total Sleep Deprivation

The effect of total sleep deprivation on physiological sleep tendency can be precisely measured by using the Multiple Sleep Latency Test. Although five nap sessions in one day is standard, additional sessions may be added. During a prolonged period of wakefulness, sleep latency can be measured every two hours around the clock. Assuming a well-rested individual has been awake throughout the day, sleep latencies will be long until bedtime approaches. At the time when a person normally expects to sleep, latencies will be short through the night as expected. During total sleep loss, however, physiological sleep tendency tends to remain strong (sleep latencies are very short) until recovery sleep is finally allowed. The usual robust 24-hour oscillation of physiological sleep tendency is greatly reduced in amplitude. Figure 3-4 shows the sleep latency measured every two hours throughout a prolonged period of total sleep deprivation. This study was actually carried out during the Spring Break 1976 at Stanford University with undergraduate volunteers as subjects.

When subjects are forced to stay up all night, they frequently experience a sense of feeling a little better after sunrise or breakfast. This is primarily a change in subjective sleepiness. Thus, an apparent daily rhythm in the ability to remain awake and "feel alert" may continue a day or two in the face of total sleep deprivation. However, physiological sleepiness persists as long as wakefulness continues. Motivation to succeed will influence how long subjects can go totally without sleep before giving up.

The Relation of Physiological Sleep Tendency in the Daytime to Different Amounts of Sleep the Night Before

The relationship of sleep tendency during the day to the amount of sleep at night has been studied. This relationship for young adults is illustrated in Figure 3-4, which shows the daytime sleep latency profiles that follow various amounts of sleep the night before. As the amount of sleep at night decreases, the strength of sleep tendency in the daytime increases. The relationship, however, is not quite linear. The effect is somewhat greater below five hours of sleep.

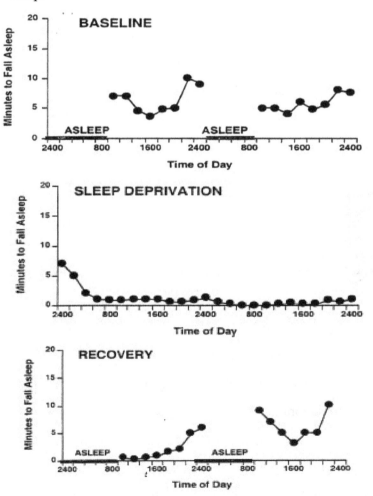

Figure 3-4: Two-night sleep deprivation study with MSLT measurements every two hours. Subjects' sleep patterns were recorded prior to the beginning of the sleep latency measurements during 8 hours in bed on two successive nights, and then subjects remained awake from 7:30 A.M. until 11:30 P.M. 64 hours later. They then slept for two consecutive recovery nights

on the same schedule: in bed from 11:30 P.M. to 7:30 A.M. Sleep latency measures were carried out throughout each day following the two recovery nights. At the first midnight test initiating the two days of prolonged wakefulness, sleep latencies began to decline and then stayed at a very low level on every test for the remainder of the period of wakefulness. The pattern of sleep latency scores after the first recovery night was completely unexpected and counterintuitive. Back then, one expected that the sleep tendency would be greatly reduced after getting up. The exact opposite was seen. Sleep latency was short after getting up and progressively lengthen during the day. After recovery night two, the typical U-shaped pattern was seen. Another noteworthy result not appreciated at the time was the overall strong sleep tendency (short sleep latencies) seen during the baseline days. We now know that undergraduates are very sleep deprived and being in bed eight hours for one night is not going to change this.

The profiles in Figure 3-5 demonstrate another important phenomenon. Sleep tendency increases in the middle of the day and decreases toward the end of the day in the absence of daytime sleep or napping. This typical U-shape pattern will be considered in more detail later.

The studies that generated these results were all performed by measuring physiological sleep tendency through the day following a nocturnal sleep period of varying amounts ranging from over nine hours to zero. The sleep latency profiles following different amounts of sleep at night illustrated in Figure 3-5 are the average results for groups of young adult subjects. Changes in physiological sleep tendency in response to sleep loss are generally consistent among subjects although there are modest individual variations. Objective measures are always more consistent than subjective measures. For example, some individual might feel nothing the day after a reduction of two hours of sleep on a single night. With five hours of sleep or less, however, individuals usually feel subjectively sleepy during much of the day, although the degree of sleepiness will fluctuate as seen in figure 3-5.

Chronic Partial Sleep Loss: The Most Important Cause of Increased Daytime Sleep Tendency

Far and away the most important aspect of the homeostatic regulation of sleep is the following simple fact: sleep loss is cumulative. When total nightly sleep is reduced by exactly the same amount each night for several consecutive nights, the tendency to fall asleep in the daytime becomes progressively stronger each day. The first study that demonstrated this effect was carried out at Stanford in 1977. Our group studied a sample of ten young adults around the clock for ten consecutive days. The subjects slept in laboratory bedrooms and all-night sleep recordings were carried out every night. Subjects were in bed from 10 P.M. to 8 A.M. on three baseline nights sleeping about 9 hours on the average. After baseline measurements, sleep in the laboratory was restricted to five hours a night for seven consecutive nights. Throughout each day, the standard Figure

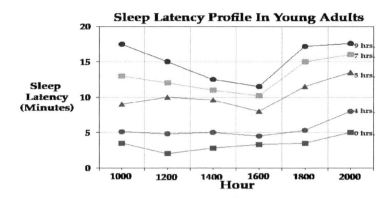

3-5: MSLT profiles after varying amounts of nocturnal sleep. This figure is based on a very large amount of data. The MSLT profiles and mean scores are directly related to the amount of sleep at night. In these tests, the scheduled time in bed varied from ten hours to zero. Generally, with ten hours in bed, total sleep time was nine hours or more. With the lower values, total sleep time closely approached total time in bed.

MSLT was administered to each subject and no napping was allowed at any time. With an identical five hours of sleep each night, the mean MSLT scores of the subjects progressively decreased on each successive day. The pooled results for the entire group are displayed in Figure 3-6a. MSLT profiles from one individual subject are presented in Figure 3-6b and 3-6c. Figure 3-6b shows a sleep latency profile from the baseline period. Figure 3-6c is the MSLT profile from the day after the fifth night of moderately restricted sleep. Comparing the profile in Figure 3-6b to the profile in Figure 3-6c demonstrates the profound build-up of physiological sleep tendency in this subject after only five nights of restricted sleep. This particular subject may have needed more sleep than some of the others and therefore, the accumulated sleep loss was effectively larger which would result in an earlier appearance of very short sleep latencies. All subjects without exception showed a progressively increasing daytime sleep tendency during the seven days of reduced sleep.

If we examine Figures 3-6a, b, and c, we will see that the MSLT has a limitation. Once an individual falls asleep in less than a minute or two on every test, there can be no further change in the MSLT though the underlying sleep drive and sleep tendency can continue to increase. This is referred to as a floor effect.

Partial sleep loss has been studied for a longer duration in the laboratory of Professor David Dinges at the University of Pennsylvania. Subjects were restricted to four hours of sleep each night for two weeks (14 consecutive nights). As in the Carskadon/Dement study, the subjects' tendency to fall asleep became progressively stronger. On the fourteenth day after the last night of restricted sleep, the impairment of the subjects had not leveled off and their sleep debt was presumably continuing to accumulate. The subjects were at their worst level of psychomotor impairment.

Although laboratory studies of large numbers of consecutive sleep restriction nights (e.g., 20 or more) have not been done, it is likely that the sleep tendency strengthens as the amount of accumulated sleep loss gets larger and larger. In the real world, many people, hospital interns and residents for example, may exhibit very severe impairment and make serious errors after prolonged partial sleep deprivation due to excessive work hours.

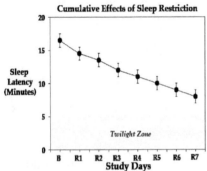

Figure 3-6a: Cumulative effects of sleep deprivation. This figure summarizes the first experiment demonstrating that sleep loss accumulates. Ten young adults participated. After three baseline measurements with nine hours in bed for three consecutive nights the MSLT was carried out on each successive day. Each point in the graph is the mean of daily MSLT scores, or the group As can be clearly seen, there is a progressive decrease in the overall sleep latency, indicating a progressive increase in daytime sleep tendency. This was the first study demostrating that sleep loss accumulates. While there is often considerable individual variation, no one has reported a single instance of the failure of sleep tendency to increase with accumulating sleep debt.

Figure 3-6b & c: These two graphs illustrate the results in a single individual during the baseline MSLT and the MSLT after only five nights of sleep restriction. The MSLT profile on Friday is well within the "twilight zone" even though the basal score is close to the level of "good alertness." Most other subjects did not show as rapid a decrease in sleep latency scores, but all showed a significant decrease.

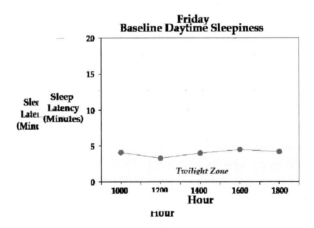

Since individuals often do not feel sleepy even when they have accumulated a very large sleep debt, they do not realize they could fall asleep at almost any moment. Such individuals have a very high risk of making a sleep-related error or having an accident.

The Flip Side of the MSLT: the MWT

After the development of the MSLT, it was apparent that the limitation of the test was not measuring the abililty to stay awake but only the ability to voluntarily fall asleep. A complimentary test was necessary. The MSTL protocol was modified to create the Maintenance of Wakefulness Test (MWT). During the MWT, a person is asked to stay awake while reclined in a quiet dark room for 20 to 40 minutes depending on the protocol. During this time, they are continuously monitored. The person is instructed to stay awake and be still (no arm flapping or lip biting allowed) These sessions are scheduled at 2-hour intervals just as an MSLT.

The MWT is now an accepted test used by the regulatory agencies such as the Federal Aviation Administration. The MSLT is often used as a diagnostic tool however the MWT is used to measure the efficacy of treatment. For example, if an airline pilot is diagnosed with sleep apnea and claims to have had successful surgical treatment, the FAA might request a MWT to prove the pilot is able to stay awake in an extended monotonous environment. Failing an MWT is a dramatic demonstration of the power of sleep even when your career depends on staying awake. Do you think you would stay awake in a quiet dark environment without falling asleep? Clearly your answer will depend on your motivation to stay awake.

Sleep Debt

From the scientific studies discribed previously in this chapter, we developed the concept of sleep debt. Every hour of sleep that an individual obtains less than his or her nightly requirement appears to accumulate by the brain as a debt. This debt appears to be precisely added up over time. It may be possible that the debt includes an hour lost a month ago or a week ago as well as an hour lost on the previous night.

The concept of sleep debt presumes that for each individual there is a specific amount of sleep which, if obtained on a nightly basis, will maintain the same level of daytime physiological alertness over successive days. It also presumes that the daily sleep need may vary somewhat from individual to individual.

In an individual who needs nine hours a night, and who sleeps six hours a night for a week, the lost sleep, or "sleep debt," would add up to a debt of 21 hours by the end of the week. In other words, the human sleep debt accumulates just like a credit card balance. This sleep debt drives the tendency of the brain to fall asleep, and the size of the debt determines the level of risk that any person operating hazardous equipment or making crucial decisions may make a disastrous error.

A large sleep debt does not go away spontaneously. It does not diffuse out of the nervous system. It can only be reduced by extra sleep. Sleep loss of only an hour or two over a number of nights can eventually lead to serious physiological sleepiness that manifests as a very strong tendency to fall asleep especially in sedentary activities such as reading this book.

Reducing the Sleep Debt and Improving Daytime Alertness

Persons who are very tired and sleepy in the daytime can reduce their sleepiness by getting extra sleep. Increasing the time spent in bed at night usually accomplishes this although sometimes naps must be added. Figure 3-7 shows the average daily MSLT profiles in subjects who were permitted to be in bed ten hours a night on four nights that followed three nights with bedtime limited to eight hours. These individuals averaged almost one and one half additional hours of sleep each extended night, and the sleep latencies are clearly increased. These results are evidence that the excessive daytime sleepiness experienced by individuals who do not have sleep disorders is primarily due to a large sleep debt as a result of insufficient sleep for multiple nights. They also demonstrate that extra sleep improves daytime alertness by reducing sleep debt. A major conclusion from these result is that 8 hours in bed does not allow the average undergraduate to obtain "enough" sleep.

Sleep researchers at the Henry Ford Hospital in Detroit have recently extended the "10 hours in bed each night" routine to two weeks (14 consecutive nights of extra sleep). In subjects who had low (sleep deprived) MSLT scores, improvement was continuous and progressive over the entire two weeks. However, even with this large "pay back" of lost sleep, the subjects still had not reached an optimal MSLT score when the experiment ended.

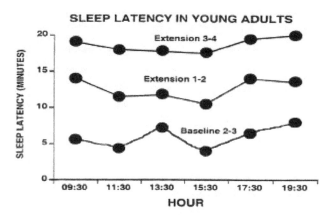

Figure 3-7: Mean multiple sleep latency profiles with sleep extension in Stanford undergraduates. MSLT scores following testing after 8 hours in bed (7 1/2 hours of sleep or more per night) generally yield lower sleep tendencies. When time in bed was extended to 10 hours generally yielding 9 1/2 hours of sleep for several successive nights, the MSLT profiles progressively improved, e.g., sleep latencies lengthened.

Further Documentation of Our Sleep Indebtedness

Two similar experiments carried out almost thirty years apart dramatically confirm the concept that human beings can unknowingly carry a large sleep debt. Neither experiment was designed by the researchers primarily to study sleep. In both experiments, brain waves were recorded continuously as one of many measurements of the effects of the experimental protocol on the volunteers.

The first experiment was carried out many years ago, at the U.S. Naval Hospital in Bethesda, Maryland. It was intended to be a test of the role of sensory deprivation on cognitive function. At that time, it was believed that a substantial reduction of sensory input could dramatically impair normal mental processes and that disorientation, hallucinations, and even psychosis might be the consequence. The study required adult subjects to lie on a comfortable cot in a cubicle where they were completely isolated from light, sound, and interactions with the outside world. The temperature was held constant, neither perceptibly cool nor perceptively warm, and the subjects wore thick gloves to minimize tactile sensations. Brain waves were recorded continuously. Subjects remained in this sensory deprivation situation for one week.

Having absolutely nothing to do, subjects generally slept a great deal throughout the first twenty-four-hour period. The mean total sleep time for the group was above 16 hours on the first day. However, the mean total sleep time declined on each successive day, and on the last (seventh) 24-hour period, the group mean was just under eight hours. The subjects in this experiment were young naval personnel. Since we now believe that nearly all young adults are chronically sleep deprived, it may be assumed that even if the subjects had maintained reasonably normal schedules prior to the experiment, they would have begun the sensory deprivation protocol carrying substantial sleep debts. We may assume that their sleep debts were caused by the enormous increase in total sleep time when there was essentially nothing else to do except sleep throughout the entire 24 hours. However, as the sleep debt was progressively reduced, the associated sleep drive progressively weakened and the total amount of sleep accordingly decreased on each subsequent day. Even with absolutely nothing to do all day long, the subjects were eventually unable to sleep more than eight hours.

This experiment can also be viewed as refuting the old theory that sensory bombardment of the brain is necessary to maintain wakefulness. From this perspective, it is an extremely negative result. Although the tremendous reduction in sensory input was maintained for the entire week, there was a large increase in wakefulness.

The second experiment, carried out by Dr. Thomas Wehr and his colleagues at the National Institutes of Health, was intended to examine the effect of different photoperiods (duration of time spent in the light each 24-hour day) on human mood and emotion. The nocturnal portion of the experiment was carried out in a laboratory setting. Sleep parameters were recorded while subjects remained continuously in bed in the dark. When they were out of bed in the light, they were allowed to go about their daily routines. The sleep of each subject was monitored for 42 consecutive nights.

During the first week, the daily photoperiod during which subjects were out of bed in the light was the conventional 16 hours, and each night they were in bed in the dark from 11 P.M. to 7 A.M. (a total of eight hours). After the one baseline week, the 16-hour photoperiod was decreased to 10 hours during which the subjects were out of bed in the light. They were consequently in bed in the dark for 14 hours every night (9 P.M. to 11 A.M.) for 5 weeks. During the six-week period, daily mood scales and a variety of other tests were administered to the subjects.

The subjects' average nightly baseline sleep time was 7 hours and 36 minutes. They were awake only 24 minutes during the 8 hours they were in bed in the dark. Such a result means that sleep efficiency (time asleep divided by time in bed) was very high (95%).

When the subjects were switched to the ten-hour photoperiod followed by 14 hours in bed in the dark, their total daily sleep times jumped to amounts above 12 hours on the first night and then gradually declined. By the fourth week of the ten-hour photoperiod schedule, total sleep time for the group had leveled off at a nightly average of eight hours and 15 minutes even though they continued to be in bed in the dark for 14 hours every night. Their mean time lying awake in the dark doing nothing was 5 hours and 45 minutes! The sleep efficiency dropped to 59% (which must have felt very inefficient to them)

The interpretation of these results is that the subjects entered the protocol carrying sizeable individual sleep debts. Of course, neither the subjects nor the researchers were aware of such a possibility. The baseline period of eight hours in bed certainly did little to reduce the subjects' sleep debts and may even have resulted in a small increase. As in the Navy hospital experiment, when the opportunity to sleep was greatly increased, their large sleep debts caused a very large increase in total daily sleep time. As the subjects' sleep debts decreased, total sleep time per day declined proportionally.

If we assume that the "steady state" value in the last week of the 10-hour photoperiod (14 hours in bed in the dark) schedule represented the real daily sleep need for this group of subjects, we may conclude that all amounts of sleep above these values represented "extra" or "make-up" sleep. Accordingly, the mean "pay back" of sleep debt for this group of volunteers averaged about thirty hours. For purposes of comparison, individuals who were not at all sleep deprived would have to spend more than three consecutive days with no sleep at all to accumulate a sleep debt of similar magnitude.

Another very important result of this experiment was a dramatic improvement in subjects' mood, energy level, and sense of wellbeing as indicated by daily checklists and questionnaires. Figure 3-8 shows the nightly amounts of sleep for one typical subject during the 14-hour in-bed schedule. These results show that although we can accumulate sleep debt, we cannot bank alertness.

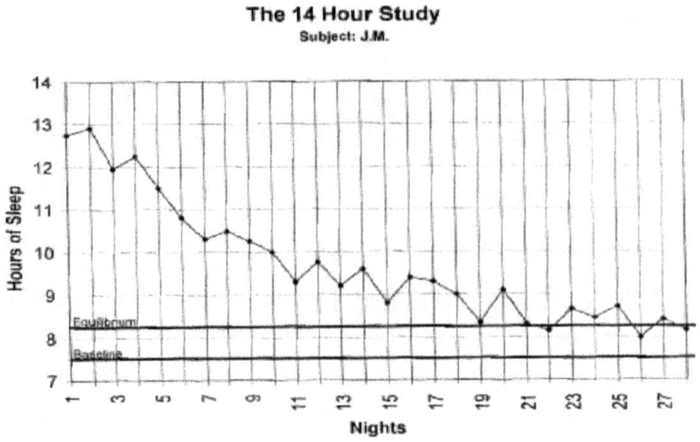

The 14 Hour Study
Subject: J.M.

Figure 3-8: Typical subject in the 14 hours in bed study. The last seven nights of the experimental period are not shown because there was no further change in total sleep time. Total sleep time on the first four nights of the 14-hour schedule is 12 hours or more. There is an asymptotic decline in total sleep time, which levels off at about night 21.

Sleep Debt Consequences

The scientific experiments described in the preceding section demonstrate that individuals, who are obtaining what most people would deem to be normal amounts of sleep, can be carrying a large accumulation of sleep indebtedness which they are not reducing. How long such indebtedness would persist without change if no extra sleep was obtained is not known. However, it is obvious from the 14 hours-in-bed, multi-week study that to be able to sleep 30 extra hours and to have accumulated such a debt in small increments means that sleep indebtedness must persist for substantial amounts of time — weeks or months at the very least. In addition, this second experiment and others like it present evidence that a large sleep debt impairs our mood, our sense of well-being, our energy, and our intellectual function. This means that the negative consequences of chronic sleep deprivation are not just nodding off, i.e., having microsleeps, at critical moments. There is also a general and unrecognized global impairment of performance, attentiveness, and mood. We have long known that sleep can affect classroom behavior and mood in children. There is solid emerging data that sleep deprivation is associated with suicide ideation and suicidal behavior. A study of over 8,000 teenagers in Korea found that about 20 percent of them felt sleepy in the daytime and were getting only seven or less hours of sleep on school nights while they habitually slept in more than two hours on the weekend. These teenagers were much more likely to have suicidal behavior. A study in 2012 of over 6,500 teenagers in the U.S. found that sleep problems predicted suicidal behavior independent of depression or substance abuse.

This effect is not limited to adolescents. A meta-analysis of over 140,000 adults and teenagers confirmed that suicidal behavior is 2-3 times more likely in people suffering from sleep problems. Again, these findings were independent of depression. A possible explanation for this problem comes from a 2013 study using fMRI technology, which demonstrated imbalances in brain regions of adolescents with sleep disturbances which were associated with more risk-taking behavior. The last thing we need is for our sleepdeprived drivers to take more risks!

In view of this, it is reasonable to hypothesize that a major improvement in our health and function could be achieved by reducing our sleep indebtedness to a very low level. With a very low sleep debt, fulfilling our sleep requirement each night will be associated with optimal alertness and energy all day every day.

Sleep deprived undergraduates who obtain extra sleep for one or two nights (a typical weekend) probably may not experience a spectacular improvement in how they feel because they will have accomplished only a small reduction of their large sleep debt. This might be very discouraging to individuals who are trying to "catch up on their sleep" and achieve "good sleep habits" unless they have a clear understanding of sleep debt, how large it can be, and how it can be reduced. Like any other debt, it is better to stick to a budget and make partial payments rather than ignoring it and letting it accumulate.

Interrupted Sleep: The Concept of "Minimal Units"

The concept of "minimal units" of sleep has evolved in recent years as a result of both clinical and experimental observations. In patients with obstructive sleep apnea, sleep is interrupted many times as patients must breathe. In severe cases, such interruptions may occur every minute. Such a patient may awaken over 500 times during an eight-hour night in bed and never sleep continuously for more than a minute or two. Even though the hundreds of brief fragments of sleep add up to nearly eight hours, such patients are so sleepy in the daytime that they are literally zombies.

Sleep specialists feel that the fragmentation accounts for the daytime sleepiness typically experienced by patients with severe sleep apnea and several other sleep disorders. To test this hypothesis, laboratory studies which replicate the sleep disturbance seen in sleep apnea patients have been carried out using normal human volunteers. In these studies, basal levels of daytime alertness were measured using the MSLT. On the following night, volunteers were awakened every minute or so. They generally returned to sleep immediately and the duration of the arousal was no more than several seconds. Even though subjects were awakened hundreds of times, usually only a few awakenings were recalled in the morning. Although very fragmented, the accumulated total sleep added up to a solid eight hours. Despite getting 8 hours of sleep, MSLTs during the days following the experimental nights indicated that the level of daytime alertness was impaired.

Interrupting sleep every minute or two throughout the entire night is a difficult experimental manipulation. Nonetheless, this research has been replicated several times and it is certain that the restorative value of sleep is severely curtailed if sleep periods are not allowed to continue for at least several minutes. When at least ten or more minutes of sleep are allowed prior to an awakening, the restorative value appears relatively maintained.

The Daily Sleep Requirement

"Sleep Need" As Measured by the MSLT

It is possible to determine the sleep requirement of a single individual with precision. We have seen that reducing nightly sleep by a relatively small amount leads to a progressive increase in physiological sleep tendency during the day. We have also seen that increasing nightly sleep by a relatively small amount leads to a progressive decrease in daytime sleepiness. Therefore, if measurements of daytime sleep tendency do not change on successive days, we may conclude that the amount of sleep at night is the precise sleep requirement of that particular individual, neither more than is needed nor less than is needed. In summary, by utilizing the MSLT and measuring the mean daily value over successive days, the specific sleep requirement of an individual can be determined.

In the sleep laboratory, the procedure begins by scheduling sufficient time in bed to allow the amount of sleep an individual thinks he or she needs. The MSLT is administered on each successive day.

- If the mean daily MSLT value does not change over a reasonable period of time (7 or 8 consecutive days), we can conclude that the amount of sleep being obtained is the amount needed, or very close.
- If the daily mean MSLT score progressively decreases, the amount of sleep being obtained is less than the individual needs.
- If the daily mean MSLT score progressively increases (the sleep latency is getting longer and the sleep tendency is getting weaker), the amount of sleep being obtained is more than the individual's daily requirement.
- If the MSLT is changing very slowly, then the amount of sleep is close to the amount of sleep the individual needs.

Figure 3-9 (a-c) illustrates the measurement process. All other things being equal, it is better to start the daily measurements with an initial MSLT value for the first day which falls within the five minute to fifteen-minute range. If the first day MSLT is well into the twilight zone, it is possible that a marked increase in nocturnal sleep depth and sleep efficiency would confound the results somewhat. If the mean MSLT score is at the ceiling (20 minutes), the individual might have some initial difficulty sleeping in a laboratory setting. However, carrying out this long-term measurement of sleep need is essentially independent of the initial level of the mean daily MSLT. It is this type of evidence that supports the assumption that each person has a specific sleep need. Theoretically this protocol can be carried on for many days, and if the daily sleep required were consistently fulfilled daily MSLT should/would be more of less the same. To carry out such prolonged and demanding lab measurements for a large number of subjects however, would be impractical.

If a person could remain in the sleep laboratory for several weeks, the sleep requirement could be determined to within at least the nearest quarter hour. It would be an amount, finally, which if obtained every day, would produce exactly the same mean MSLT score over time. That is, there would be no significant change. A small day-to-day variability might be seen, but a running average of seven or eight days should show no change at all. Newer technology has become available to carry out sleep studies in a person's home. Also, the introduction of smart phone apps that measure sleep could be a value in these situations.

Is Sleep Vitally Necessary?

It should also be very clear that the homeostatic need for sleep is not necessarily the same as the vital need. The latter refers to a hypothetical function or functions which can only take place during sleep and which presumably must take place during sleep to preserve our lives. However, such hypothetical vital functions may require much less time to be accomplished than the homeostatic requirement. If this is true, why would the homeostatic requirement for sleep be a much larger amount than the amount required to fulfill the vital need? A possible answer to this question may be found by invoking evolution and survival. It is adaptively advantageous to sleep a great deal because the need for calories is least during sleep, and less food is therefore required. It is also adaptively advantageous to have a drive or motivation to find and stay in a safe place during the dark phase of the earth's rotation. Human beings and other primates depend almost entirely on their visual senses to guide them around in their environment.

(Consequences of sleep debt and falling asleep at the wheel.)

As we have seen, losing sleep increases the strength of the overall tendency to fall asleep, and obtaining extra sleep decreases the tendency to fall asleep. If someone's homeostatic need for sleep is eight hours, then sleeping only six hours for many nights would inevitably result in a very large sleep debt with severe tiredness and impaired function in the daytime. If this person obtained an hour of extra sleep per night for many nights, the daytime tiredness and impairment would finally disappear. Figure 3-9 (a-c) illustrates the measurement process. All other things being equal, it is better to start the daily measurements with an initial MSLT value for the first day which falls within the five minute to fifteen-minute range. If the first day MSLT is well into the twilight zone, it is possible that a marked increase in nocturnal sleep depth and sleep efficiency would confound the results somewhat. If the mean MSLT score is at the ceiling (20 minutes), the individual might have some initial difficulty sleeping in a laboratory setting. However, carrying out this long-term measurement of sleep need is essentially independent of the initial level of the mean daily MSLT. It is this type of evidence that supports the assumption that each person has a specific sleep need. Theoretically this protocol can be carried on for many days, and if the daily sleep required were consistently fulfilled daily MSLT should/would be more of less the same. To carry out such prolonged and demanding lab measurements for a large number of subjects however, would be impractical.

Optimal Sleep/Wake Schedule It should be very clear that individuals with a large sleep debt will continue to feel fatigued each day even when they are obtaining their nightly sleep requirement. Moreover, they will feel fatigued and sleepy the next day even when extra sleep was obtained the night before. To re-emphasize, getting the amount of sleep one needs each night does not ensure optimal alertness in the daytime. Thus, when a person who is in the "twilight zone" because of prior sleep deprivation begins fulfilling his or her daily sleep requirement, optimal daytime alertness is not likely to be achieved because there is no "pay back." The sleep debt is not being reduced. Enough extra sleep must first be obtained so that the sleep debt is very small. Once this debt reduction is accomplished, obtaining the sleep one needs each night will maintain high daytime alertness; this would be the optimal sleep/wake schedule. Figure 3-10 recapitulates the discussion of our daily sleep requirement and how obtaining it might not necessarily lead to progressively greater daytime alertness. Clearly, if the amount of sleep at night is more than the daily requirement, the tendency of fall asleep in the daytime will progressively decrease (sleep latency progressively increases). It should be very clear that it is not necessary to feel fully alert in the daytime for the amount of sleep at night to be the amount needed by individual. However, if one obtained extra sleep over a period of time and the daytime tendency to fall asleep became very weak or not existent, the optimal alertness could be maintained with nightly amount of sleep somewhat less; i.e., no extra sleep. The standard MSLT is terminated at 20 minutes, which is good enough for most situations. However, if the sleep debt is very small, latency of 30 min or more would presumably be obtained if the length of the test were increased.

In summary, achieving an optimal sleep/wake schedule requires two steps: (1) first, your current sleep debt must be reduced to as low a level as possible, ideally several hours from zero; (2) second, you must maintain a bedtime schedule which will allow you to fulfill your personal sleep requirement every night, or almost every night.

a.

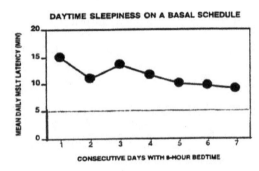

Figure 3-9 abc: These graphs depict mean MSLT scores each day following seven successive nights on a basal sleep/wake schedule.

Time in bed is generally 9 hours with total sleep. Accordingly, the MSLT score on day 1 is generally between 5 and 15 minutes.

In the result depicted in a., the mean MSLT score tends to shorten suggesting that the amount of sleep obtained is less than the b. amount needed by this individual.

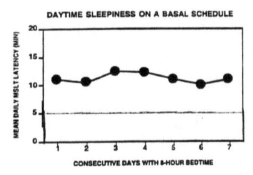

In graph b., the MSLT score does not show a statistically significant change over time. This suggests that the amount of sleep obtained each night is very close to the daily sleep requirement of this individual. c.

mean MSLT score progressively lengthens individual's daily sleep time in this situation is requirement, hence alertness is improving. It is that supports the assumption that each person has a

The Sleep Homeostat Revisited

We now know that accumulating a

In graph c., the daily suggesting that this more than the daily this type of evidence specific sleep need.

large sleep debt is a by-product of the fundamental biological mechanism of sleep homeostasis. Figure 3-11 depicts the concept of a sleep homeostat in the brain. We are certain that the brain houses the neuronal machinery for carrying out the homeostatic regulation of sleep. The precise mechanism for this is unknown. A number of years ago, scientists were in an analogous situation with regard to the biological clock. Everyone knew there was a time keeping mechanism in the brain, but no one knew where it was located. All sorts of possibilities were proposed until the specific location of the biological clock was finally identified. It is housed in the suprachiasmatic nuclei within the hypothalamus.

Not surprisingly, the homeostatic process that we are discussing is continuously active. There have been numerous studies showing that sleep tendency increases progressively throughout any period of wakefulness. This means that all wakefulness is sleep deprivation and is continuously adding to the ongoing sleep debt. During sleep, the homeostatic process continuously and progressively reduces sleep tendency.

If this is true, why would the homeostatic requirement for sleep be a much larger amount than the amount required to fulfill the vital need? A possible answer to this question may be found by invoking evolution and survival. It is adaptively advantageous to sleep a great deal because the need for calories is least during sleep, and less food is therefore required. It is also adaptively advantageous to have a drive or motivation to find and stay in a safe place during the dark phase of the earth's rotation. Human beings and other primates depend almost entirely on their visual senses to guide them around in their environment. Obviously, we are speaking about the distant evolutionary past, perhaps even before human beings could control fire.

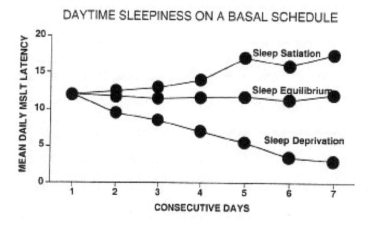

Figure 3-10: A schematic representation of the operational definition of daily sleep requirement based upon the knowledge that sleep drive is regulated.

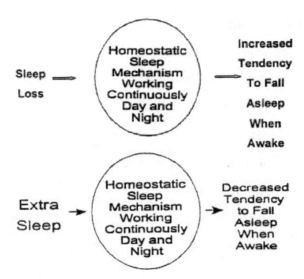

Figure 3-11: The sleep homeostat. This figure is a schematic of the hypothetical machinery underlying the homeostatic regulation of sleep and wakefulness. The model requires the following assumptions: (a) All wakefulness is associated with accumulating sleep debt. (b) The homeostatic process must operate continuously. (c) It must have some sort of time keeping mechanism. (d) Each individual's sleep homeostat has its own "set point" which is the amount of sleep required each night to pay back the sleep debt accumulated during the previous day's wakefulness.

In spite of the widespread belief that people can get too much sleep, there is also no evidence whatsoever that the effects of sleep deprivation are not always reversed by extra sleep. In other words, no sleep-deprived human being has ever been found whose alertness and energy could not be restored by obtaining extra sleep. The sleep homeostat, wherever its location, must be an extremely robust and stable neural process. The potential implications for research should the location of the sleep homeostat finally be discovered are very exciting.

Summary

The material in this chapter has covered sleep homeostasis. It has introduced the concept of sleep debt to account for the consequences of multiple nights of inadequate sleep. It is supremely important that most people do not understand why they are tired all the time. They may think they are depressed, anemic, or hypothyroid. Or, they may be misdiagnosed with "chronic fatigue syndrome". Such erroneous thinking means that there is no strong motivation for students and other individuals to determine their personal sleep requirement and by consistently fulfilling it, to experience the wonderful benefits of maintaining a small sleep debt. These benefits are peak performance, high energy, optimism, and being wide-awake and fully alert all day long.

CHAPTER 3 Sleep Quality & Sleep Debt

"What wakes you up may not be what keeps you awake." - Dr. Christian Guilleminault

CHAPTER 4 Alert Wakefulness from Inside Ourselves: The Biological Clock

"...the rules we postulate today may be revised tomorrow, always based on data."- Franz Halberg

Vignette #1 : When would you like you parents to visit you?

Some years ago, I had the chance to observe a completely unambiguous and beautiful episode of clock-dependent alerting at work on mood and energy. It occurred while I was visiting my daughter Cathy who was then a freshman student at (sic) the Stanford of the East. We were sitting in her dormitory room after a late lunch. I was attempting to satisfy my parental curiosity about how she was doing, what classes were most interesting, how her social life was going, and so on. Her monosyllabic, almost sullen answers, conveyed no worthwhile information. Indeed, she was almost alarmingly apathetic and seemed to be totally uninterested in my visit. I was even beginning to ask myself why I had even bothered to make the trip. Since anything would be better than the non-conversation we were having, I suggested we take a walk. It was a sunny May afternoon and now about 4 P.M.

After we had strolled silently along the banks of the Charles River for about half an hour, Cathy began to speak a little. Over the course of perhaps 20 to 30 minutes, she metamorphosed into a talkative, informative, smiling, completely vivacious person. This continued through dinner until I departed at about 8 p.m. We had a great time, and I learned all I wanted to know about how things were going and much more. Had I departed at 4 P.M., it would have been a lousy visit. But Cathy was still at an age when even a severely sleep deprived college student will have such a strong late afternoon/evening period of clock-dependent alerting, that it was able to abolish her fatigue completely. This marvelous transformation occurred in the complete absence of any outside stimulation; it was entirely internal and spontaneous.

-William C. Dement, The Promise of Sleep, 1999, pp 279-280

Introduction to Clock-Dependent Alerting

Many individuals have read or heard about "circadian rhythms" and the "biological clock," but relatively few have a clear idea about how this physiological system actually works. This ignorance influences how we handle things such as jet lag and shift work. The main thing to understand is that your brain houses a very precise internal timekeeping mechanism whose function is to make certain things happen at the same specific time in each successive 24-hour period. For example, certain hormones are secreted at approximately the same time each day and these secretory episodes are mediated by the biological clock. Another example of a circadian rhythms is the daily oscillation in body temperature. In the absence of other influences on body temperature, the peaks and troughs of core body temperature will occur at the same time each day.

In a regular environment, the circadian oscillation has a period length of 24 hours. This is because it is coupled to the light/dark cycle created by the earth's rotation. If a human being is completely isolated from the environment and has no cues whatsoever about the time of day (as in an underground bunker or cave without clocks, radios, or TV), the period of the circadian oscillation will usually lengthen slightly. When this happens, the circadian rhythm is called **free-running**. The free-running period may be slightly different from person to person. In this chapter, we will deal mainly with the role of the biological clock when its daily oscillation is precisely coupled to the 24-hour periodicity of our environment which is full of cues about time of day. The most important environmental factor in entraining our biological rhythms is the daily cycle of light and darkness.

Although our sleep debt and our drive to sleep can be opposed by many factors, the main opposition comes directly from the biological clock in our brains. It comes in the form of a daily alerting process which maintains and consolidates daytime wakefulness in humans and other animals, and enhances our level of alertness while we are awake. We have named this alerting process **clock-dependent alerting**. If our sleep debt is steadily increasing throughout the daytime and is not entirely eliminated by a single night of sleep, why do we wake up? It is also clear that the sleep drive dominates the biological clock during the night. This is why it is so difficult to stay awake all night long.

At the beginning of the 1960s, a London based

organization called "the Association to Preserve the Flat Earth" was still active. This society was officially disbanded when photographs of the earth from space were made public unequivocally showing the Earth to be round. Many photos show both day and night on Earth. The fact that the environment changes continuously and rhythmically from light to dark, noisy to quiet, and hot to cold every 24 hours has biological consequences. The behavior of nearly all species is organized on a 24-hour basis. Primates who depend disproportionately on vision for directing their behavior are relatively helpless in the dark. Other species, e.g., rodents, which depend less on vision and more on smell, hearing, and touch, are safer in the dark. Remaining in a safe place during the night was adaptively advantageous for humans who would be easy prey for predators if they wandered about in the dark. In addition, remaining quiet in a safe place for sizeable portions of time each day had the advantage of conserving calories. In addition, there is a consistent daily pattern in which an increased tendency to fall asleep occurs at about 2 P.M. to 3 P.M. in the afternoon often accompanied by increased fatigue and waves of drowsiness. Since this dip in alertness occurs after the conventional lunchtime, the increased sleepiness is wrongly assumed to be caused by food. Remember as we mentioned earlier, breakfast doesn't make you sleep, and we perk up after dinner. Research studies have conclusively shown that the afternoon increase in sleep tendency will occur whether or not food has been ingested.

The Location of the Biological Clock

The biological clock, the fundamental pacemaker of the entire circadian timing system has been localized to two tiny bilateral groups of nerve cells in the brain named the **suprachiasmatic nuclei (SCN)**. The SCN is located within the hypothalamus at the base of the brain just above the point where the optic nerves cross on their way to the visual thalamus. Figure 4-1 is a schematic sagittal view of the anatomical landmarks. The clock is located in the perfect position to get info from both eyes. Identifying the brain locus of the biological clock was a major breakthrough in sleep and biological rhythm research.

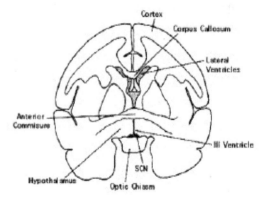

Figure 4-1: The SCN is located in the brain just above the point where the optic nerves from the eyes cross en-route to deeper brain structures. This cross section of the brain shows the location

This knowledge has allowed the SCN to be selectively lesioned in research animals. When the SCN is selectively lesioned, all circadian rhythms completely disappear. Clockless animals appear otherwise normal, and indeed rats and mice in whom the SCN has been removed have lived out

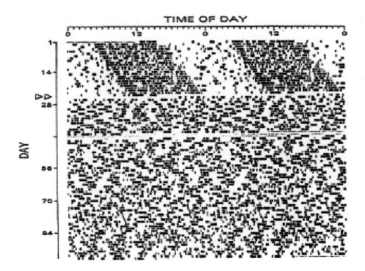

Figure 4-2: Raster plot of a rodent's sleep and wakefulness during constant conditions. It is a convention in biological rhythm research to plot data (usually wheel-running) in successive 48hr bouts. This is also called a "double plot." The dark bars are wakefulness. A predictable circadian periodicity is present in days 1-26. At the point indicated by the triangles, the SCN is eliminated.

normal lifespan in the laboratory. Figure 4-2 illustrates the profound changes in daily sleep and wakefulness patterns seen before and after the SCN is destroyed.

The SCN can also be completely removed from the brain as a tissue slice preparation. When bathed with the proper nutrients in a petri dish, this "clock in a dish" preparation maintains its normal circadian oscillation for several 24-hour cycles. More than that, the isolated clock continues to show a characteristic "re-setting" response to certain neurotransmitters dripped into the petri dish at the specific times when the phase shifting mechanism is responsive. This research will be further described in this chapter.

Vignette #2: I've had many assistants.. I recently observed another dramatic example of reversing sleepiness. It occurred in my own office. I had just hired a young Stanford graduate as my number one assistant. Our desks were at opposite ends of a single large room. During the afternoon, he often looked very tired. He maintained a conventional schedule, arriving at 8 A.M. and leaving around 5 P.M. each day. On this particular occasion, the time was around 4 or 4:30 P.M. and I was very irritated by the disorganization and clutter in the office. I was pushing my young assistant hard to go through piles and get things filed out of sight when it became clear that I was pushing too hard. He looked so tired that I said, "Never mind," and returned to my desk. About thirty minutes later, he suddenly was standing over me, demanding to know where to file three or four documents. "We've got to get these things done," he muttered. He was moving quickly around the office and almost quivering with energy. His eyes were bright, he was talking faster and more precisely than usual, and of course it dawned on me. "Do you realize what has happened?!" I asked. "Your clock-dependent alerting has finally kicked in." He hadn't taken a cold shower, he had drunk no coffee, he hadn't taken amphetamines. Responding to an unseen force, he had become more wide awake and alert than I had ever seen him. I suppose this change ordinarily took place while he was driving home, and had probably prevented an accident many times. Those of us who hire students do not realize our loss. Just as their brains are starting to work at peak alertness, office hours are over. I wonder how many of my fellow professors have any clue that this is happening. Since this revelation, I tried to have my assistant spend more time in the office after 5 P.M. We arranged a schedule in which he would come to work at 10 A.M., thus catching up on his sleep in the morning, and I had him do personal errands and things requiring very little energy and intelligence in the late morning and after lunch. Then as my day was winding up, his late afternoon enthusiasm motivated me to work another hour and accomplish quite a few additional tasks. This situation proved to be an exceptional schedule for us, but I was left with a feeling of how conventional schedules, sleep debt, and clock-dependent alerting deprive other employers or cheat them from getting their money's worth.

- William C. Dement, The Promise of Sleep, 1999, pp 86-87

What Keeps Us Awake?

In experimental animals, whose suprachiasmatic nuclei have been completely eliminated, the daily circadian rhythm of sleep and wakefulness disappears. It is replaced by many short periods of wakefulness and many short periods of sleep with no internally driven tendency to cluster at any time of the day or night (see Figure 4-2). It is extremely important, however, to point out that research animals without a clock remain able to transition from wake to sleep and from sleep to wake in a completely normal fashion. In addition, if clockless animals are continuously and vigorously stimulated, they can be kept awake for much longer periods of time than usually occur when they are left alone.

The arousing effect of external sensory stimulation is entirely separate from clock-dependent alerting. In marked contrast to clock-dependent alerting, however, the effects of external stimulation are usually quite transitory. If we are being kept awake in the middle of the night, as soon as the stimulation is removed we will immediately fall asleep. Similarly, external stimulation can keep clockless animals awake for longer durations than occur spontaneously, but as soon as the stimulation is removed, they will immediately return to sleep. Once asleep, they can, of course, be reawakened by sufficiently intense stimulation.

"The Biological Clock is an Alarm Clock"

In 1984, Dr. Dale Edgar, a sleep researcher at Stanford University, obtained results following elimination of the biological clock (SCN) that no one else had ever reported or emphasized in any way. Several previous studies of sleep and wakefulness had been carried out in nocturnal rodents with suprachiasmatic lesions, and no significant changes in total amount of sleep were seen. Dr. Edgar's research utilized a diurnal (awake in the day, asleep at night) primate, the squirrel monkey. Day/night differences in sleep/wake consolidation are generally very much greater in primate species than in rodents. Indeed, in squirrel monkeys almost the entire amount of wakefulness is consolidated into the daytime, and conversely most of the sleep is consolidated into the night. This high degree of sleep and wake consolidation also persists in normal monkeys in an environment where influences related to time of day are entirely eliminated (see Figure 4-3).

In Edgar's study, the monkeys were housed individually in sound insulated cubicles maintained at a constant temperature and with continuous dim light illumination. Food and water were always available. Under these constant conditions, the free-running circadian period was a few minutes longer than 24 hours.

Following the SCN lesions, the circadian alternation of consolidated sleep and consolidated wakefulness completely disappeared giving way to numerous brief bouts of wakefulness alternating with equally brief bouts of sleep (see Figure 4-4). In addition, there was a very large increase in total daily sleep time which was four to five hours in a single circadian period (see Figure 4-5). Furthermore, the marked reduction in total wake time (increase in sleep time) and wake consolidation in the activity phase of the circadian cycle was in no way compensated by an increase in time awake in the remainder of the cycle. This dramatic reduction in total wakefulness persisted for days in the constant dim light environment.

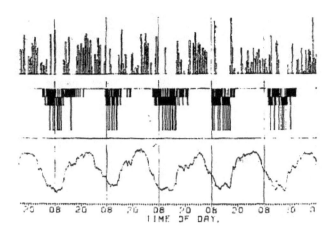

Figure 4-3: Normal squirrel monkey in constant dim light. Data from one squirrel over 5-6 successive days are depicted in this figure. A robust circadian (near 24 hour) fluctuation of body temperature is obvious and the plot of sleep stages (vertical bars) is highly consolidated. The absence of sleep (wakefulness) is also highly consolidated

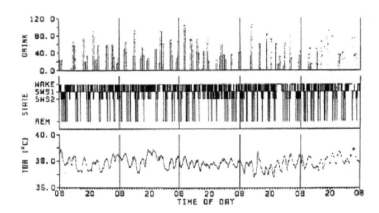

Figure 4-4: The same squirrel monkey without a biological clock. There is a very obvious disappearance of consolidated sleep and wakefulness. The monkey keeps falling asleep. The circadian oscillation of body temperature is also eliminated

The foregoing studies on squirrel monkeys suggest that the major effect of the biological clock on the regulation of sleep and wakefulness is the promotion and consolidation of the waking phase. As noted earlier, this effect is called "clock-dependent alerting." The biological clock is an alarm clock; its sole or major function in the regulation of sleep and wakefulness is to induce and maintain the waking state. For social animals, it is highly essential we predict when we will be awake. Students should ponder the effect on their daily lives if they had to sleep 12 hours every day. Further, falling asleep every hour would fragment one's daily existence; for all intents and purposes it would be no existence at all.

Figure 4-6 shows the pooled data plotted in terms of percentage of wakefulness in each successive 30-minute interval. This figure should be examined carefully with the phenomenon of clock-dependent alerting in mind. This internal alerting process forces the monkey to remain awake for large amounts of time although there is absolutely no stimulation in the environment and there is absolutely nothing for the monkey to do. Then the awake phase gives way to a similarly consolidated sleep phase. The sleep debt is reduced, and as the clock-dependent alerting turns on, there is no sleep debt (or a very small sleep debt) to oppose its action.

F i g u r e : 4 - 5:
Total sleep time. This figure shows the total sleep time in normal monkeys and in clockless monkeys. There is a consistent and dramatic increase in sleep time of about 4 hours per day. LL is light light, which means constant light exposure without dark phase.

What Makes The Clock Tick?

How does the SCN actually take light and transforms it to a change in our behavior? The molecular basis of circadian rhythm generation been revealed to involve an intricate interaction between light and our DNA. This interaction is based on transcription-translation feedback loops comprising a number of "clock genes" and their protein products. The clock genes have quasi-descriptive names such as Period (per) and Timeless (tim). The protein products of these genes are called PER and TIM. Perhaps the most whimsically name circadian gene is **CLOCK**, which stand for **Circadian Locomotor Output Cycles Kaput.** The ticking of our biological clock is based on the amount of time it takes for the DNA in these genes to make messenger RNA. The messenger RNA is translated into proteins in the cell cytoplasm. These protein products reenter the cell nucleus and feedback on the DNA to regulate the clock genes. Variability in these genes influences our individual differences in our circadian rhythms. The biological clock synchronizes to environmental light–dark cycles through direct retinal projections via the optic nerves to the SCN. The retina has a subset of photoreceptors, which have been identified as having unique characteristics for the circadian system. This subset of receptors contains the photopigment melanopsin.

Most, if not all, tissues and organs of the body appear to contain the core molecular circadian clock machinery, such that the molecular clock in the SCN entrains similar molecular clocks throughout the body to ensure overall internal temporal organization. With hundreds and even thousands of genes exhibiting rhythmic expression in any given organ system, under the control of both the central SCN and local self-sustained circadian oscillations, the circadian clock system must have a central role in regulating cellular function.

Absent Biological Clock Has No Effect on Sleep Homeostasis

Though it mediates the daily circadian timing of sleep and wakefulness, the biological clock does not play a necessary role in the occurrence of sleep onset or in waking up. We have noted that animals with complete SCN lesions are able to fall asleep and awaken from sleep in an apparently normal fashion. If the clockless animals are kept awake by artificial stimulation for long periods, they get very sleepy and the subsequent period of sleep is lengthened. In addition, the increase in sleep appears to be about the same as the induced increase in prior wakefulness. We may conclude that sleep homeostasis, the process that we learned keeps track of the amount of sleep and regulates the drive for sleep, remains completely normal in animals without a clock. This means the biological process which mediates homeostatic regulation of sleep is separate from the area which mediates clock-dependent alerting. When the biological clock is successfully eliminated in an animal, the homeostatic sleep process remains normal and can then be studied in isolation.

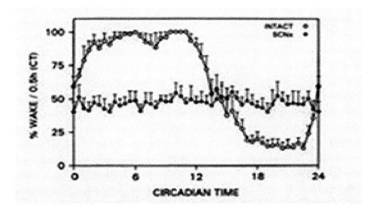

Figure 4-6: Distribution of wake time in successive 30 minute epochs throughout the circadian cycle. The pooled data for intact squirrel monkeys and for the same monkeys after discrete lesions of suprachiasmatic nuclei are plotted in this figure.

The Extended MSLT Demonstrates the Clock

In the standard MSLT (recall Chapter 3) utilized in sleep clinics and laboratories, individual sleep latency measures are carried out only in the daytime. Obviously, sleep latency can be measured at any time of the day in non-standard protocols if required by the experimental design. Non-standard MSLTs have yielded important information which has advanced our understanding of the daily cycle of sleep and wakefulness in humans. In one such non-standard MSLT protocol, the standard five tests are carried out, but two additional sleep latency measures are added at 8 P.M. and 10 P.M. When this is done, we have an extended MSLT consisting of seven individual sleep latency measures in a single day which evaluates sleep tendency in the evening as well as the daytime. The extended MSLT has yielded extremely important results in young adults. Recall that in the research version of the MSLT, all individual tests are terminated the instant the subject falls asleep. In nearly every young subject (puberty to early 20s), sleep tendency is found to decrease in the evening without additional sleep. Be very sure you understand this. Figure 4-7 is a plot of an extended MSLT constructed from the pooled data of ten subjects. Their individual results were quite similar. The data shows that the tendency to fall asleep decreases dramatically in the later tests even though no sleep occurs. This is a highly consistent result in normal young adults and is robust evidence for clock-dependent alerting opposing and reversing the strong sleep latency that is seen in the 2 P.M. and 4 P.M. tests. This result is counterintuitive for the unenlightened in every respect. Clearly, if sleep debt is large enough to produce sleep latencies near or in the twilight zone at 4 P.M., the process that reverses this tendency later in the day must be very strong.

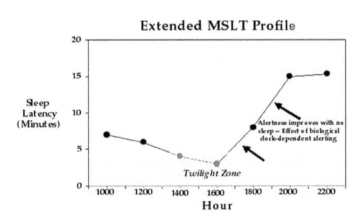

Figure 4-7: Clock-dependent alerting. This figure depicted the pool results of standard MSLTs with two additional measurements (at 8 P.M. and 10 P.M.) As in all other instances, the individual sleep latency measurement is terminated as soon as the onset of sleep is detected. Alertness improves with no sleep. This is the effect of biological clock dependent alerting.

Since there is no sleep we must assume that the ongoing sleep debt is unaffected and continues to build up while individuals are awake in the evening. Most people think their fatigue and sleepiness has dissipated. However, around midnight or later when clock-dependent alerting subsides, sleep can come very quickly and without warning. If the sleep debt is very large, the biological clock dependent alerting may be overwhelmed and a short sleep latency may be obtained even in the evening. The opponent process model explains the apparent contradiction that individuals can feel more alert late in the day even when they have had no daytime sleep whatsoever.

Figure 4-8 is a plot of an extended MSLT from a single individual. We now know that the decrease in sleep tendency (increase in alertness) in the late afternoon and evening is due to strong clock-dependent alerting. The sleep drive has been steadily increasing during the day and has reached a high level in the evening. Nonetheless, the internal stimulation from the biological clock successfully opposes this increased sleep tendency. Since we know that the sleep debt is continuing to build up even while sleep tendency is decreasing, we can infer that the evening clock-dependent alerting is very strong because the interaction with the increasing drive to sleep nonetheless produces a net decrease in the tendency to fall asleep.

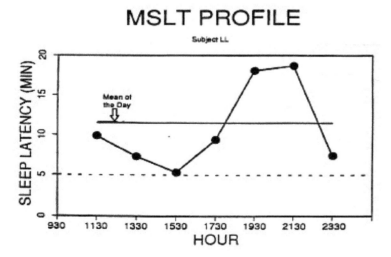

Figure 4-8: An extended MSLT from one individual on one day. This graph shows the typical decrease in sleep latency late in the day, but also shows the rapid increase in sleep latency when clock-dependent alerting begins to subside and the even larger

sleep debt is unopposed. Readers should reflect that one might feel wide awake and alert (as in the graph) at 7:30 P.M. and be unable to stay awake at 11:30 P.M. while driving home.

Summary

The biological clock is an alarm clock located inside our brains. It directly promotes alert wakefulness. It does not directly promote sleep. Since sleep homeostasis is maintained following elimination of the biological clock, homeostatic regulation is clearly a completely independent neural process located somewhere else in the brain. The biological clock fosters nocturnal sleep indirectly by causing or promoting the long period of daytime sleep deprivation which would not occur in the absence of a normal biological clock and clock-dependent alerting. If the circadian system was not promoting wakefulness at specific times each day, it would be difficult to understand why jet lag and shift work would result in difficulty sleeping despite prior sleep loss. The clock-dependent arousal process drives the state of the organism in the direction of consolidated wakefulness and increased alertness.

> *"Why does the eye seem more clearly when asleep than the imagination when awake?"*
> *Leonardo Da Vinci*

CHAPTER 5 The Opponent Process Model

"I love sleep. My life has the tendency to fall apart when I'm awake, you know?" - Ernest Hemingway

The brain of every human being houses a process whose function is to induce and maintain sleep. This process is sleep homeostasis. In addition, the brain of every human being possesses a second and independent process whose function is to induce and maintain alert wakefulness. This second process is called **clock-dependent alerting**. The ongoing interaction of these two processes which produces the daily cycle of sleep and wakefulness creates the **opponent process model**. See figure 5-1.

The homeostatic process in our brains that causes us to fall asleep and to remain asleep through the night is also continuously active throughout our waking day. All during our waking hours, we have some tendency to fall asleep. If our sleep debt is very small, the tendency to fall asleep will be weak and unnoticeable, but it is still present. When the tendency to fall asleep is strong, we may experience repeated waves of drowsiness — we feel sleepy.

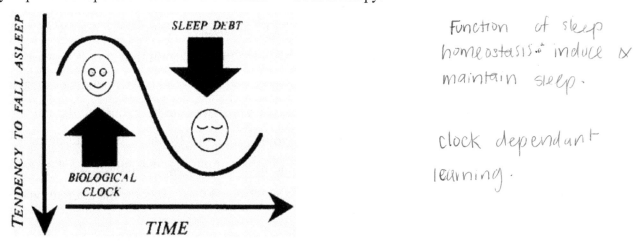

Figure 5-1: The Opponent Process Model. This drawing reminds us that there are two crucial, completely internal processes that regulate our sleep and wakefulness: an alerting influence from the biological clock and a continuously active drive to sleep whose strength depends upon the size of the sleep debt.

As stated, the homeostatic regulation of sleep tendency is one of the two crucial components of the opponent process model of sleep and wakefulness. When total daily sleep time is reduced, there is a homeostatically mediated increase in the tendency to fall asleep in the daytime. Conversely, when total daily sleep time is increased, the tendency to fall asleep in the daytime decreases.

The second crucial component in the regulation of sleep and wakefulness is the biological clock. The homeostatic drive to sleep in response to not sleeping is opposed by periods of clock-dependent alerting which normally occur at the same time each day. In other words, the main reason that we do not fall asleep as soon as we have been awake a few hours is that the homeostatic sleep inducing process is held at bay by an independent internal alerting process. As just stated, clock-dependent alerting is not active all the time — it is only active during specific portions of the day. At these times, we may feel wide-awake even in the presence of a strong sleep drive.

When we are awake, the underlying homeostatic drive to sleep is continuously increasing. It builds up steadily all day long. The sleep drive becomes unopposed when the clock-dependent alerting process subsides at the end of the day. This now maximally strong sleep drive enables us to sleep through the night. By morning, the sleep drive is sufficiently weak so that we can easily awaken. In summary, the sleep drive increases continuously during wakefulness and decreases continuously during sleep.

In contrast to the homeostatic sleep drive, clock-dependent alerting is not readily influenced by variations in the amount of prior time spent sleeping or not sleeping. It simply consists of lengthy intervals of stimulation that alert the brain and opposes the sleep tendency at specific times during the 24-hour cycle. Clock-dependent alerting could be likened to the effect of ingesting a stimulant in the morning which would last about 16 hours and, as its activity subsided at bedtime, the sleep drive which had been building up throughout the day would then be unopposed. The opponent process model satisfactorily accounts for the alternation of sleep and wakefulness in each successive daily cycle.

To foster an optimal daily cycle of human existence, the opponent process model requires (a) the correct timing of clock-dependent alerting, and (b) avoiding the accumulation of an overly large sleep debt. The latter can be accomplished by habitually obtaining adequate sleep at night. To sum up, except for the effect of transient stimulation, an individual's ability to stay awake at any time during the day or night is strongly influenced by the ongoing interaction of clock-dependent alerting and homeostatic sleep drive.

The Biological Clock and Homeostatic Determinants of Alertness and Sleepiness

CHAPTER 5 The Opponent Process Model

The Tendency to Fall Asleep at Any Time, Day or Night At bedtime, the physiological sleep tendency is strong because the sleep drive is high and clock-dependent alerting is subsiding or entirely absent. Once an individual has gone to bed and has fallen asleep, there is a tendency to remain asleep. We may assume that during the night, the sleep drive is essentially unopposed. Accordingly, the size of the sleep debt at bedtime will, to a large extent, determine how long an individual will sleep at night.

Keeping in mind the opponent process model, we should revisit the experiments in chapter 3. The following experiment was designed to evaluate the entire 24-hour day. Sleep latency was measured every 2 hours throughout the daytime in subjects who then went to bed at 11:30 P.M. and got out of bed at 8 A.M. Every 2 hours during the night, the subjects were fully awakened for a brief time and then requested to return to sleep while the sleep latency was measured in a standard manner. In this way, the 24-hour pattern of sleep tendency was measured without seriously interfering with nocturnal sleep. The results of this experiment are illustrated in Figure 5-2.

In general, it was found that sleep latencies were short during much of the conventional sleep period. When awakened at night, subjects returned to sleep fairly quickly. This is similar to what happens in real life; that is, individuals normally awaken a number of times during the night (though they may not be aware of the awakenings due to retrograde amnesia). In the morning, with declining physiological sleep tendency, normal awakenings are more frequent and extend for longer durations until individuals no longer desire to remain in bed for as long as it might take to fall asleep again.

 In Figure 5-2, which illustrates sleep latency testing every two hours over a 24-hour period, counter-intuitive results are depicted.   The sleep tendency is not maximal in the early sleep period. (Recall that the MSLT is terminated at the onset of sleep so that no sleep accumulates.) In Figure 5-2, it can also be seen that sleepiness decreases (alertness improves) from about 2:30 P.M. in the afternoon until about 9 P.M. If the only mechanism involved were the homeostatic regulation of sleep, one would expect physiological sleep tendency to increase progressively across the entire daytime waking period, and to decrease steadily across the entire

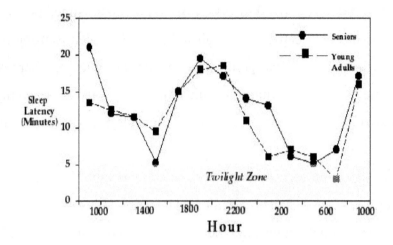

Figure 5-2: 24-Hour Sleep Latency. This figure depicts the results of measuring sleep latency every two hours around the clock in young adults and senior citizens. During the daytime, the sleep latency measurements are carried out in the conventional fashion, every two hours from getting up in the morning until going to bed at night. While subjects are in bed at night they are awakened every two hours and the time it takes to return to sleep is carefully recorded. However, when the subject has fallen asleep, they are allowed to remain asleep until the next test. In this protocol there is a clear bi-aphasic pattern with longer tendencies in the morning, a midday trough with a subsequent increase, and then during the period in bed, a progressive increase in sleep latencies until about 6 A.M. The result at the beginning of the night is counterintuitive, but reflects that clock-dependent alerting has not fully subsided at this time.

sleep period. Clock-dependent alerting is the only possible explanation for the decrease in sleepiness in the latter half of the ordinary waking hours, e.g., after the trough in alertness at 2:30 P.M.

The tendency for sleep latencies to decrease to the lowest point as the night progresses (shortest latency) around 5 A.M. to 6 A.M. is also counterintuitive. It can only be explained by assuming that the period of clockdependent alerting that produces long sleep latencies in the late afternoon and evening is still operating at the beginning of the sleep period, although decreasing. This is consistent with the commonplace observation that staying awake all night is most difficult in the hours just before dawn.

The U-shaped MSLT profile that is characteristic of daytime sleep tendency in human adults is explained by assuming that clock-dependent alerting is divided into two independent episodes — one in the morning which begins to subside before noon, and one later in the day beginning in mid to late afternoon and persisting into late evening.

This model of the 24-hour manifestations of sleep tendency is illustrated in Figure 5-3. Careful study of this figure is recommended. Primates and rodents whose biological clock has been

CHAPTER 5 The Opponent Process Model

eliminated continue to fall asleep normally and show normal sleep. They can also wake normally and can be kept awake for fairly long periods of time by external stimulation after which they will sleep for longer periods of time. The homeostatic sleep mechanism seems completely unimpaired. The biological clock independently fosters continuous, alert wakefulness during specific periods of the nychthemeron that are advantageous for the species. The opponent process model includes timed periods of clock-dependent alerting opposed by wake-related build up in homeostatic sleep tendency or drive to sleep. The interaction of these processes determines the degree of the sleep latency at any particular moment. External stimulation will also induce waking but unless the stimulation is relatively intense or prolonged, the effect in terms of the 24-hour day is negligible.

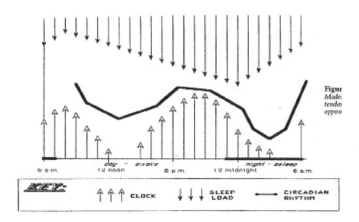

Figure 5-3: The opponent process model. This figure is a detailed schematic illustration of the opposing process that regulates the daily cycle of sleep and wakefulness.

If, for the sake of discussion, we postulate that the mechanism of clock-dependent alerting is the timed release from the SCN of a specific stimulating substances, then we must assume that more of these substance are released in the evening because in order to achieve the high level of alertness that is typically seen in the evening in young adults, they must overcome the sleep deprivation and associated "build-up" of the sleep drive during the waking day. A substance fitting this description is the neuropeptide, **hypocretin**, which was discovered at Stanford and will be discussed later this chapter. This timing is well illustrated in the extended MSLT plot in the chapter three (see Figure 5-7). The hypothetical morning release seems to have a half-life of about four hours; the evening's release appears somewhat slower in its rise time and faster in the fall off, though the faster fall off may be to some extent a function of a strong sleep drive in the late evening.

Humans can rarely maintain an absolutely regular daily schedule. Therefore, it is advantageous to possess a mechanism that responds to reductions in sleep time by increasing the tendency for more sleep to occur. In addition, something must oppose this sleep drive so that a relatively small sleep debt does not overwhelm us. The opponent processes ensure that the total needs of the human organism are met in an optimal fashion. Humans and other primates need to be awake, alert, and active in the daytime and they need to obtain a certain amount of sleep each night. Rodents need to be awake, alert, and active at night and to be safely sleeping and out of harm's way in the daytime. The occurrence of clock-dependent alerting typically during the day or typically during the night favors the specific ecological requirements of diurnal and nocturnal species. It is not entirely clear why a night of sleep is terminated at any particular moment. Why does a sleeper awaken and what causes the decision to get out of bed for the day? The cause of arousal could be as trivial as thirst, hunger, or a full bladder. Discomfort could cause spontaneous movement leading to an arousal. One definite process may be the first wake-up signal from the biological clock. The underlying sleep tendency also probably makes a difference; a large carry-over sleep debt will foster the continuation of sleep. It is clear that high school and college students have a large carry-over sleep debt as they frequently sleep well into the afternoon on weekends, or on any day when nothing is scheduled; however, even with a large carry-over sleep debt, it remains nearly impossible to sleep through the period of strong clock-dependent alerting in the late afternoon and evening. This is something that needs to be taken into account, for example works shifts or study schedules.

Demonstration of the Opponent Process Model

Ultrashort Sleep Wake Schedule/ The 90 min day

If MSLT were administered in its pure form every two hours around the clock even at night, and subjects were aroused as soon as they fell asleep, they have no opportunity at all to accumulate sleep. The result would resemble the total sleep deprivation study described in Chapter 3. In an approach to measuring 24-hour physiological sleep tendency and the strength of clock-dependent alerting, a special experimental schedule was created that permitted repeated measurement of sleep latency in the face of partial, but not total, sleep deprivation. Carried out in the Stanford Summer Sleep Camp (Jerry House), this experiment was designed to measure sleep tendency every 90 minutes around the clock see Figure 5-4. The "90-minute day" was organized with the 24-hour day being divided into 16 individual 90-minute periods.

Subjects were required to be in bed for the first 30 minutes of each 90-minute interval, or 16 times during each successive 24-hour day. The duration of the time in bed was exactly 30 minutes whether or not subjects fell asleep. At the end of each 30 minutes in bed, subjects were required to get out of bed and stay out of bed for 60 minutes. Accordingly, each couplet of sleep and wake constituted a "90-minute day." During the 60-minute period out of bed, subjects were required to carry out simple performance tests and make subjective ratings of sleepiness. Adequate nourishment was

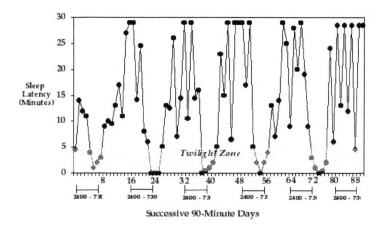

Figure 5-4: The "90-minute day." This figure represents one reprasentative volunteer who lived for five consecutive 24-hour periods on a schedule which required going to bed every 90 minutes and getting out of bed after only 30 minutes had elapsed. The light grey shade corresponds to the "twilight zone"

provided and free time was permitted for meeting personal hygiene and bathroom needs.

If the subjects had fallen asleep immediately on every test, they would have accumulated 480 (16 x 30) minutes of sleep each day, or eight hours. If the subjects did not fall asleep immediately, the total sleep during any given interval in bed would be 30 minutes minus the sleep latency time. The same overall results were obtained from every participant. Figure 6-4 displays data from one representative subject. The plot shows the five consecutive 24-hour days during which subjects went to bed and had their sleep latency measurement every 90 minutes. Over the 5-day period, there were 90 consecutive sleep latency tests. The important results were as follows:

(1) A very robust 24-hour rhythm in sleep latency values was apparent.

(2)   The bi-phasic pattern of longer sleep latencies disappeared. What remained was a single unambiguous period of alertness during the evening hours (the subjects were mainly college student volunteers). It is important to emphasize that this single pronounced period of clock-dependent alerting occurs in the late evening at about the same clock time that college students and young adults tend to be very alert.

(3)   The total amount of sleep obtained each successive 24-hour day was about four-and-a-half-hours, considerably less than the presumed daily need. In spite of the very substantial sleep deprivation, the remaining single daily period of clock-dependent alerting did not change across the five-day period of chronic sleep loss. Does this mean that the daily need for sleep is actually reduced in the 90-minute day schedule? We don't really know. When the experimental schedule ended and the subjects were allowed to sleep as long as they wanted, the total amount of sleep in the single recovery period monitored by polysomnography ranged from 16 to 20 hours-unambiguous evidence of prior sleep deprivation. Over the five days on the 90-minute day schedule, the total amount of sleep lost was approximately 17 to 18 hours, or between three and four hours per day, assuming an eight-hour need.

The results of the "90-minute day" experiment depicted in Figure 5-4 may be readily understood in terms of the opponent process model. Since there is only one obvious period of alertness in each successive 24-hour period, and since this peak occurs late in the day with reference to regular clock time, it is clear that the weak morning period of clock-dependent alerting was rapidly overwhelmed. On the other hand, the strong evening peak is maintained. Although an overall change in sleep latencies over the five experimental days is not apparent, a substantial sleep debt must has accumulated which in turn caused the greatly increased amount of recovery sleep.

Had the "90-minute day" schedule been maintained for many more days, it is likely that the partial sleep deprivation and accumulating sleep debt would have finally suppressed the evening period of strong clock-dependent alerting. However, such an experiment has never been carried out. Until it is done, we cannot rule out the alternate possibility that the opponent process system might make a fundamental readjustment in response to this unique schedule in order that at least one period of daily alertness can be maintained indefinitely. For example, the biological clock might be able to increase the strength of the late-in-the-day alerting period. In view of the fact that human beings in the modern world may often undergo weeks or even months of reduced daily sleep time, gaining more knowledge about what, if anything, happens to the opponent processes during such long intervals is very important.

Sleep Deprivation: Daytime Alertness Following Recovery Sleep

The discovery of clock-dependent alerting has made it possible to understand a research result which long appeared to be completely at odds with common sense notions. The study in question was carried out more than 25 years ago, at Stanford University. In the experimental protocol, subjects were required to go to bed at 10 P.M. and stay in bed until 8 A.M. for several baseline nights. They were then totally sleep deprived for two days. This was followed by two recovery nights with the same in-bed schedule, as the baseline, 10 P.M. to 8 A.M.

The investigators expected that subjects would be much more alert when they woke up on the morning after their first recovery night than they were at the same clock time on the day following a night of no sleep. The opposite result occurred. The MSLT measurements showed that the subjects had a very strong physiological sleep tendency on the first test after more than 9 hours of recovery sleep. They then gradually became more alert through the day without additional sleep (recall that the sleep latency test is terminated at the onset of sleep and essentially no sleep is allowed). The entire experiment was depicted earlier in Figure 4-8. The unexpected recovery day MSLT pattern is repeated in Figure 5-5. To re-emphasize, the recovery of alertness toward the end of the day occurred without any sleep at all!

We now know that morning alertness is absent on the day after the first night of recovery sleep because the sleep debt has only been partially reduced and continues to overwhelm the morning period of clock-dependent alerting.

The later decrease in physiological sleep tendency, we now know, is caused by the second stronger evening period of clock-dependent alerting.

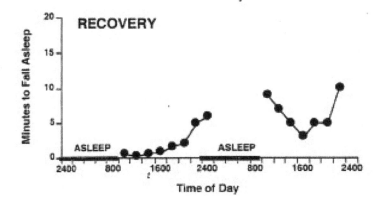

Figure 5-5: Sleep latency measurements every two hours on the day after the first night of recovery sleep. This experiment was described in Chapter Four. The sleep debt from two consecutive nights without sleep is not sufficiently "paid back" after one night of recovery sleep so it continues to suppress the weaker morning period of clock-dependent alerting.

Principles of Optimal Alertness

Optimal Alertness is More Than the Absence of Sleep Indebtedness

In the past, the varying strength of the daytime sleep tendency was viewed by some as a unitary process. In this formulation, the weaker the tendency to fall asleep, the greater the state of alertness. However, a very important principle derived from the opponent process model is that optimal alertness is not only the result of an absent or very low sleep debt. Optimal alertness may also require stimulation either from the clock or from external sensory input or both. Obviously, optimal alertness will always be more readily achieved when the sleep debt is small such that the drive to sleep is weak. A very large sleep debt, on the other hand, can overcome or overwhelm clock-dependent alerting even when strong external sensory stimulation is also present. Thus, for each individual, there is an amount of sleep which, if habitually obtained, will reduce the sleep debt to a very low level by morning, and will permit an optimal influence of clock-dependent alerting in the morning as well as in the evening.

Napping

The opponent process model explains the characteristic early-afternoon sleepiness seen in adults as well as the daily siesta that is still scheduled in a few parts of the world. In the case of the siesta, the biological tendency for afternoon sleepiness has evolved into a cultural phenomenon. This biological tendency is readily seen in Figure 5-2 earlier in this chapter in which physiological sleep tendency is plotted across 24 hours. A marked midday increase in sleep tendency is evident. The best time to take a nap is obviously during this afternoon dip in alertness.

What happens when we take a nap? In the first place, the ability to take a daytime nap suggests that total nocturnal sleep time has not been optimal and that a carryover sleep debt remained after the morning arousal. This does not mean, however, that a napping schedule is not optimal. Indeed, depending on an individual's life-style, it may be much more efficient and effective to spend a little less time sleeping at night and to take a nap in the daytime. Napping in the daytime will increase overall alertness unless any associated reduction in nighttime sleep is too large.

If one intends to take a nap, it should not be postponed. If the attempt to nap is delayed until after the midday physiological sleep tendency reverses, it will be difficult to fall asleep. Accordingly, the time spent trying unsuccessfully to nap will be completely wasted. The daily pattern of sleep tendency varies somewhat from person to person. Thus, the midday increase in sleep tendency may occur at noon in one person, at 3 P.M. in another, and even in the evening in another. The typical time for most people, however, is in the early to mid-afternoon. To re-emphasize, the best and most efficient time to take a nap is at the onset of the midday dip in daytime alertness.

Must the Sleep Debt Be Fully Repaid?

Must we "make up" lost sleep with the same precision that we must repay borrowed money? If we assume a specific individual's sleep requirement is eight hours, this means that for every two hours of being awake this individual must pay with one hour of sleep to satisfy the biological ratio of sixteen hours awake-eight hours asleep. If the individual in this hypothetical example habitually sleeps eight hours and then one night sleeps four hours, must the amount of sleep be twelve hours on the next night to set things right? Such a precise pay-back has not been conclusively demonstrated. However, the evidence to date suggests that all lost sleep must be made up to totally cancel the sleep debt. There has been no laboratory research to confirm or deny that this principle applies equally to very long periods of partial sleep loss (the longest laboratory study that has been conducted up until now is fourteen consecutive nights). However, there is enough evidence to warrant the assumption that we must make up all or most of the lost sleep that we have accumulated at least over several weeks. In addition, we know that most people carry a fairly large sleep debt and that obtaining extra sleep on several consecutive nights is not guaranteed to return them to optimal daytime alertness. The study by sleep researchers at the Henry Ford Hospital in Detroit previously mentioned, provides strong support for the principle that lost sleep must be fully made up. In this study, subjects were totally sleep deprived for one night. Starting the next night, they were required to remain in bed 24 consecutive hours. When confined to bed and requested to sleep as much as possible, they made up the full amount of sleep they had lost the night before by sleeping 16 hours or very close to this amount.

Clock-Schedule Mismatch

It should be obvious by now that a harmonious relationship between the homeostatic sleep process and the clock-dependent alerting process is highly desirable. In other words, the best sleep occurs when the biological clock and your personal schedule are in synchrony. The strong afternoon and evening period of clock-dependent alerting should be ending or in sharp decline at your scheduled bedtime. If clock-dependent alerting continues past the scheduled bedtime, you will then be trying to go to sleep when your biological clock is working to keep you awake. Also, if sleep onset is delayed, most individuals are likely to become severely sleep deprived because their class and/ or work schedules are usually inflexible and require getting up and out of bed at a relatively early hour. In such a circumstance, there will be sleep deprivation with increasing sleep debt every day until an individual sleeps through the alarm clock. A large mismatch of the scheduled bedtime and the occurrence of clock-dependent alerting can cause prolonged and severe difficulty falling asleep and all the consequences of severe sleep deprivation.

The effect of a lack of a good matchup between the timing of clock-dependent alerting and the desired schedule of hours in bed is shown in Figure 5-6. This figure depicts a situation, very common in adolesence and young adults in which the circadian rhythm is running later than desirable. As a consequence, the daily period of strong clock-dependent alerting is inappropriately delayed into the hours that are usually scheduled for sleeping. Students and others who are unaware of this role of the biological clock will erroneously attribute the inability to fall asleep to other things, typically worry and anxiety or being "unable to turn off my mind." Such students are likely to complain that they suffer from insomnia.

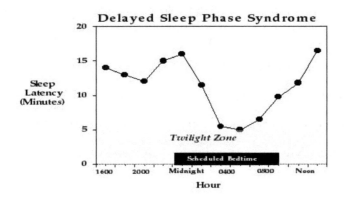

Figure 5-6: Delayed Sleep Phase Syndrome. This figure depicts a hypothetical 24-hour MSLT when the circadian rhythm is delayed with respect to the scheduled bedtime. Thus, clock-dependent alerting is strong when at the conventional bedtime and at the conventional wake up time, sleep latencies are in the twilight zone.

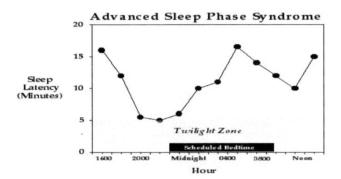

Figure 5-7: Advanced Sleep Phase Syndrome. This figure depicts a hypothetical 24-hour MSLT in a person whose rhythm is advanced with respect to the scheduled bedtime. This individual is in the twilight zone well before the scheduled bedtime and clock-dependent alerting occurs well before the scheduled waking time.

Another way this sleep onset insomnia typically occurs is when a person who lives in San Francisco flies to New York. His or her biological clock will continue to run on San Francisco time. As a result, it will be particularly difficult to go to sleep at bedtime in New York because bedtime in New York is three hours earlier with respect to the biological clock. Thus, it will be time to go to bed when clock-dependent alerting is strongest. Due to both the biological rhythm and sleep loss, it will then be very difficult to get up in the morning, (this, by the way, is called jet lag, which will be discussed later). If the individual remains in New York, an adjustment will take place with the clock advancing itself about an hour a day until synchrony between the circadian sleep/wake tendency and the desired schedule is re-established.

The reverse example, that of the biological clock running earlier than is desirable, is depicted in Figure 5-7. This situation occurs when a person who lives in New York flies to San Francisco. The periods of clock-dependent alerting will occur three hours earlier with respect to clock time in the new time zone. The person is likely to feel sleepy in the early evening and to wake up too early in the morning, perhaps at 4 A.M. Since we generally eat dinner and socialize in the evening, it is usually impossible to go to bed three hours early. Sleep onset will therefore be postponed. This and waking up three hours early will result in sleep deprivation. Whenever the circadian period of clock-dependent alerting and the desired schedule of sleep and wakefulness are not in the proper relationship, sleep deprivation is inevitable.

Putting it all together

Physiological sleep tendency varies throughout the day. This variation is the result of the interaction of two opposing processes: (1) sleep homeostasis, and (2) clock-dependent alerting. Sleep homeostasis is the process by which so many hours of wakefulness produce a need for so many hours of sleep. Clock-dependent alerting is a process which functions to induce and maintain alert wakefulness.

The interaction of these two factors is not well understood at the molecular/cellular level because we do not yet know the exact neural mechanisms by which they produce their effects. However, at the physiological/behavioral level, the operation and interaction of sleep homeostasis and clockdependent alerting have been accurately described and understood.

The homeostatic regulation of sleep and the mediated drive to sleep underlie feelings of fatigue and sleepiness as well as the desire to go to sleep. Although the mediated sleep promoting process is operating continuously whether we are awake or asleep, fatigue and feelings of sleepiness refer to feelings that can only be consciously experienced in the waking state. They are always associated with an increased internal tendency for an individual to fall asleep. As with the homeostatic response to fluid reduction in which thirst is the subjective aspect, seeking water is the behavioral aspect, and dehydration causes the physiological drive state; analogously, the feelings of fatigue and sleepiness are the psychological side of the coin, reduced sleep latency is the behavioral component, and the strong physiological sleep drive occurs in response to the sleep-deprived state of the brain.

The exact brain locus of the biological clock is known. The biological clock is situated in the suprachiasmatic nuclei (SCN). Because the location is known, it is possible to eliminate the clock in experimental animals by electrolytic lesion. When this is done, the animals no longer show any circadian periodicity in wake or sleep in constant conditions. In primates, the long daily episodes of consolidated alert wakefulness are eliminated. It must be strongly emphasized, however, that the ability to wake up and remain awake for brief periods is fully maintained in the absence of the biological clock.

The major change in clockless animals is that wakefulness cannot be sustained in the absence of continual intense external stimulation. The homeostatic regulation of sleep is unimpaired in such animals and if an animal is kept awake for sufficiently long periods of time by external stimulation (at least one or two hours), a compensatory increase in amount of sleep is always seen. Thus, animals without a biological clock appear to stay awake only until sufficient homeostatic sleep drive has built up to facilitate the occurrence of sleep. This indicates that the drive to sleep is building up whenever the organism is awake. In other words, in terms of the homeostatic process, all wakefulness is sleep deprivation. Although sleep tendency builds up during any period of wakefulness, in common parlance the term, "sleep deprivation," is never applied to the ordinary waking day, though it is technically incorrect. It is typically used only when sleep at night is substantially below the amount of an individual's typical time in bed. For ethical and humanitarian reasons, we will never undertake to destroy the clock in human beings. Therefore, we will never be able to study the subjective correlates of the clockless human condition. However, it may be imagined that animals or humans without a clock would experience feeling sleepy after they had been awake one or two hours and this would motivate them to go to sleep. After they had slept one to two hours, they would presumably wake up feeling less sleepy. We do not know if normal levels of physiological and psychological alertness and mental performance could be achieved in the complete absence of clock-dependent alerting.

Most people think that drowsiness is a transient phenomenon easily reversed and of no serious consequence. They also think that such things as a warm room, a heavy meal, boredom, monotony and alcohol directly cause drowsiness. The general public does not know that besides strong drowsiness, there are many levels of sleepiness and alertness that affect waking function. At an optimal level of alertness, mental powers are generally at their peak, concentration is optimal, and the sense of being fatigued is completely absent. A fully alert individual has high energy and clarity of mind that fosters peak performance.

A sleepy individual has some degree of mental impairment and is generally less energetic and motivated. The alertness/ sleepiness dimension of waking function is a product of two factors: the size of the sleep debt and the time of occurrence and strength of alerting influences arising from the biological clock.

Peak performance requires adequate sleep. It now appears that peak performance is most readily achieved when the sleep tendency is low. Unfortunately, nearly all adults are chronically sleep deprived. Many individuals carry a very large sleep debt which undermines their ability to achieve optimal alertness and performance in the daytime. We also know that adults can be chronically sleep-deprived and impaired without being fully aware of their condition. Some people assume they are performing at their full capacity when they are far from it. In addition to the overall impairment of cognitive processes, chronic sleep deprivation leads to increasing feelings of fatigue and increasing vulnerability to attacks of severe drowsiness and unintended sleep episodes. This, in turn, is very hazardous because it tremendously increases the risk that individuals will injure themselves and/ or others through an error of impaired judgment, inattention, or actually falling asleep. One of the most common of such events is falling asleep while driving.

The internal brain process which causes people to fall asleep and to remain asleep is always active. This process is active 24 hours a day. If the intensity of the sleep-inducing process is high, daytime sleep tendency is strong. A strong sleep tendency impairs our creativity and our cognitive ability even when we are not consciously drowsy. Furthermore, any moment we relax or let down, we can experience an unintended sleep episode.

Drowsiness is RED Alert! Drowsiness may mean you are seconds from a disaster. If we could respond as to an emergency when we experience the first wave of drowsiness, many catastrophic events and errors and the attendant human suffering could be avoided (figure 5-8.)

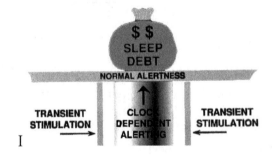

CHAPTER 5 The Opponent Process Model

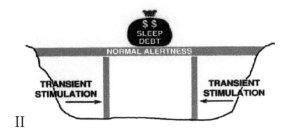

II

III

Figure 5-8: The opponent process model and safety I. This schematic figure shows a large sleep debt resting on the bridge of normal alertness. The bridge is strongly supported by clock-dependent alerting. Two other weak supports symbolize transient stimulation. When strong clock-dependent alerting is present, normal alertness can be maintained. The opponent process model and safety II. When the sleep debt is small, normal alertness can be maintained with relative ease even in the absence of strong clock-dependent alerting. The opponent process model and safety III. This schematic suggests that transient stimulation alone cannot maintain normal alertness when the sleep debt is large.

"You had everything sewed up tight. How come you lie awake all night long?" -Sugaree

CHAPTER 6 Drowsiness RED Alert!!

"My mind clicks on and off... I try letting one eyelid close at a time while I prop the other open with my will. But the effort's too much. Sleep is winning. My whole body argues dully that nothing life can attain is quite so desirable as sleep. My mind is losing resolution and control." - Charles A. Lindbergh

Feeling Sleepy

The above quote was brought to our attention by Honorable Mark R. Rosekind. This Stanford and Sleep and Dreams alumni is the first sleep specialist to serve on the National Transportation Safety Board.

Until now, we have covered mainly those aspects of sleep and wakefulness that can be measured objectively. This chapter will explore how you feel when you are awake, specifically, your subjective levels of alertness and other feelings that are the direct result of the amount and quality of your sleep. The most important concepts to understand clearly are (a) the amount and quality of your sleep determines the way you feel when you are awake in class, studying and driving, (b) you must have a sleep debt if you feel sleepy, and (c) the way to communicate how you feel to yourself and others.

There often comes a moment in a dull conference, reading this textbook or very late at night when something happens. Most people know exactly what it is. We feel sleepy. Our eyelids want to close and they would very much like to go to sleep. It must be clear that such a wave of sleepiness is at the extreme end of the subjective scale of sleepiness and alertness. If we consider the full range of **subjective sleepiness** and alertness during the entire day, we find that most people are not particularly sensitive to the different levels of sleepiness and alertness. People almost never say they are "just a tiny bit sleepy." In addition, a documented increase in the objective physiological tendency to fall asleep is not always associated with a parallel change in the way we feel. People may deny feeling sleepy despite the evidence of the contrary. Have you ever done this?

Finally, under ordinary circumstances we only become sleepy or drowsy when we are physically at rest, lying or sitting, and external stimulation is at a minimum. For all intents and purposes, people almost never feel sleepy when engaging in vigorous activity.

CHAPTER 6 Drowsiness RED Alert!!

All normal individuals have frequently experienced sleepiness and drowsiness in their daily lives. For example, when it is necessary to stay awake far beyond the customary bedtime and then get out of bed early the next morning to go to work, to catch a plane, or to participate in some other specific occasion, most people will experience waves of sleepiness as they go through the day. These feelings of sleepiness may persist throughout the day, or occur intermittently.

Some people will use words like fatigue, tiredness, zoning out, bushed and weariness instead of sleepiness. However, people also use the word fatigue to describe the muscular burn that comes with working out. The distinction between the two is underscored by the fact that most people experience feelings they call tiredness, fatigue, and sleepiness in the complete absence of vigorous physical activity. In addition, the feeling of sleepiness that precedes falling asleep for the night usually bears no obvious relationship to the activities of the preceding day. Rather, it is as if the brain has simply decided it is time to go to bed, and creates the feeling of sleepiness to motivate the appropriate "going to bed" behavior.

Obviously, the way one feels cannot be objectively measured. Therefore, it is extremely important to be as precise as possible in using words that describe our subjective levels of alertness. A fair number of individuals seem to have difficulty describing the feeling of sleepiness. Nonetheless, it is an elemental and unique feeling. The feeling of sleepiness is also typically associated with a constellation of physical sensations that can be described. In this sense, it is not purely subjective. These sensations, particularly when we are trying to resist falling asleep, as Charles Lindberg describes, often involve the eyes.

A mild itching or burning usually accompanied by an increasing heaviness of the eyelids that requires a conscious effort to hold them open against a strong inclination to close them. The limbs and body may feel heavy, and a reluctance to move develops. We rub our eyes and stretch our limbs to produce momentary relief, yet the desire to sleep grows slowly but persistently. The desire to close our eyes and lie down in a resting position becomes overwhelming. At this point, almost everyone would admit they were very sleepy or drowsy.

A number of situations encourage the unmasking of subjective sleepiness. If the book we are reading is overly repetitive, ponderous, or filled with minutiae, sleepiness quickly takes over. If the material is exciting, or if we are physically active, sleepiness can often be held at bay for long periods of time. Maybe, if we could remain continuously active, such as dancing, to encourage us not to sit down, we could resist sleepiness for two or three days. If we were alone, however, we would soon yield to the impulse to close our eyes or rest and, even without intending, we would soon fall asleep. In a person who has been wide awake throughout the day and has arrived at bedtime, sleepiness occurs spontaneously, growing inexorably from an indiscernible beginning to overwhelming strength in a relatively short time. A person may be caught by surprise and fall asleep while reading or watching television.

The Total Range of Sleepiness and Alertness

An interesting relationship between sleep and wakefulness is the wide fluctuation in daytime sleepiness and alertness that follows a similarly wide fluctuation in overall daily amount of sleep. This subjective dimension of the waking state encompasses the entire range of levels of alertness — the continuum that extends from optimal alertness and energy to extreme drowsiness and actively struggling to stay awake. Optimal alertness is generally associated with feeling energetic, motivated, and desiring to do things and to be active. One feels at the peak of one's creative and intellectual powers. Severe sleepiness is associated with diminished intellectual powers, apathy, a desire to do nothing, to lie down, or to go to bed. A severely sleepy person is unable to function mentally at more than a rudimentary level.

One of the most striking aspects of increasing sleepiness is the way it changes our motivation toward what we are doing — we lose interest, feel dull, become indifferent and we crash. Whether reading a book, watching television, participating in a conversation, doing a crossword puzzle, or studying, our interest wanes. As sleepiness increases, concentrating become increasingly difficult, and interest in what we are doing decreases. In much the same way that individuals might begin to fantasize about a cool drink of water as thirst increases, they will begin to think about a comfortable bed as sleepiness increases, and the idea of taking a nap becomes more and more attractive.

Sleepiness Indicates a Behavioral Drive State

The motivation and desire to sleep that are associated with feelings of sleepiness or drowsiness are the behavioral and psychological manifestations of a drive state. The importance of this drive is that it compels the human organism eventually to sleep which may ensure that other vital functions associated with sleep are carried out. It also motivates various animals to seek a place to sleep which, in terms of their ecological niche, is usually also a safe place — for example, a burrow for a mouse or a tree for a baboon. Sleepiness or drowsiness can be very pleasant or very unpleasant depending largely on the availability of the consummation response, i.e., being able to go to bed and go to sleep. The parallel with thirst and hunger becomes even more obvious. Intense thirst makes the consummation response, drinking, exceedingly pleasurable whereas the absence of water makes thirst an ever-mounting torture. Analogously, a very sleepy person finds the act of getting into bed, snuggling down, relaxing, letting the mind drift very pleasant. It is very unpleasant if this same person is required to maintain vigilant behavior although very sleepy. If the duration of wakefulness is unusually prolonged as in sleep deprivation experiments, the intense subjective sleepiness can become so painful that a person will do almost anything to end it.

Experiencing an unambiguous wave of sleepiness is the ultimate manifestation of a strong sleep tendency for most adults. However, this subjective awareness cannot always be taken for granted. Studies of sleep loss in children (ages 6 to 8) have shown that there is difficulty in verbalizing sleepiness at this age level. These children cannot seem to compare subjective levels of alertness after adequate sleep versus inadequate sleep even though their behavior may suggest strong drowsiness.

At its peak, the biological drive to sleep may well be the most powerful in the human repertoire. When the sleep debt reaches a certain size, falling asleep cannot be prevented. While very thirsty individuals can still refuse to drink and very hungry individuals can still refuse to eat, a very sleepy human being cannot avoid falling asleep. It is the consequences of such irresistible and unintended sleep episodes that make understanding sleepiness extremely important to both individuals and society. When human beings are asleep, they cannot receive and process sensory information nor rapidly elaborate appropriate motor responses. If a person is in a situation where alertness is crucial, an unintended sleep episode can lead to a catastrophic accident.

Thus, far in our text, the word "alertness" has been used essentially as the reciprocal of sleepiness. In common parlance, it has other connotations such as being more intelligent, more interested in things, quicker to comprehend, etc. In the context of the impact of nocturnal sleep on daytime function, full alertness is the complete absence of sleepiness and/or tiredness and the optimization of energy, motivation, and intellectual capacity. In our analogy with thirst, the drinking behavior that subjective thirst motivates can very quickly reverse the thirsty state. Similarly, a brief nap can often reverse subjective sleepiness. However, it must be very clear that substantially or totally reversing subjective sleepiness in this manner is not at all the same as substantially or totally reversing an underlying physiological drive to fall asleep. In other words, optimal alertness is more than a minimal or absent tendency to fall asleep.

Why People Rarely Complain Specifically About Sleepiness

Introspectionism is a historical school of thought within psychology. This psychological discipline failed to precisely define subjective sleepiness because it was not regarded as an elemental feeling. Some psychologists had asserted that sleepiness was actually a combination of muscular fatigue and depression. However, there is absolutely no doubt that a sleepy person can also be contented and happy.

Although the reality of subjective sleepiness seems straight forward, it is actually very common for obviously sleep people to deny feeling sleepy. There are some people who claim they are never sleepy. In the general population, subjective sleepiness is definitely a somewhat slippery quality. If people were asked what they would experience if they were not allowed any fluids for one whole day, almost everyone would answer, "I would feel thirsty." On the other hand, if people were asked how they would feel after having no sleep at all for one whole night, some would say, "I would feel sleepy," but many would say, "I would feel tired," "I would feel fatigued," "I would feel irritable," and a few would say, "I would feel nothing." In American society, the association between being sleep deprived and sleepy, and the words used to describe this state, are somewhat ambiguous. The global failure to deal effectively with sleep deprivation and subjective sleepiness has created a pervasive insensitivity to their psychological, behavioral and safety consequences in every component of society.

Finally, being a sleepy person has a negative image so that there is a tendency to exclude the word "sleepy" from everyday language. For a young child, admitting to being sleepy usually means being hustled off for an unwanted nap or a too early bedtime. For an older person, being sleepy has a negative connotation of being lazy, lethargic, and in some way, subnormal. "Tiredness" and "fatigue," the two words most commonly used in place of sleepiness may at least carry the positive implication that they are the result of hard work or extra effort. Thus, we find that huge numbers of patients complain about and are being evaluated for "fatigue" while it is rare to hear patients complain that they are "excessively sleepy."

Sleepy Behavior

Cardinal signs of sleepiness are glazed expressions and a marked reduction of spontaneous activity and animation. Individuals may rub their eyes and show periods of increased blinking. The eyelids may close briefly. Headdrop and the sudden jerking up with a startle indicate the occurrence of a "**microsleep**." As the tendency to fall asleep becomes more and more irresistible, prolonged eye closure will occur.

Yawning

A widely accepted behavioral sign of feeling sleepy is yawning. However, yawning is not exclusively associated with maximum sleep propensity. For example, because yawning seems contagious, people who may not feel sleepy will begin to yawn if they are with another person who is yawning. On the other hand, fully alert people generally do not appear to yawn. Therefore, we may attribute at least some degree of sleep propensity to a person who is yawning. There is very little research on yawning. Although, sleep researchers are often asked, why is yawning contagous and why do we yawn. It has never been a funding priority of the NIH.

Diminution of Maximum Performance

Over many years, a major approach to evaluate the effect of sleep deprivation has been performance testing through a wide variety of tests. The results generally show an impairment of performance in association with sleep loss, particularly when a feeling of drowsiness is present. Performance testing, however, does not yield precise and consistent results when the effect of relatively small amounts of sleep loss are being evaluated. In general, the most sensitive performance tests are those which involve prolonged periods of carrying out simple, repetitive tasks, such as adding two figure digits for one or more hours, and where motivational and situational effects are less important.

Quantifying Sleepy Behavior

Behavioral measures of sleepiness are sometimes used by investigators. Ratings are made by trained observers who judge the behavior of the subject. In this approach, observers look for the behaviors associated with sleepiness such as yawning, head nodding, and so forth. Some sleep deprivation studies have required subjects to rate one another during the course of the study. Although behavioral ratings are often used to evaluate sleepiness, a widely accepted standard rating scale for behavioral sleepiness/alertness has not been universally developed or validated. Two scales that have had some acceptance are the he Stanford Sleepiness Scale Figure 6-1 and Epworth Sleepiness Scale Figure 6-2.

Quantifying Subjective Sleepiness

It is obvious that an individual may be able to say, "I am sleepy" or "I am not sleepy." In order for subjective measurements to be more useful, however, more precise gradations beyond "yes or no" are required. Self-rating scales are very common in psychological measurement. A widely used scale for measuring subjective sleepiness is the **Stanford Sleepiness Scale** (SSS). The SSS consists of statements that describe a range of feeling states associated with seven levels of sleepiness/alertness. A patient or subject is given a form upon which the statements are printed and is asked to choose the one that best describes his or her feelings at that moment. A typical use of the SSS requires making a rating every hour throughout an entire day.

After the Stanford Sleepiness Scale was devised, it was validated by asking volunteers to make ratings every hour throughout several baseline days, during one night of sleep deprivation, and during the day following recovery sleep. The ratings were generally worse during sleep deprivation and improved following recovery sleep. The Stanford Sleepiness Scale lead to the development of other self-rating scales.

The **Epworth Sleepiness Scale** has been very useful (figure 6-2.) The Epworth sleepiness scale has been validated primarily in obstructive sleep apnea and narcolepsy. The maximum score is 24 and the scores below 10 are consistent with excessive sleepiness. What is your score?

Another way to measure sleepiness is by means of a simple linear rating scale. Such a scale might consist of a horizontal line 100 millimeters in length on which the right extreme is labeled "Very Sleepy" and the left extreme, is labeled "Very Wide Awake." Subjects can be oriented to using this scale by asking them to recall the sleepiest they have ever been and the most alert they have ever been and to consider these their personal extremes. They are requested to draw a vertical mark through the line that best approximates the way they feel at the time of the rating with regard to their recalled extremes.

Measuring the distance (in millimeters) of the vertical mark from the left extreme yields the score. Thus, a score of 0 would correspond to maximum alertness, and a score of 10 would correspond to maximum sleepiness.

The Stanford Sleepiness Scale

Degree of Sleepiness	Scale Rating
Feeling active, vital, alert, or wide awake	1
Functioning at high levels, but not at peak; able to concentrate	2
Awake, but relaxed; responsive but not fully alert	3
Somewhat foggy, let down	4
Foggy; losing interest in remaining awake; slowed down	5
Sleepy, woozy, fighting sleep; prefer to lie down	6
No longer fighting sleep, sleep onset soon; having dream-like thoughts	7
Asleep	X

Figure 6.1: Stanford Sleepiness Scale

Epworth Sleepiness Scale

How likely are you to doze off or fall asleep in the following situations?
Answer considering how you have felt over the past week or so.

0 = Would never doze
1 = Slight chance of dozing
2 = Moderate chance of dozing
3 = High chance of dozing

1. Sitting and reading

2. Watching TV

3. Sitting inactive in a public place (e.g., theater or meeting)

4. As a passenger in a car for an hour without a break

5. Lying down to rest in the afternoon when able

6. Sitting and talking to someone

7. Sitting quietly after a lunch without alcohol

8. In a car while stopped for a few minutes in traffic

Figure 6.2: Epworth Sleepiness Scale

Unfortunately, subjective rating scales are susceptible to bias. For example, if a person overheard a conversation in which someone was described as sleepy and lazy with the implication that sleepiness is a negative personal characteristic, then that person would be more likely to deny feeling sleepy. In addition, there appear to be wide individual variations in introspective ability. Although quantifying the subjective feeling of sleepiness is important, it is rarely the approach used to diagnose excessive daytime sleepiness in sleep clinics. Much greater emphasis is placed upon objective measures.

While a single measurement of objective sleep latency can vary somewhat unpredictably, the Multiple Sleep Latency Test (MSLT) score, which represents the sleep tendency of an entire day, is very stable and can be regarded as a continuous linear variable when carried out on successive days. The overall daytime objective sleep latency changes in a linear fashion in relation to the amount of sleep loss or extra sleep. This is not true of the subjective feeling of sleepiness.

How do you feel right now?

A. Wide Awake — 18%
B. A Little Tired — 45%
C. Definitely tired, Not Sleepy — 25%
D. Sleepy — 12%

Figure 6-3

Before the development of a scientific field of sleep research and particularly research on daytime sleep tendency, people were simply sleepy or not sleepy. There was no impetus to quantify the feeling of sleepiness or even to define it. Moreover, the feeling of sleepiness was hardly ever studied either quantitatively or qualitatively as an outcome measure of sleep deprivation. Rather, measuring the impact of sleep deprivation on a variety of performance tests was the standard approach.

When a professional clinical interest in sleep disorders developed, it became necessary to confront the issue of daytime sleepiness as the major complaint. For example, in treating a patient with narcolepsy the crucial outcome is an improvement in daytime alertness and a recurring question is how much improvement has been achieved. At this point in time, that is prior to the development of the MSLT, the focus was on the subjective response of patients and experimental subjects.

The Stanford Sleepiness Scale (SSS) and other scales that have been devised for rating subjective sleepiness assume without much evidence that the range from feeling fully alert and most wide awake to extremely sleepy is smoothly linear in the same way that we regard the MSLT. However, this does not seem to be the case. Figure 6-4 shows several examples of daytime subjective sleepiness in terms of SSS ratings. They appear to be extremely variable. In addition, often often sees a succession of alert ratings (1 or 2) made by individuals who are, in fact, actually falling asleep.

Historically, the development of the Multiple Sleep Latency Test arose out of the study described on page 118 called the "90-minute day". In addition to measuring their objective sleep latency and total sleep time during the successive 30-minute periods they were in bed, the subjects were required to make subjective ratings using the Stanford Sleepiness Scale at the time of getting into bed.

Nearly 1,000 pairs of subjective sleepiness ratings and objective sleep latencies were accumulated from more than 50 volunteers in this study. The results are displayed in Figure 6-5 and yield considerable insight into the difficulties of accurate and predictable subjective ratings.

The vertical bars in the plot represent mean sleep objective sleep latencies associated with each of the seven subjective Stanford Sleepiness Scale values. The vertical lines show standard deviation of the means. When individuals rated themselves as very wide-awake (SSS=1) or very sleepy (SSS=7), their subjective sleepiness predicted the speed of falling asleep quite well. When individuals gave themselves ratings that were ambiguous i.e., in the middle of the Stanford Sleepiness Scale, the subjective ratings were very poor predictors of the speed of falling asleep and hence the

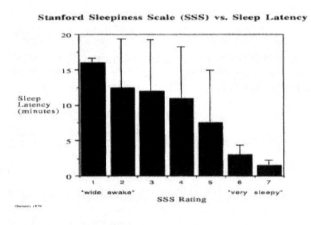

Firgure 6-5: Stanford Sleepiness Scale (SSS) versus Sleep Latency

standard deviations are very large. The subjective ratings of SSS=3 or SSS=4 were followed by an extremely variable sleep latency ranging from half a minute to twenty minutes. Due to this problem, the use of the SSS has decreased in research. The Epworth Scale is currently more commonly used.

Getting In Touch With Our Feelings of Alertness: A New Introspection

Given these results, the authors of this textbook have completely abandoned the arbitrary attempt to differentiate feeling sleepy and feeling drowsy. Rather, it has been assumed there is one internal level of subjective alertness which is labeled sleepiness and/or drowsiness by most people. In order to buttress this assumption, we need to conduct a vigorous research on the subjective dimension of sleepiness/alertness.

Individuals are readily able to state that they are not sleepy or drowsy. What enables them to make such unambiguous responses? In order to understand this, we designed a survey which asked the question: "Which of the following statements do you associate with feeling drowsy?" Figure 6-6 is the list that was developed.

The results of the survey were quite clear. The vast majority of individuals associated the feeling of drowsiness with symptoms involving their eyes, i.e., eyelids are heavy, eyelids want to close, and/or eyes burning. Difficulty concentrating, staying focused, and holding the head up were less common symptoms selected by respondents. Almost every other symptom on the list was only rarely chosen. Using the same list of statements, a survey was carried out asking the question, "Which of the following symptoms or statements do you associate with feeling sleepy or sleepiness?" The results were essentially identical with the drowsiness survey. Taking the survey results into account, it is possible to define drowsiness as the state prior to falling asleep during which a conscious effort is required to remain attentive, to stay focused, and/or to keep the eyelids from closing.

Anecdotal results strongly suggest that most individuals can recognize the onset of this state which we now call the "moment of the onset of drowsiness."

The bimodal relationship between SSS ratings and objective sleep latency measurements (e.g., good at the extremes, bad in the middle) have yielded a more precise understanding and communication about the way we feel in relation to our sleep as well as our tendency to fall asleep. We can divide the subjective continuum that ranges from peak alertness to extreme sleepiness into three discrete levels. These three internal levels of subjective alertness can be readily introspected, identified, and communicated with the desired degree of accuracy.

Three Readily Identifiable Levels of Subjective Alertness/Sleepiness

The first or top discrete level of subjective alertness is the level at which we feel completely wide awake, alert, and able to maintain a high level of attentiveness or concentration without any effort. This is the level of peak alertness. At this level, we feel energetic and we want to do things—we want to "get going." There is no impairment of motivation. When we are at this level, we would always say we are definitely not tired, and definitely not fatigued.

The middle or second internal level of subjective alertness is defined to include all the periods of wakefulness when we are neither unambiguously wide-awake and at peak alertness nor unambiguously sleepy. During these periods, people may say they are feeling tired, fatigued, lethargic, or unmotivated. Or they may simply say they are not quite at their peak of alertness. They never or almost never say they are sleepy or drowsy. People almost never feel intensely energetic or eager to do things at this level, - there is little desire to "get going." It is very important to specify that even if we have lost some sleep, at this level of alertness no specific effort should be required to maintain attentiveness or focus while we are in class, studying, or driving. To restate this important point, at the "tired" level, a conscious effort is not required to remain continuously attentive.

The bottom or third subjective level is feeling sleepy—or the state of being sleepy. This is the lowest of the three levels of alertness. To speak generally about this level of alertness, it is necessary to define "drowsy" and "sleepy" as meaning exactly the same thing. Since many people feel this way and the remainder are equally divided on which is more intense, the best decision is to declare that the words sleepy and drowsy are exactly equal. In all of the following text, any time the words sleepy and sleepiness are encountered, the words drowsy and drowsiness can be substituted and vice versa.

The sleepy state is the period of wakefulness immediately prior to falling asleep. It is the level where you will soon fall asleep unless you make a conscious effort to stay awake, e.g., to actively avoid falling asleep. Symptoms involving the eyes are almost always present at the sleepy level. One or more of the other symptoms in Figure 6-6 may also be present. If the sleepy state is uninterrupted, there will usually be a head drop unless the individual is lying down or the head is being supported.

The precise onset of the sleepy state can therefore be defined as the moment of the first awareness that the eyelids feel heavy—that they want to close and that a conscious effort is required to keep them open and to stay attentive. Although the majority of individuals might refer to heavy eyelids, the real essence of the onset of sleepiness is the unequivocal awareness that remaining alert and attentive is no longer effortless. This principle is so supremely important that it will be repeated—the onset of the sleepy state is the moment we become aware that remaining alert and attentive is no longer effortless --- that is, a conscious effort of will is required to remain attentive.

Although they are very simple, differentiating these three relatively easily recognized levels of subjective alertness is constant with the fundamental principle underlying the Stanford Sleepiness Scale. This principle states that the subjective domain of sleepiness and alertness is a continuum that runs all the way from most alert to most sleepy. What is most crucial, however, is that distinguishing these three subjective states allows us to define and specify a detectable internal warning of impending disaster—a signal that absolutely requires an immediate response. We are referring to the moment of the onset of drowsiness— the moment we have defined when you must abruptly make a conscious effort to stay awake—to not allow your eyes to close. So—drowsiness is red alert! Sleepiness is red alert! Get out of harm's way. Do it instantly! Do not delay even a few seconds! If you are driving, this is an absolute imperative! Do not end up a statistic.

Levels of Alertness Are Both Dynamic and Static

Given a certain objective tendency to fall asleep and a specific amount of internal and/or external stimulation opposing this tendency, one can imagine a situation in which the manifestations of the ongoing drive to sleep are held at exactly the same level by the opposing stimulation. In such a case, it might be assumed that the level of subjective alertness would also remain the same. Of course, in light of the principal that all wakefulness is sleep deprivation, the opposing stimulation would have to slowly increase to maintain the same objective sleep tendency. In this hypothetical situation, the level of alertness would appear to be a static and relatively stable state.

Level One - Wide awake and energetic, peak alertness, high motivation; absolutely no feeling of tiredness.

Level Two - Definitely not at peak; low motivation, may feel tired, fatigued; remaining attentive and focused does not require conscious effort.

Level Three - Conscious effort required to remain attentive and focused; drowsy, sleepy, unmotivated; must resist ever more strongly or sleep will occur.

On the other hand, we can imagine a situation where someone with a very large sleep debt becomes tremendously excited, frightened, angry or engages in vigorous exercise and at that moment feels he or she is very wide awake and fully alert (Level One). If this person then sits on a comfortable chair in a warm and quiet room, we may assume that the subjective feeling of being wide awake and fully alert would fairly quickly give way to feeling tired, perhaps very tired (Level Two). This would soon be succeeded by the moment of sleepiness (a.k.a. the moment of drowsiness) (Level Three). Once again, the state of drowsiness (a.k.a. sleepiness) would be quickly reversed by engaging in vigorous exercise or by strong emotion. In this hypothetical situation, the subjective alertness would appear to be very dynamic and changeable.

Why Students Are Tired Much, Most, or All of the Time

Close-up

The following is an example indicating that feeling tired is more pervasive than feeling sleepy. In October 2001, I gave a lecture to an Introductory Psychology class at Stanford University. The time of the lecture was 11:00 A.M. and all of Jordan Hall's 400 seats appeared to be occupied. I distributed the survey that I have been using to ask about subjective feelings and requested that the students to fill out the questionnaires immediately and then at the end of class leave them on the table by the door. To emphasize my point about tiredness even though the paper and pencil survey had been distributed, I asked all students who felt tired to hold up a hand. It appeared that every hand was in the air so I asked them to lower their hands. Then I requested any student who felt absolutely wide-awake and fully alert to hold up a hand. Exactly one hand was raised. "Do we believe her?" I asked the class. There were shouts of "No, No, No!" Then I asked, "How many students feel sleepy right now?" As I recall, approximately 20 hands were raised.

Given that these students had almost certainly gotten out of bed only a few hours earlier, probably had not eaten a large breakfast, and most probably had not engaged in any vigorous

physical exercise prior to the class, there can be only one explanation for several hundred healthy, young adults feeling tired. Sleep Debt! It is further likely that the percentage of students responding that they feel sleepy would steadily increase if the room were warm and the lecture extremely dull. However, the degree of stimulation at that time (an interesting lecture plus some clock-dependent alerting and the small amount of stimulation from the seats) was sufficient to forestall a decrease in alertness to the point of the moment of sleepiness and entry into the sleepy state. There are many more consequences of a large sleep debt than feeling sleepy which will be discussed and cognitive impairment is perhaps under ordinary circumstances the most important. Were the students in Introductory Psychology on that occasion cognitively impaired? Of the several hundred surveys, I distributed, only 23 students remembered to leave them on the table by the door! Need I say more?
- William C. Dement

Are There Causes of Tiredness and Fatigue in Addition to Sleep Debt?

In the survey mentioned earlier, Figure 6-3, slightly more than 90 percent of all respondents answered "yes" to the question, "Do you believe that tiredness can be caused by anything other than losing sleep or an untreated sleep disorder? If yes, what?" The "other causes of tiredness" that were cited by those who were surveyed fell into two general categories: (a) decreased stimulation and (b) physical exertion. It is well established that such things as a big lunch, boredom, a warm room, sexual activity, and alcohol are basically reducing the stimulation that opposes sleep debt and its tendency to cause tiredness and sleepiness. Thus, these various things that are believed to cause tiredness merely unmask an existing sleep debt. When individuals lower their sleep debt by obtaining substantial amounts of extra sleep, their tendency to feel tired in these circumstances usually disappears. In the absence of education about sleep debt, most people erroneously assume that whatever is followed by tiredness is the direct cause of the tiredness.

It is less certain that physical activity does not independently cause feelings that resemble the feelings of tiredness caused by sleep debt. If the average person were to run around a 440-yard track without stopping, a point would inevitably be reached at which the individual would not be able to continue. It is quite likely that "I am too tired to keep going" would seem to be an appropriate and meaningful explanation. One might assume, however, that in this case, if the individual had a low sleep debt, sitting quietly for a while would relieve the feeling of tiredness and indeed the individual could start running again. However, most people feel quite certain that if they had strenuous physical activity for a very long period of time, at least several hours to most of the day, this would directly and unambiguously cause them to feel tired. We prefer an interpretation in which tiredness after strenuous exercise could be related, at least in part to sleep debt - the reduction of muscle tone and spontaneous movement could permit sleep debt to express itself.

The importance of understanding these relationships is based on the fact that as long as individuals believe tiredness has many other causes, their motivation to get extra sleep and reduce the sleep debt is definitely undermined. As far as we know, at the present time no one has done a largescale survey in which the question has been asked, "Do you believe that sleepiness (not tiredness) can be caused by anything other than losing sleep, a sleep debt, or an untreated sleep disorder?" We hypothesize that the percentage of individuals who would answer yes to this question would probably be less than 90 percent and perhaps substantially less. However, we feel certain it would not be zero. Perhaps the most important reason for the belief that sleepiness can be directly caused by factors other than sleep debt is the tendency for students to experience drowsiness after the noon meal. This is associated, as we saw in previous chapters, with a mid-day decrease in sleep latency which in turn is due to carry over sleep debt and a mid-day reduction in clock-dependent alerting. In a classic example of a *post hoc ergo propter hoc* erroneous conclusion, people assume the post-prandial drowsiness to be caused directly by the food eaten at lunch.

What is Fatigue?

As was noted earlier, the term **fatigue** is commonly applied in the transportation industry as well as other components of society to account for a wide variety of accidents, errors, and other untoward results. In a variety of studies and applications, the term fatigue has been used more or less interchangeably with such descriptors as tiredness, exhaustion, sleepiness, drowsiness, boredom, apathy, and illness. Fatigue has been used one time or another in place of all the words that any of us may use to describe how we feel when we are not fully energetic, not totally healthy, and/ or not fully alert. When a highway accident is said to be caused by fatigue, it can mean falling asleep at the wheel, or it can mean being inattentive, or it can mean merely that the driver made a clear-cut error.

Early attempts to define fatigue stated that it is primarily the result of time on task (for example, driving long hours) and is accentuated when work is monotonous and repetitive such as assembly line work. Fatigue can occur when a task that is very demanding and stressful continues too long. Finally, those who feel that fatigue is a real and precise phenomenon often invoke unproven physical and biological processes to account for inattentiveness and crashes. So, clearly the term, fatigue, has too many meanings and too many implications to foster precise understanding of our subjective feelings particularly as they are related to inadequate or unhealthy sleep.

Over the years, many scientists have entered the sleep research field and we are now decades down the road. We have a huge amount of knowledge based on scientific experimentation. It is a firm belief of most sleep scientists that 95% of what is labeled fatigue is caused either by sleep deprivation or undiagnosed, untreated sleep disorders, or both. It is possible the reader already knows some of this, but we absolutely must take a moment to ask ourselves "What are all the different things that I believe can make me feel tired and fatigued." At the risk of repetition, remember if we become tired, fatigued, or sleepy in the early afternoon following a heavy meal, do we assume that the direct cause of the way we feel is the heavy meal?

We will not try to create a definition for the term, "fatigue." In relation to our new more precise terminology, if a driver uses the word fatigue casually, it should mean the same as feeling tired, but not the same as feeling sleepy or drowsy or having to make a conscious effort to remain attentive. In addition, it should be stated that a disinclination to continue a task that is unpleasant because it is extremely boring or very stressful should not be called fatigue. In the first place, neither boredom nor stress has a direct relation to sleep deprivation or the sleepiness/alertness continuum. In the second place, why not say, "I am bored," or "I am stressed." These words are much clearer and much more meaningful.

Now, in all this, our greatest concern must be focused on the moment when even a microscopic effort is required to remain attentive. Every individual, without exception, must learn to recognize instantly the moment when remaining attentive is no longer effortless. This moment must be immediately recognized inside himself or herself by everyone. Until this sensitivity is established, learning to recognize the moment when remaining attentive is no longer effortless must be a major focus of our attention and practice. Each one of you now reading this text must be able to state with total honesty, assurance, and certainty that you know exactly within a second or two when this clear change in the way you feel takes place. Practically, it is the signal that will sooner or later save your life.

Summary

In summary, the tendency to fall asleep is part of a drive state which increases in response to sleep deprivation. A strong drive to sleep gives rise to subjective sleepiness and motivates sleep-seeking behavior and is always associated with an increased tendency to fall asleep. If sleep loss is very pronounced, the tendency to fall asleep will become overwhelming. Subjective sleepiness is an elemental feeling often associated with heavy eyelids and indicates a readiness of the brain to fall asleep due to accumulated sleep debt. If we consider its full range, the dimension of sleepiness/alertness is the most important determinant of levels of human mental function and ability to pay attention. In summary, three subjective levels of alertness are defined and characterized that all students can easily recognize and communicate.

Level One: The peak of alertness—a complete absence of any slowness or tiredness.

Level Two: Not at your peak, but remaining attentive does not require a conscious effort.

Level Three: The sleepy/drowsy state when a conscious effort, a struggle against the desire to close your eyes is required to stay attentive.

Finally, we exhort readers, especially students, to always consider sleep deprivation and sleep debt whenever they feel tired regardless of any other possibility, particularly when they have not recently engaged in vigorous physical activity. A number of suggestions for improving one's sleep and thus reducing one's sleep debt will be described in the next chapter.

CHAPTER 6 Drowsiness RED Alert!!

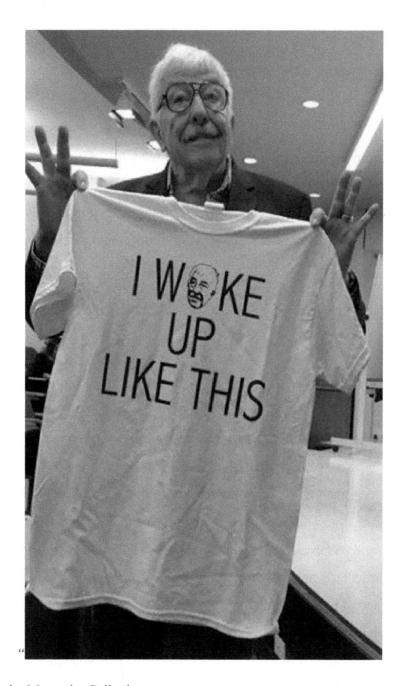

Photo by Margarita Gallardo

CHAPTER 7 Exercise, Nutrition & Sleep: The Triumvirate of Health

"I want to die in my sleep like my grandfather. Not screaming and yelling like the passengers in his car." - Emo Phillips

Our society has transformed without taking sleep requirements and biological rhythms sufficiently into account. Sleep is often treated like an inconvenience in our lives. Accordingly, we are now seeing numerous problems due mainly to chronic sleep loss and its negative impact on human performance. This including a predisposition for accidents and other serious consequences.

Sleep and Alertness in Our Daily Lives

Self-awareness of Sleepiness

The most important recommendation to the public is to become much more aware of the strength of their ongoing tendency to fall asleep. The strength of your tendency to fall asleep is, of course, an indication of the size of your underlying sleep debt.

It is not uncommon to hear people say "I can fall asleep anywhere anytime," but being able to sleep anywhere at any time means you probably have a significant sleep debt. You should not be able to fall asleep so easily. The best way to evaluate whether or not you are significantly sleep deprived and carrying a large sleep debt is to pay close attention to how you feel during sedentary situations particularly at times when clock-dependent alerting is not strong. For most people, the greatest vulnerability is during the early afternoon. In a Gallup poll, 82% of American adults responded "yes" to the question, "Do you typically experience drowsiness after lunch or early in the afternoon?" It must be clearly understood that you can be obtain your daily requirement of sleep and still have a troublesome amount of tiredness and sleepiness during the day. This is because you have a large sleep debt and substantial extra sleep payback will be required to get out of the "twilight zone," or more obviously, you have a sleep disorder. If you are tired, no matter how much sleep you get, then the problem is the quality not the quantity of sleep. If the quality of sleep is of concern, then an evaluation by a sleep medicine physician and possibly a sleep study may be indicated.

How To Determine The Sleep You Need

One important area of sleep and its impact on wakefulness that has not been well studied is the range of individual differences in average amount of sleep needed each day. From anecdotal accounts, we should expect the range of individual differences to be fairly large. However, documenting the individual differences in sleep need of several hundred representative human volunteers utilizing all night PSG and the MSLT method described in Chapter 3 has yet to be done, and may never be done. Such an approach would require individuals to spend many days and nights in a sleep laboratory undergoing nocturnal polysomnography and daytime MSLT's. The time required and the costs are simply prohibitive.

Another method called actigraphy (Figure 7-1) could provide a reasonable approximation. In this approach, movements are continuously monitored wearing a "wristwatch" motion sensor. A number of validation studies have yielded a very high correlation of amount of movement

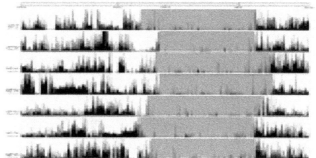

Figure 7-2

Figure 7-1

of the non-dominant wrist with the presence of sleep or wakefulness. This approach is now widely accepted when the nightly or daily amount of sleep must be monitored over several or many nycthemerons. Depending on the desired resolution, the data can be downloaded from the motion sensor every day, or at much less frequent intervals. The small loss in precision of detecting sleep and wakefulness is more than compensated by the increased convenience and greatly lowered costs. Actigraphy is used in sleep medicine and in various setting to determine sleep patterns over time.

(Figure 7-2)

It is possible, however, to make an adequate estimate of your sleep need without either laboratory testing or actigraphy. Sleep management systems such as Zeo and LARK, and smart phone apps are available for tracking the quality of your sleep and then giving you personalized advice to help improve it. You can measure how much restorative deep or REM sleep you are getting, which are critical for mental clarity, physical well-being, and performance. It is strongly recommended that everyone seriously attempt to determine his or her personal sleep requirement at least once during once in their life. As we get older, the need for sleep does not change. It is the ability to sleep that changes. In light of this you should assume that your individual sleep need will tend to remain essentially the same for your life. Determining your personal sleep requirement should be attempted only when you have a chance to maintain a fairly regular schedule for several weeks, including weekends. This will be difficult for the average college student while school is in session. College life is punctuated by varied social and academic demands on your time. If at least a semblance of regularity cannot be maintained, there is little point in beginning such an effort. It would be like starting a diet on the day before Thanksgiving. Nonetheless, what follows is the recommended approach to determining your personal sleep requirement.

Step One. Most people have some notion about the amount of sleep they need. The following example will involve a hypothetical person who says, "I believe that I need eight hours of sleep to feel energetic and alert all day." This person should select a bedtime when falling asleep is likely to occur relatively promptly. The getting up time should be 10 or 15 minutes more than eight hours after the bedtime so that the estimated eight-hour requirement is likely to be fulfilled each night. Given that there are no early morning commitments, the selected bedtime might be midnight in which case the wake-up alarm should be set for 8:15 a.m. After the first night, our hypothetical subject should ask the question, did I fall asleep promptly? If the answer is no, the midnight bedtime may be too early. If sleep onset occurs promptly but getting up in the morning was still very difficult, our hypothetical person likely has a very large carryover sleep debt. If an adjustment of bedtime seems desirable, it would be wise to do it after the first night in order to get on with the rest of the program as soon as possible.

Step Two. Once you have selected your initial sleep and bedtime schedule, you must pay careful attention to the tendency for sleepiness to occur in non-stimulating, sedentary situations throughout the entire day. How often do waves of drowsiness occur in the daytime, in afternoon classes, in morning classes, while studying, after lunch, or while driving a car? Do waves of drowsiness persist? How long? If daytime sleepiness is very troublesome even with eight hours of sleep, the time in bed should be extended 30 or 60 minutes. If daytime drowsiness is not very troublesome, continue the original schedule and note carefully any occurrence of drowsiness and the situation in which it occurs. In other words, you must seek out boredom.

If possible, the selected schedule should be maintained for at least two weeks and very close attention should be paid to daytime function. If you begin to note an increasing tendency for sleepiness to occur during the day in soporific situations, then the scheduled eight hours of sleep is less than your daily requirement. In this case, the procedure should be repeated extending the time in bed by about one hour.

On the other hand, if the time scheduled for sleep exceeds the daily amount of sleep that you need, your waves of daytime drowsiness should unambiguously begin to diminish in strength. In addition, it should become more difficult to fall asleep at night. There might also be an increasing tendency to awaken too early. These changes would indicate that your individual sleep requirement is less than eight hours. As your sleep debt is reduced, the diminishing sleep drive will no longer keep you asleep most of the time you are in bed.

By repeating this procedure, we should be able to "zero in" on our personal daily sleep requirement. Once your personal requirement has been determined, you can use the information to reduce your sleep debt and lead a more enjoyable and fruitful life.

Sleep Debt and Alcohol Do Not Mix!

In addition to determining their personal sleep requirement, readers should also try to estimate as accurately as possible the amount of sleep debt they can tolerate without experiencing truly overwhelming attacks of drowsiness and sleep in class, while driving, or in any sedentary, soporific situation. Once this self-knowledge has been acquired, people should avoid sleeping below this minimum amount as much as possible, except in very rare and true emergencies, or when their waking activities involve no risk to themselves and others. Finally, it is important to emphasize that alcohol and other sedative drugs are greatly potentate by a sizable sleep debt. With regard to the use of alcoholic beverages, our clock-dependent evening alertness can be very misleading. A person can remain alert at a party in the evening after consuming an alcoholic beverage, but can almost instantaneously become dangerously sleepy while driving home if this is when the biological clock stops opposing the alcohol sleep drive.

Reducing Your Sleep Debt

We have learned that individuals can accumulate a very large sleep debt without realizing it. Sleep and Dreams student surveys have shown that some very sleep deprived students can feel "tired" most of the day but only feel sleepy during brief periods. This is why we emphasize being exquisitely sensitive to the moment of drowsiness when suddenly a conscious effort must be made to stay attentive, keep the eyes open, and to hold your head up. Poor understanding about how sleep deprivation works also fosters insensitivity to our sleep indebtedness. For example, someone with a very large sleep debt might have extreme difficulty getting out of bed in the morning because of sleepiness and grogginess. "I am not a morning person" is a common phrase, and such an individual typically thinks this is an intractable personal characteristic rather than the weight of a huge sleep debt. When the strong clock-dependent alerting in the late afternoon and early evening reverses the sleepiness, it is interpreted as "I am definitely a night-owl."

Since night owls are likely to have trouble going to bed early, they should reduce their sleep debt by sleeping late. Larks, on the other hand (people who are wide awake in the morning), would be better off obtaining extra sleep and reducing their sleep debt by going to bed early. Relatively few students are unambiguous larks. We may have tendencies to be night owls or morning larks. It should be noted that your tendencies are not your destiny. Sometimes we need to work harder to counter our tendencies to improve our health. Finally, students must be aware sleepiness might be caused by a sleep disorder. In an undergraduate population, these are most likely stress-related insomnia, delayed sleep phase syndrome, obstructive sleep apnea, and restless legs syndrome - all of which will be described in later chapters. The likelihood that a sleep disorder is the cause of tiredness and/or sleepiness in the daytime is greatly increased if individuals feel that the amount of their nightly sleep is more than adequate. After many years, student health centers, such as Stanford's Vaden, now recognize sleep disorders. For example, victims of obstructive sleep apnea who awaken hundreds of times every night to breathe may not recall even one such awakening. Rather, they may spend more hours per night in bed than most healthy individuals. They feel their sleep is lengthy and continuous and they therefore cannot understand why they feel so tired in the daytime. Any student or other person who is definitely excessively sleepy and claims to spend adequate or more than adequate time in bed sleeping should be promptly referred to a sleep specialist for diagnostic evaluation see figure 7-3.

Can We Get Too Much Sleep?

Sometimes when students have an unusually long sleep period, they are surprised to find that they wake up feeling groggy, achy, and listless. They are puzzled that they did not wake up feeling "great" and fully restored. Some students report that they feel worse than when they went to bed. This experience is fairly common particularly among young adults and has given rise to the widespread but entirely erroneous belief that it is possible to get "too much sleep." The feeling of over sleeping most likely arises from **sleep inertia** (see glossary.)

There are several reasons for this response to an unusually long night of sleep. In the first place, the ability to sleep 12 or more hours or into the late afternoon means that one has a very large sleep debt. Most nights of prolonged sleep occur following the bouts of severe sleep loss that may occur during final exams. Several hours of extra sleep is almost certainly a negligible reduction of a very large sleep debt. A large sleep debt simply cannot be eliminated by one extended sleep period. In the second place, feeling groggy and unrested when awakening contrasts shockingly with what was expected by the naive individual, although it simply reflects a continuing strong physiological sleep tendency. Thirdly, when sleep is markedly extended, individuals are likely to awaken close to the midday dip in alertness, much like being jet lagged. Finally, it is likely that any muscular aches, pains, and/or stiffness are the possible result of the unusually protracted period of immobility, not specifically the result of "too much sleep."

Figure 7-3: As the sleep debt gets larger and larger, it is harder and harder to cope. At some point, no matter what a person does, sleep will seize the brain.

There is no evidence whatsoever that there is a true phenomenon of getting "too much sleep." On the contrary, the evidence suggests that when the sleep debt is near zero, one cannot sleep more than the daily requirement. This also means that we cannot store up sleep. The best result in terms of individuals sleeping several hours more than their usual amount is achieved when such a schedule is maintained much longer than just a day or two. It may take more than a week to produce a noticeable reduction in the sleep debt. When such a reduction has been accomplished, individuals will begin to experience increased energy, improved motivation, peak performance, feeling "great" throughout the entire day and will never think they had "too much sleep."

Sleep Hygiene: Healthy Sleep Habits

Just as good physical health implies lack of disease, healthy sleep implies lack of specific disorders of sleep. A student who has good physical health can undermine his or her health by inadequate sleep. Hygiene generally stands for everyday habits that promote good health, and **sleep hygiene** emphasizes proper habits to promote healthy sleep. The goal of good sleep hygiene is to foster refreshing deep, continuous sleep at night and optimal daytime alertness. Sleep hygiene guidelines include:

1) Create an optimal environment for sleeping.

It is not unusual for students to convert their bedrooms into mini studio apartments. Many high school students may spend more time awake in their bedrooms than sleeping in them. Your own bedroom may be one of the worst places for sleeping. In many home environments, the bedroom receives the lowest priority in the quality of its furnishings. The following suggestions may be difficult to implement; however, readers should do their best.

Quiet, dark, secure surroundings are desirable. Some people can, of course, be champion sleepers, and possibly due to severe chronic sleep deprivation, can often sleep anywhere. As one gets older, environmental factors become more and more important. A comfortable bed, the right degree of firmness, and the right amount of covers should receive meticulous attention. The place you sleep should be your sanctuary.

When possible, the bedroom should be used primarily for sleeping. The more the bedroom and the bed are used for other activities, the weaker the association will be between the bedroom and sleep. This may allow distracting thoughts and activities to invade an individual's "sleepy" phase, prolonging wakefulness and reducing both sleep efficiency and total sleep time. "The bedroom is for sleeping" is a fundamental principal of optimal sleep hygiene. It should be remembered, however, that as the sleep debt is reduced, the depth of sleep at night will also be reduced.

2) Predictability. Maintain a regular sleep schedule. If people are truly serious about their sleep, they will consistently schedule predictable hours in bed. In young children, the day of the week usually makes no difference: they go to bed at the same time on Saturday nights as on Tuesday nights; they get up at the same time on Sunday morning as they get up on Wednesday morning. Students, as well as most other adults, shatter this principle by staying up late on Friday and Saturday nights, sleeping late Saturday and Sunday mornings, and changing bedtime hours throughout the week. They also continually disturb sleep by trans meridian travel, changing work schedules, staying up late on-line. In general, it is undesirable to take naps in the evening. If sleepiness becomes overwhelming, limit naps to 30 minutes or less and take them before 3 P.M. Designate a specific personal bedtime for yourself so at least you know when you depart from it. Furthermore, you should rarely need an alarm clock to get up in the morning. If a person must be dragged out of bed by another person, insufficient sleep, delayed sleep phase syndrome or both, are implicated (see Chapters 4, 6, and 21). Allow an adequate time to unwind before going to bed.

Avoid all substances that adversely affect the sleep/wake cycle. The most common drugs used by adults are caffeine and alcohol. Too much caffeine late in the day interferes with sleep. Most people have learned not to drink coffee before going to bed, but some have not. Chronic use of alcohol also appears to impair the sleep mechanism. One of the most common causes of insomnia in the United States is consumption of alcohol at bedtime; as the alcohol wears off, it often produces middle-of-the-night arousals. Alcohol may also lead to fragmented sleep.

3) Good general health favors optimal sleep. Exercise, proper diet, and regular timing of meals promotes optimal sleep and optimal alertness. These activities, when performed regularly, may act as social time queues (**zeitgebers** -see glossary), further strengthening the circadian rhythm of sleep and wakefulness. Readers should exercise regularly (daily), preferably in the morning or afternoon (about 3-5 hours before bed time). Try eating a light snack, drinking herbal tea, or warm milk. Also, avoid spicy foods and heavy eating in the evening (i.e., large meals, large quantities of liquids). Do not go to bed with hunger pangs. Above all, avoid beverages and food containing caffeine.

A Lifetime Philosophy: Triumvirate of Health

The Most Important Rule: Drowsiness is RED Alert! The reader is once again reminded that the terms sleepy and drowsy are defined to mean exactly the same thing. In any situation where falling asleep is dangerous, the moment of drowsiness must always be responded to as a "red alert." We must respond in the same manner as we respond to the burglar alarm, a siren, or the continuous honking of "red alert" signals that Star Trek fans heard so often on the Enterprise. The appropriate response is to get out of harm's way immediately. If one is driving, this means getting off the road without delay and taking a nap. If one is performing important duties and cannot be relieved, this means recognizing that one is impaired and very error prone, and that one must be extremely careful and obtain sleep as soon as possible. Any maneuver other than obtaining sleep only postpones serious drowsiness. If the drowsiness occurs during the mid-afternoon dip in alertness, clock-dependent alerting in an hour or so may reverse it. This does not mean one should keep driving until clock-dependent alerting kicks in! Mid-afternoon drowsiness is also extremely dangerous when performing other potentially hazardous tasks.

There are three general areas of health and healthy behavior that apply to all human beings and the way they conduct their lives. They are: good nutrition, physical fitness, and healthy sleep. The first two, nutrition and physical fitness, are not relevant to this discussion. However, it is likely that the motivation to exercise and the general level of waking activity are related to the level of alertness. In addition, these first two areas of general health maintenance are very well established in the consciousness of the American public.

While all human beings do not necessarily exercise properly or eat a healthy diet, there can be no doubt that everyone is aware that this is desirable. In addition, the promotion of good nutrition and physical fitness are well established in the educational system. From kindergarten on, the school lunches, the education materials, and regularly scheduled and purposefully organized exercise are part of the school day for almost everyone. Classes in health and physical education of one sort or another are required through the grading system, and in some instances even in college.

On the other hand, only occasional lip service is paid to getting enough sleep and napping stops after kindergarten, if not preschool. Sleep need is essentially ignored. The cycle of ignorance continues generation after generation with parents not knowing exactly what to do in terms of structuring the bedtime hours at home, and children learning relatively little about this very important aspect of their lives. It will be extremely important for newly aware individuals to be proactive in promoting healthy sleep not only with regard to themselves and their friends, but in their community as well. Hopefully, these habits will be lasting.

Although a great deal of knowledge about sleep, sleep deprivation, and sleep disorders has been accumulated, this knowledge has not been communicated effectively at any educational system. As a result, the American public and education and health professionals have a pervasive ignorance about the determinants and consequences of sleep deprivation and the associated behavioral and cognitive impairment. It is often argued that even if knowledge about sleepiness, alertness, and the need for sleep were effectively communicated to all individuals in society, they would still behave in a way that maintained chronic sleep deprivation. This may or may not be true, but at the very least, individuals could make informed choices about the amount of sleep they would try to obtain, and they would be forced to accept responsibility for any consequences due to not obtaining enough.

In recent years, anecdotal evidence has accumulated that children in the elementary grades often appear to be sleep deprived, and as a consequence, are sleepy and irritable in class. This is generally attributed to late night television, the Internet, and lax parental control. On the other hand, if parental control is adequate, it is likely that children will usually obtain adequate sleep. In addition, the alerting influence of the biological clock subsides early enough for the child to fall asleep at an appropriate bedtime. However, as adolescence occurs, there is a tremendous increase in the pressures to stay up late. These include homework, part time jobs, Internet and increasing social activities.

Two Ways to Live Your Life

When information about sleep need, sleep deprivation, and sleep debt is presented superficially or ineffectively, most people come away with the idea that they are being urged to spend several additional hours in bed every night for the rest of their lives. This is absolutely not what is being recommended. The many nights of extra-long sleep which are required to pay back a very large sleep debt should not be equated with a permanent, lifelong need for this amount of sleep. Recall the results of the 14 hours in bed study described in chapter 3, Figure 3-8. This study clearly showed that eventually extra hours of sleep are unnecessary and, in fact, without a sleep debt one cannot sleep extra hours above his or her daily requirement.

We have learned that all wakefulness is sleep deprivation, and that sleep deprivation is required to empower the sleep homeostat to induce sleep. Accordingly, each reader should settle on the duration of daytime wakefulness that fosters sound and adequate sleep at night. They must also determine as precisely as possible the size of a small and manageable carryover sleep debt that will do this, but will be easily opposed and overcome by clock-dependent alerting in the daytime, and will therefore not undermine optimal daytime alertness.

If individuals reduce their sleep debt even below the aforementioned small carryover sleep debt, they may begin to have difficulty sleeping at night. For those whose clock-dependent alerting is late in the day, the first sign of an overshoot in reducing the sleep debt is the development of difficulty falling asleep at the accustomed bedtime (sleep onset insomnia). Researchers hope to be able to provide more precise values in the future, but at the present time, individuals must determine their optimal carryover sleep debt by trial and error. A fundamental guideline may be derived from our vast experience in exhorting students to lower their sleep debt. Assume that the ideal carryover sleep debt is between the amount that exists when afternoon drowsiness disappears and the amount that exists when sleep onset insomnia appears.

As a hypothetical example, imagine a woman who needs exactly eight hours of sleep per day and who is in the process of diligently reducing her sleep debt. On the previous night, instead of falling asleep in 5-10 minutes at her usual 12:00 bedtime, she did not fall asleep until 1:00 A.M. She then deprives herself of one hour of sleep each night by going to bed at one o'clock and getting up with an alarm clock exactly seven hours later. She pays close attention to how she feels in her 2:00 P.M. class. Assume that after she accumulates exactly twelve hours of additional sleep debt, her afternoon drowsiness begins to reappear. Accordingly, her optimal carryover sleep debt would be the amount that existed on the first night she experienced sleep onset insomnia plus an amount less than twelve hours and more than zero. The only way to know the size of the remaining invisible carryover sleep debt when sleep onset insomnia developed would be to lie in bed, in the dark, for 14 or more hours each night until no extra sleep occurred, and then measure the total amount of prior extra sleep (which would equal the minimal carryover sleep debt). This is absolutely not practical!

Figure 7-4 shows this philosophy. Two individuals are shown, both of whom are spending an average of eight hours in bed every night. The man on the left in Figure 7-4 (a hypothetical truck driver) is going through life with a small and manageable sleep debt, able to function at the peak of his abilities throughout the entire day, energetic, cheerful, highly motivated, and able to be very creative and clever. Thus, a low sleep debt optimizes performance, sound sleep is fostered, emergencies (when adequate sleep cannot be attained) are easier to handle, and in general, life is a bowl of cherries.

Figure 7-4: Which lifestyle will you choose?

The truck driver on the right spends the same amount of time sleeping each night, but he leads a very different life. He carries a very large sleep debt, which impairs his performance during the day. He tends to be lethargic and not very interested in doing things. His memory and his ability to perform complex mental tasks are far from optimal. He is frequently drowsy while driving. Finally, he tends to be pessimistic and thinks he might have depression or chronic fatigue syndrome and feels sexually inadequate. It is also likely that he cannot handle nocturnal emergencies because the sudden addition of substantial sleep debt to an already large amount will lead to catastrophic sleep episodes.

Managing sleep is analogous to managing money. One can be very prudent about accumulating debt and be considered a good risk; to have monthly payments, which are very easy to handle; and to be able to deal easily with financial emergencies. The imprudent financial manager will have huge debts; will always be on the verge of bankruptcy; will have large interest payments each month; and will be considered a very poor risk any time a loan might be needed. Why would anyone choose to live one's life with a huge sleep debt? You have been given a choice about how to live your life. Which way will you choose?

Summary

Although the phenomenon of drowsiness is ignored by almost everyone, it is always in the wings as we act out our entire lives. The daily fluctuation in sleep tendency organizes and limits our activities, determining whether we work and perform with efficiency and whether we sleep with efficiency. The continuum of sleepiness/alertness extends from very drowsy to peak alertness. Obviously, mental function is essentially impossible during the intense sleepiness that immediately precedes falling asleep. It is amazing, therefore, that the waking dimension of sleepiness/ alertness has not received attention, ignored by cognitive and clinical psychologists as well as many other health professionals. Extreme sleepiness can, in effect, be likened to sleep since an extremely sleepy person is essentially a walking zombie.

In order to lead a healthy and productive life, everyone should cultivate a daily level of alertness that fosters his or her life objectives. We will accordingly list the fundamental facts and principles that students should know and remember for the rest of their lives in order to manage their sleep and wakefulness in an optimal manner.

(1) There is a necessary amount of sleep for each individual and any amount less than this accumulates as a sleep debt.

(2) The sleep debt can only be reduced by extra sleep.

(3) The amount of sleep we need is almost always more than we think, and assessing this need is obscured by external and internal stimulation.

(4) The amount of sleep needed may stay the same or even increase in the second decade of life over the amount needed in late childhood. It may then decrease very slowly, if at all, as the years go by.

(5) Taking a nap is often a very good strategy to avoid a disastrous sleep episode.

(6) Breaks and coffee provide only temporary relief, and over a longer period increase the danger that an unintended sleep episode will occur.

(7) Unintended or unwanted sleep episodes, in addition to severe sleep deprivation, may occasionally be due to a specific sleep disorder such as obstructive sleep apnea or severe insomnia.

(8) "Drowsiness is red alert!" More than anything else, feeling drowsy in a situation which has any hazard at all if an unintended sleep episode or error of inattention occurs, should be a very strong warning signal to get out of harm's way. The best strategy at this point is to stop whatever you are doing and take a nap.

"Sleep is the best meditation." - Dalai Lama

CHAPTER 8 Sleep Deprivation: Consequences & Counter Strategies

"We have zero tolerance of sleeping on the job..." - Ray LaHood, former Secretary of Transportation

A tendency for students to fall asleep is present to a greater or lesser degree during every minute of the waking day. From the perspective of motivation, the tendency to fall asleep is a part of homeostatic drive. The subjective manifestation of the drive to sleep is progressive tiredness giving way to sleepiness and a desire to sleep. The drive to sleep can be very strong or very weak depending upon the prior history of the individual. As such, it competes successfully or unsuccessfully with other drives and other demands upon the student. It is clear that the drive to sleep and the neurobiological processes that mediate this drive have the potential to undermine all other cognitive activities and behaviors. When very strong, the drive to sleep can compete successfully with even very basic drives such as reproduction, hunger, and even thirst. The stronger the drive to sleep (the larger the sleep debt), the greater the negative impact on every aspect of our waking life.

Waking Functions Affected by Sleep Loss

Task Performance

There is an enormous amount of research that documents impairment in a wide variety of simple performance tasks, e.g., reaction time as a function of sleep loss. Initially, researchers had difficulty showing consistent effects. However, it was soon determined that longer tasks such as adding columns of numbers for an hour were much more sensitive to sleep loss than shorter tasks. The interpretation of this is simply that sleep deprived subjects can usually summon a burst of energy and motivation for a very short period of time even when very sleepy. In simple performance tasks, impairment increases progressively with accumulating sleep loss. Errors increase and performance slows.

Complex Tasks

Very complex tasks are difficult to standardize. Nonetheless, sooner or later in a sleep loss paradigm, a subject's ability to solve difficult puzzles, achieve special insights, and complete all sorts of complex tasks is always impaired. In general, subjects do better when they have as much time as they need to complete a task than when the time allowed for a task is strictly limited. Some say, "Sleep deprivation makes us stupid." All of the functions and behaviors that have been measured deteriorate as sleep loss continues. This is one of the most firmly established facts in biology. The rate of deterioration of different functions varies and there are sizable individual differences, but sooner or later everything becomes impaired.

Mood

In general, mood is negatively affected by sleep loss. The sense of optimism, joy, and happiness, all disappear as sleep debt increases. The relation of sleep loss to clinical depression is unclear. Most people with depression complain of poor sleep but many sleep-deprived subjects say they experience a depressed mood. Paradoxically, a depressed person may have brief improvements in their mood by being REM deprived. In addition, emotional responses appear to become unstable. Subjects undergoing experimental sleep loss may have periods of uncontrollable laughter, giddiness, or crying. Some individuals become silly and euphoric.
Others sob and cry without being able to give a reason. On the whole, there appear to be enormous individual differences in emotional responses to sleep loss.

Motivation

For many individuals and experimental subjects, the most sensitive response to sleep loss is in the area of motivation. The desire to do things and the ability to maintain interest in various activities and subjects decreases remarkably. As sleep debt increases, there is a progressive apathy--a tendency to sit and do nothing. Often people do not understand that sleep debt is responsible for their high levels of lethargy and apathy.

Self-Control and Impulsiveness

There is plentiful anecdotal evidence that sleep deprived subjects are more readily provoked to anger and rage. Pediatricians have testified at hearings of the National Commission on Sleep Disorders Research that the leading cause of infant abuse is the sleep deprivation of parents and, in this state, there is a greater likelihood they can be provoked to harm a continuously crying infant. Sleep deprived subjects are often cranky, grumpy, and extremely irritable. The casino industry is well aware that risk taking behavior and impulsiveness increase with sleep derivation. Gambling casinos rarely have windows or clocks that would help people know how long they have been awake. Although we have long known that risk taking behaviors increase with sleep deprivation, recent neuroimaging studies in 2013 have localized specific brain regions that may be involved. Using fMRI imaging, sleep deprived adolescences were confirmed to exhibit greater risk taking behavior. These results suggest that poor sleep creates an imbalance between affective and cognitive control systems in the brain leading to greater risk taking in adolescence.

High Level Cognitive Function: Memory and Creativity

Numerous studies have documented impairment of memory with increasing sleep loss. These studies have tested recall and/or recognition of nonsense syllables, simple words, stories, complex phrases, and so forth. The time interval between presentation of material and testing has been varied. Impairment is consistent and is related to the amount of sleep loss. Creativity is a more difficult area of functioning to examine. However, given the overall impairment of so many functions, it is likely that creativity would be very negatively affected by sleep loss. The few studies reported support a role for sleep and specifically REM sleep to play a role in creativity. Loose associations have been documented to increase in REM sleep. This may play a role in creativity and also explain the bizarre nature of our dreams.

Lapses and Micro-sleeps

As was pointed out earlier, when individuals are running, walking, or even standing without support, they rarely fall asleep. When sitting and in situations where posture does not have to be consciously or semi-consciously maintained, a sleep deprived individual is likely to fall asleep. Recall that the Multiple Sleep Latency test has a clear lower limit. When subjects fall asleep in less than a minute on all five tests, no additional change in test scores can occur as a result of additional sleep loss.

Laboratory based sleep deprivation studies have documented the occurrence of **micro-sleeps**. Micro-sleeps are very brief periods of sleep usually consisting of NREM sleep of which a person is not aware and denies being asleep. The typical approach to document micro-sleeps and lapses is requiring a subject who is being partially or totally sleep deprived to sit in front of a computer monitor and respond to signals while brain waves are being recorded. In addition to an overall decrease in response time, there will be increasing numbers of response failures called "lapses." Such lapses are associated both with micro-sleeps (periods in which brain waves change) or complete inattention though apparently awake. In one of the longest partial sleep loss studies to date, subjects were limited to four hours of sleep per night for two weeks (14 nychthemerons).

In this well-designed study, the subjects were tested many times each day. The number of lapses increased progressively. Not only was there no evidence of plateau in this measure, but also during the second week there was an acceleration in the rate of increasing impairment!

Psychomotor Vigilance Task

In recent years, there has been a consistent effort of the sleep research community to devise a standard measure of fitness for duty. The outstanding candidate in this regard is the **Psychomotor Vigilance Task** (PVT) developed by Dr. David Dinges at the University of Pennsylvania. One of his goals was for the task to be convenient and practical. The PVT is limited to ten minutes. The subject watches a small screen for flashes of light. The task is to press a button as quickly as possible in response to each flash. The reaction time in milliseconds appears on the screen. The individual taking the test must try them to react as quickly as possible. Subjects who are not sleep deprived typically perform at a certain level. There is a characteristic lengthening of reaction times during the ten-minute test. Experimentally, sleep deprived subjects will start off with a slower reaction time than normal subjects and typically show a steeper negative slope. This test has been very widely used, appears to be very sensitive to sleep loss, and has a very large database. Compared to the MSLT, it is certainly much more practical.

Brain Activity

Studies that directly observe brain activity as a function of sleep deprivation (with the exception of examining brain waves) are quite rare because they are typically difficult, expensive, and cannot readily employ repeated measures. More recently there has been research utilizing functional magnetic resonance imaging technology to monitor activity in the brain of sleep-deprived subjects performing simple learning tasks. Certain areas that were activated in controls failed to become activated in sleep-deprived subjects when both groups performed identical simple arithmetic tasks. This is evidence that during sleep deprivation, the function of specific brain areas may diminish.

Variability and Overlap

It should also be clear that most areas of impairment described in this text are not completely isolated from one another. Performance on any kind of task is affected by the level of motivation. If memory is impaired, complex tasks are more difficult. Impaired reaction time affects the number of problems that can be solved. Negative mood impairs motivation.

Managing Sleep Crises

One of the most convincing studies involved several hundred adults who were recruited on the basis of saying they were fully alert in the daytime. After undergoing the MSLT testing, only ten percent of the individuals in this subjectively "most alert" cohort were in the optimally alert range of MSLT scores (15 to 20). One may assume that the population of all other adults who do not deny feelings of sleepiness in the daytime would have overall less alert MSLT scores. There are also numerous surveys showing a high population prevalence of fatigue and sleepiness.

In spite of this impressive documentation, the application of the scientific understanding of sleep debt has not been effectively transferred to areas of everyday activities in our society where it would be most crucial. The leading issue is to understand the danger of sleep deprivation when people are engaged in activities that are hazardous. The overall decrement in function plus the dynamic tendency of sleep to occur in sedentary situations makes driving one of the areas of greatest concern. Other areas abound where impaired performance can have catastrophic results but the opportunities for disaster occur less frequently than drowsy driving.

The grounding of the giant oil tanker, Exxon Valdez (1989) was catastrophic. There have also been significant air transportation disasters not to forget the nuclear plant accident; Three Mile Island near meltdown. One can imagine a sleepy person failing to recognize a blip on a radar screen which indicates an incoming ballistic missile. It has been speculated by some newscasters that on the morning of September 11, 2001 the sleep deprivation of air traffic controllers may have played a role in their not responding to the bizarre course changes of the hijacked American Airlines 767s on their radar screens.

When a developing situation reaches a decisive turning point, the moment is termed a "crisis." Applying this term to sleep-related situations may seem a little too dramatic, but it is certainly a life or death crisis when a sleep deprived college student is struggling to stay awake behind the wheel while driving home for vacation after finals week. Anyone who is severely sleep-deprived is likely to confront this crisis sooner or later. This includes airline pilots, truck drivers, shift workers and many others who find it difficult to obtain adequate sleep because of their work schedules.

Another variation of a **sleep crisis** is a sudden or unexpected summons to stay up all night for some critical task or emergency. Firemen, police, paramedics and physicians are very familiar with being on-call or on duty for long hours. Railroad engineers are essentially on-call all the time. Finally, a sleep disorder may have worsened to a crisis point when serious damage to health or safety is so likely that the victim must do something immediately to improve the situation or suffer serious consequences.

When the moment of crisis arrives, actions taken can either make things better or make them worse. Therefore, it is important for students and everyone to learn the correct strategies in order to manage sleep crises with minimum collateral damage to health and professional work. As in any crisis, the keys to getting through a sleep crisis are knowledge, preparation, and intelligent use of the tools at your disposal to cope with specific situations. Your tools can range from napping to medication. The potential crises are precipitated by one or more of the sleep-unfriendly conventions of our 24-hour society among which are long drives, jet travel, surfing the Internet, shift work, and emergencies among others.

It cannot be emphasized often enough that if you are driving or doing some other potentially dangerous activity and you feel sleep tugging your eyelids down, you absolutely must stop what you are doing. The life and death sleep crisis you are most likely to face will occur when you are driving. You are seconds from tragedy and should hear a warning alarm blaring in your head. You should feel and act as you would if you heard a police siren and saw flashing lights right behind you. Pull over immediately! No job, no vacation, no deadline, no goal or test, is worth putting your life, or the lives of others, in extreme jeopardy.

The first and foremost key to handling any sleep crisis is to master the basic knowledge about sleep, deprivation, and biological rhythms. You must know your own sleep needs and your best times of day to perform or to sleep. You may feel alert when you start out on a long road trip in the morning or begin a tour of duty, but surprisingly drowsy in the early afternoon. You must understand why you will start feeling better at the end of the afternoon when clock-dependent alerting once again pushes your fatigue aside. If you completely understand that sleep debt is additive and can be very large, you will know that one night of extra sleep before a crisis situation is likely to be insufficient. In this circumstance, any crisis situation will continue to be fraught with danger.

Finally, you must avoid alcohol in a sleep crisis because you know now how it multiplies the effects of sleep debt; and you must urge your family, friends, and fellow students to do likewise. Even a very small amount of alcohol that otherwise might do nothing can precipitate overwhelming drowsiness when you have a large sleep debt.

The Size of Your Sleep Debt: Implications

Is there any limit to the size of a sleep debt? When is a sleep debt truly dangerous? Before we address these questions, we should point out that we can now explain certain puzzling individual differences by invoking the concept of sleep debt. For example, in the absence of a particular sleep disorder, if someone says that they never take naps in the afternoon because then they cannot fall asleep at night, we may assume that their carryover sleep debt is small and they consequently require the uninterrupted (no nap) build-up of sleep debt over the entire day to develop a strong sleep tendency at bedtime. Conversely, a person who can take an afternoon nap and then have no trouble at all falling asleep at the usual bedtime probably has a much larger sleep debt.

We can further conclude that a carryover sleep debt in the ballpark of 30 hours may not cause severe or even obvious impairment in healthy young adults. However, we must also conclude that the subjects were nonetheless impaired at the beginning of the study because they experienced very substantial improvements in mood and energy by the time the study ended. We may regard these improvements as the benefit of getting rid of 25 to 35 hours of carryover sleep debt.

Upper Limits of Sleep Debt Revisited

Until proven otherwise, we will propose that starting the waking day with 50 hours of carryover sleep debt will constitute an upper limit for the average, healthy adult. If the sleep debt is more than 50 hours, getting out of bed in the morning will be extremely difficult and waking functions throughout the day will be severely impaired. People at this level of sleep debt are in the twilight zone of sleepiness.

Once Again, Drowsiness is Red Alert!

The paramount piece of knowledge that everyone must take to heart is that drowsiness can become sleep in an instant! There will be no further warning. Once your eyelids start feeling heavy, you are only a few seconds from sleep. For an analogy, imagine the cliffs overlooking the Pacific Ocean 30 miles west of Stanford University. Every once in a while, people ignore the warning signs and walk along the edge of the cliffs. They think they can keep from going over by staying a step or two back from the edge and watching their footing. But then, without warning, the soil gives out from under their feet and they plunge to the rocks below. Too many people are tempted to think they know exactly when they are in danger of falling asleep and that they can safely skirt the abyss. When wakefulness suddenly slides out from underneath them, it is too late to save themselves. No one should die, or be horribly injured because of a sleep-related accident due to drowsy driving. The feeling of drowsiness is the warning sign at the edge of the cliff. Feeling drowsy is red alert!

Coping with Sleep Crises

Prevention: Planning Ahead

Anyone should be able to prevent many sleep difficulties from becoming crises through prudent preparation. The simplest way to protect yourself is to make sleep a priority in your life. Then, if you lose sleep because of an emergency or a particularly demanding schedule, it won't hit you nearly as hard. If you know you are going to have a difficult schedule in a week or two, prepare for it by lowering your sleep debt. Get extra sleep in advance of the crisis.

Even when people have obtained lots of extra sleep and feel very alert all day, they will still have some sleep debt, but it will be much smaller than usual. Keeping your baseline sleep debt low can make the difference in a crisis between being sharp enough to function and being too fuzzy-headed to think, between feeling a little less energetic and feeling really awful. Some people may wish to think of getting more sleep-in advance as "banking" sleep, but technically you can't save sleep because if your sleep debt is zero you will be unable to sleep.

Finally, decide when, how, and where you will get what sleep you can during the difficult period coming up. Make a plan. If you were the coach of a basketball team, you'd plan to rest some starting players during the third quarter to make sure that they are not totally wiped out during the final minutes of the game. Make a contingency sleep crisis plan in case the first plan falls through. Don't just assume you can push yourself through the crisis and keep yourself alert and productive simply because you're facing an important deadline.

Napping

Napping is by far the most important and effective tool for coping with sleep crises. For a few people, adult napping has a negative connotation. People may think of naps as something for children, the sick, the lazy, and the elderly. The phrase "caught napping" reflects the belief that for healthy adults, napping is a blatant manifestation of a sloth.

On the contrary, taking naps is an excellent and respectable strategy for sleep management. Naps can make you smarter, faster, and safer than you would be without them. They should be widely recognized as a powerful tool in battling fatigue and the person who chooses to nap should be regarded as sensible and responsible, if not actually heroic.

While the number of people who nap every day is relatively low, several national surveys including the 1999 "Sleep In America" annual omnibus poll carried out by the National Sleep Foundation reported that only 20 percent of adults never nap. A 2006 NSF study reported that napping improved performance of shift workers. Remember that people cannot nap in the daytime unless they have a sufficiently large sleep debt and the biological clock is not strongly alerting the brain. A 2013 NSF survey reported that 50% of Americans age 25-55 nap twice a month. From annual classroom surveys, it has become clear that Stanford undergraduates are champion nappers. Fewer than five percent say they never nap and well over 50 percent nap frequently or every day.

There is no official definition of what constitutes a nap. For most people, naps should be less than four hours, and if the sleep episode is less than a minute or two, it probably should be called a micro-sleep. For ordinary people, only daytime sleep is called napping, and then only if the individual intended to sleep. On the other hand, it is likely that many of the high school and college students sleeping in the library or passengers sleeping on a plane fell asleep without intending to, but they would probably say they had been taking a nap. From the point of view of managing life's sleep and alertness domain, naps can be roughly placed in three categories. The emergency nap is used to cope with drowsiness in hazardous situations. The preventive nap is taken when an individual has to stay up all night or might be taken in the afternoon to be more alert in the evening. Finally, there is the habitual nap. Some habitual nappers take a nap every day at exactly the same time, usually after lunch. Other habitual nappers are less regular in their daily nap times.

Some formidable historical figures understood the power of planned napping. Winston Churchill took naps during the day so he could work late into the night, a skill that was useful during the Battle of Britain in World War II. President Lyndon Johnson was reportedly a clandestine napper. He actually donned his pajamas in the middle of the day and slept for 30 minutes, giving him the stamina to work longer hours.

Regular napping is a completely natural and correct answer to a biological call. The opponent process model allows us to interpreted the midday dip in alertness as a slight lull between the morning period of clock-dependent alerting and the evening period. Although body temperature generally is lower at night and higher during the day, body temperature and other biological rhythms dip slightly in the afternoon-usually about 12 hours after the middle of the nightly sleep cycle, thereby promoting the likelihood of falling asleep at this time. Accordingly, our minds and bodies are more inclined to sleep after lunch and mid-afternoon than during any other period of the day. Napping in the evening is generally a bad idea for young people because once the midday dip in alertness is past, it becomes difficult to fall asleep, or you may not be able to nap at all. Time spent making an unsuccessful attempt to nap is a complete waste. Daily napping in the early afternoon (the so-called siesta) was once very common in many tropical and Mediterranean countries; midday sleep is a good way to escape the hottest part of the day, and then to be able to utilize the cooler evening hours more efficiently. However, the siesta has become much less common in those countries caught up in the modern global economy. Nonetheless, daily naps can be an excellent strategy for many people, for example, those who work in the evening (also known as the swing shift), but have to get up early to get their children off to school.

Researchers have shown in laboratory experiments that selective, strategic naps can improve performance and measurably decrease sleep tendency. Generally, the longer the nap, the greater the benefit in subsequently increased alertness. Moreover, the benefits seem to be long lasting. It has been found that a 45-minute nap improved alertness for six hours after the nap and an improvement in alertness has been demonstrated for ten hours after a one-hour nap.

Research on napping should be a high priority for our society for all the reasons that have been discussed. There is one aspect of napping that remains somewhat mysterious. Many people claim to be completely refreshed by a five or ten-minute nap. This is difficult to explain in terms of our current knowledge. A study has been reported comparing the benefits from 10-minute naps versus 30-minute naps. Following the naps, performance tests were administered plus several sleep latency measures, though not a complete standard MSLT. The study found the 10-minute naps to be more restorative than the 30-minute naps. It was suggested that the unexpected result was obtained because the shorter nap minimized the occurrence of **sleep inertia.** Laboratory studies and anecdotes raise the question of how naps affect people in the real world. A study on naps in the cockpit was carried out by researchers at the National Aeronautics and Space Administration (NASA). Flight crews flying a four-leg, transpacific route racked up a fair amount of sleep debt over three or four days of flight. The result was an increased number of micro-sleeps during the last 90 minutes of the flight, including some in the ten minutes during the plane's approach and landing. Microsleeps were most common at the end of flights, during night flights, and throughout the last leg of the four-leg route.

The NASA research team then gave one group of flight crews a planned, 40-minute period to nap during each flight. The crew's average sleep latency (the time it took them to fall asleep) was 5.6 minutes, and they got an average of 25.8 minutes of sleep. A control group of flight crews were allowed no sleep during flights.

There was a surprising improvement in the level of alertness and performance in the nap groups. The napping crews had a median 16 percent improvement in reaction times and a 34 percent decrease in lapses of awareness during the flight. The no-nap groups showed a combined total of 120 micro-sleeps during the last 90 minutes of flights (the descent and landing phase), including 22 during the last 30 minutes. In contrast, the napping groups recorded only 34 micro-sleeps during the last 90 minutes of flight, including zero micro-sleeps during the last 30 minutes.

Since the NASA report was released in 1994, British Airways, Air New Zealand, Lufthansa, Swissair, and Finnair have implemented planned cockpit nap periods on their long-haul flights. Other airlines are looking at such plans. Regrettably, United States airline pilots remain at this writing relegated to napping "off the record." Regulations limiting flight time and pilot rest were first issued in 1940s.

A series of well publicized events forced changes in transportation regulations. Due to increased awareness about public safety and sleep, regulations are being continuously reviewed. In 2011 the D.O.T and FAA announced new rule to keep fatigued pilots out of the cockpit. Among these new rules is a minimum rest period for pilots to be 10 hours between and 8 hours of uninterrupted sleep during their rest period. Current FAA regulations for domestic flights generally limit pilots to eight to nine hours of flight time during a 24-hour period. This limit may be extended provided the pilot receives additional rest at the end of the flight. However, a pilot is not allowed to accept, nor is an airline allowed to assign, a flight if the pilot has not had at least eight continuous hours of rest during the 24-hour period. Flight time and rest rules for U.S. air carrier international flights are different from the rules for domestic flights. International flights can involve more than the standard two-pilot crew and are more complex due to the scope of the operations. For international flights that require more than 12 hours of flight time, air carriers must establish rest periods and provide adequate sleeping facilities outside of the cockpit for in-flight rest. Although pilot naps, unintended or otherwise, already occur on U.S. flights, it has proven politically impossible to pass the proposal that specifically permit napping in the cockpit. There is still the unfortunate perception that allowing pilots to sleep in the cockpit is very dangerous — even when another cockpit crew member remains fully awake — despite studies that show that planned napping is safer than not napping. These brief naps did not totally eliminate the cumulative sleep debt of crews, but they did provide relief from the in-flight fatigue. Despite these new rules, fatigue incidents still occur. Most recently, two Air Indian pilots were suspended after they left the aircraft in the hands of flight attendants - while on auto pilots- to take a nap in business class. The flight attendants turned off the autopilot allowing a sudden scare, which fortunately interrupted the pilots nap. There was no crash. However, in 2008, 158 people died when an aircraft crashed. Listening to the cockpit voice recorder, the pilot can be heard heavily snoring before the crash. Concerns, about air safety does not only include pilots but, air controllers and mechanics. The FAA has known for years that the air controllers take naps while at work, but ignored the issue. In 1983, the NTSB urged air safety controllers to take naps but the advice fell on "deaf ears" despite 95% of air safety controllers feel tired or sleepy at work. In addition, 53% admit to taking naps at work. Many say they fall asleep unintentionally.

An interesting aspect of the napping studies is that naps improve objective performance more than subjective performance. In other words, even though subjects often don't feel any better after the nap and don't believe that their performance has improved, objective measurements prove that it has. Just as we are not very good at perceiving how badly we are affected by sleep deprivation, we don't seem to be very good at perceiving the benefits of a nap. The cause of this misconception may be the occasional sleep inertia, which causes some people to feel groggy and to perform poorly for 15 minutes or so after a nap. Anyone who experiences sleep inertia should count on spending this quarter hour getting up to speed after a nap.

Surviving Driving

The following email was sent to Dr. Dement by a Stanford undergraduate who had taken *Sleep and Dreams*. "In the past week, I have heard of four fatal or near-fatal car accidents that happened to people I know or knew about. All of them were a result of falling asleep at the wheel.

1) My friend's next-door neighbor was killed while driving home from the Jersey shore at 10 P.M. When the police arrived on the scene, she was still conscious, and she told them, "I only closed my eyes for a second..." She was 20 years old.

2) A friend's cousin was driving to work one morning after getting only a few hours of sleep the night before. He hit a telephone pole, and now he is on a respirator, and we're all praying he pulls through. He is 24 years old.

3) My brother's friend was driving home late at night. She drifted off the road and hit three mailboxes, narrowly missing a telephone pole. She is 17.

4) I was telling a co-worker about these incidents and what I learned in "Sleep and Dreams," and she said, 'I wish my son had taken that class. He was driving through the night five years ago, fell asleep, and was killed at the age of 19.

These four incidents together make a very sobering reminder that drowsy driving can be lethal. It really woke me up, and I wanted to send it along to keep us all alert."

Although the variety of ways in which human error can lead to tragedy is almost infinite, the number one hazard for college students in general is surely drowsy driving. Driving is monotonous, it's not very challenging mentally, and it doesn't involve much physical effort. All of these factors relax drivers, diminish psychological alerting, and uncover the sleep debt lurking in the brain. Driving provides a terrible combination of increased risk of falling asleep and increased risk of injury or death to self and/or others, as a consequence.

If you, the reader, and your family or your fellow students are planning a long drive, think about how your sleep debt is going to affect you. Prepare for your trip by reducing your sleep debt and plan to drive while your circadian clock is making you most alert. If your peak alertness occurs during the dark hours of the evening, you must factor in the added danger of driving in darkness, particularly if your night-vision is impaired. Driving alone increases the tendency to fall asleep; a good strategy is to have someone in the car with you. Most of all, resolve beforehand that you will absolutely not continue to drive if you begin to feel drowsy.

Do Something!

If, despite all your planning, you find yourself feeling drowsy on the road and have no one else to take the wheel, you must immediately find a safe place to take a nap. Coffee can help you stay alert if sleep debt is low, but at high debt levels there is no substitute for sleep. A nap of 15 to 30 minutes can move you back from the edge of the abyss. You must get off the road and take a nap when drowsiness hits. But where do you get off the road? The concern has been frequently expressed that rest areas are not safe. Drivers feel uncomfortable and vulnerable taking a nap in a rest area, particularly at night. Parking on the side of the highway seems even more unsafe because as the huge 18-wheelers come roaring by, the car shakes, and you feel that you could easily be crashed into by a sleepy truck driver. The unusually frequent instances when a truck has managed to crash into a car on the side of the road suggest that sleepy truckers wrongly assume that the parked car is actually moving on the highway ahead.

The first law to specifically criminalize drowsy driving was enacted by the state of New Jersey in 2003. Maggie's Law defines fatigue as being without sleep for more than 24 consecutive hours and makes driving while fatigued a criminal offense. Maggie's Law states that a sleep-deprived driver qualifies as a reckless driver who can be convicted of vehicular homicide. It's named in honor of a 20-year-old college student, Maggie McDonnell, who was killed when a driver -- who admitted he hadn't slept for 30 hours and had been using drugs -- crossed three lanes of traffic and struck her car head-on in 1997.

Drowsiness is RED Alert!

A study from the United Kingdom suggests a new twist — a strategic combination of coffee and a nap. Researchers from the University of Loughborough put volunteers in a driving simulator during the usual afternoon dip in alertness. One group of volunteers spent two hours in the driving simulator with no rest and no coffee. Another group had coffee with 150 milligrams of caffeine (about two cups), but no nap. A third group had the same dose of coffee followed by a 30-minute nap before their two hours in the simulator. Caffeine alone significantly reduced the number of incidents in which drivers drifted out of their lane on the simulator, but caffeine plus a nap reduced incidents much more dramatically-by almost a factor of four. Although this is just one study done in a simulator instead of on the road, its conclusions are intriguing. Because individuals have different amounts of sleep debt and vary in their reaction to caffeine, you will have to do your own self-testing to find what combination works best for you.

The beauty of the Loughborough method is the synergy of drinking coffee **before** taking a nap. Consider the alternatives. With coffee alone, there is a 15 to 30-minute window of time before the caffeine takes effect, and during this period you will continue to be dangerously sleepy. With a nap alone, there may be a 15-minute period of sleep inertia after you wake where you feel just as sleepy-if not more so-than before the nap. (The period of sleep inertia is also dangerous for driving.) And if you drink coffee after the nap, the sleep inertia and pre-caffeine window occur at the same time-again, not a good condition for driving. Even if you are extremely sensitive to caffeine, the coffee before the nap won't keep you from falling asleep, and you can probably wake fairly easily after 15 to 30 minutes as the coffee kicks in.

Even the synergy of naps and caffeine is no match for a huge sleep debt combined with monotonous activity. It can't be emphasized enough that if your sleep debt is high, the only real solution is to lower it. Remember, it is the sleep debt that causes your drowsiness, not the monotony of driving! If you have any flexibility in your schedule and find yourself unbearably drowsy, consider taking a four-hour nap or checking into a motel to get a full night of sleep. Better yet, work at reducing your total sleep debt in the week leading up to the road trip. Reducing all or most of your sleep debt will render you far less likely to feel drowsy, even while driving monotonous roads for long hours. A lower sleep debt also will make naps and caffeine much more effective.

Work Deadlines, Final Examinations and Breaks: Beware

Most working people and students say that their greatest sleep loss occurs before deadlines and during final exams. Because deadlines and final exams often precede holiday breaks, drivers are usually on the road or in the air to their homes or other vacation destinations the instant they have completed their last project or examination. Also, because breaks are not that long, they want to arrive at the destination as soon as possible. For those who are not thoroughly aware that all lost sleep accumulates as a debt, there is often an erroneous decision to drive straight through. The files of the National Commission on Sleep Disorders Research are full of newspaper clippings from all over the United States documenting the deaths of student drivers on their way to a joyous destination at which they never arrived. Another common erroneous decision made by the unaware and already sleep-deprived person is to cram as much visiting and partying as possible into the holiday. Here again, the toll is enormous.

Special Situations in the Modern World

Jet Lag

When people traveled by horse or ship over long distances, they changed time zones too gradually to create much of a lag between their body's clock and the motion of the sun. However, as transportation sped up, it created unforeseen problems for passengers. Railroad trains were able to travel at fairly high speed for long distances-far and fast enough to shift the biological clock out of synchrony with the sun. Once we could travel faster than the biological clock could adjust, the circadian rhythm would foster sleepiness when we should be wide-awake and alertness when we should be sound asleep.

Plane and jet travel gave birth to a particular modern world commonly known as jet lag. In a nutshell, jet lag is feeling sleepy when everyone else is wide-awake and having insomnia when everyone else is sleeping due to an abrupt change in time zone. About half of all individuals experiencing jet lag also feel nauseated, and nearly all feel hungry at odd hours or not hungry at mealtimes. People with jet lag are distracted and fuzzy-headed, so out of sync with their normal body rhythms that they can feel as if they were ill with the flu.

On average, the body needs about one day for every time zone crossed in order to adjust to jet lag. So for a trip from New York to London, which is five time zones away, it takes at least five days for the body's temperature and hormonal rhythms to adjust to the new time. Yet many people don't feel the symptoms of jet lag that long. Some people actually feel better in a day or two, while others feel lousy for a week. Susceptibility to the symptoms of jet lag is determined partly by the peculiarities of an individual's circadian cycle and partly by the amount of his or her sleep debt at the time of travel.

It is easier for most people to adjust when flying west than when flying east. It has been hypothesized that this is because the "natural" circadian rhythm overshoots the Earth's 24-hour rhythm. Consequently, getting to bed later (flying west) works with our natural rhythm and getting to bed earlier (flying east) works against it. Thus, we can see that it is easier to fly westward at least partly because it is easier to go to sleep when clock-dependent alerting is in decline than to go to sleep when the clock-dependent alerting is strong. This is summarized in the common saying, "East is beast, west is best."

Strategies for Dealing with Jet Lag

Most frequent travelers have tried one or other systems to beat jet lag, but no one beats jet lag completely. There is no practical way to adapt to a new time zone instantaneously, but there are some things you can do to make jet travel easier.

Work Around Your Clock-Dependent Alerting

The first principle is the same as it is for all sleep crises: understanding. Knowledge about how the interaction among sleep debt, biological alerting, and light's effect on the biological clock are critical for understanding how you can best manage your jet lag.

Remember that it is the mismatch between your internal clock and the external (local) clock that is the main cause of jet lag. Accordingly, the first step is to calculate when your personal strong clock-dependent alerting will be occurring in the new time zone. For instance, the time difference between San Francisco and Paris is nine hours. If people fly from San Francisco to Paris and arrive in the late evening, their biological clock will tell them that it is early afternoon. Our travelers may even feel a little bit tired, because the mid-afternoon dip in alerting is just happening. They have dinner and go to bed at 10 P.M. but can't stay asleep. They are trying to sleep just as their biological clocks are swinging into the strong evening (back home) alerting period. Let's assume our travelers have biological alerting that is very strong from 6 P.M. to 9 P.M. at home. This "forbidden zone," when sleep is almost impossible now extends from 3 A.M. to 6 A.M. in Paris, which is why they fall asleep briefly in the hotel room but wake shortly afterward and are unable to return to sleep for hours.

These travelers should plan to stay up, tour Paris, and see the sights at night. They should go to bed in the early morning, and plan to sleep until after lunch. Getting immediately out into the bright light of early afternoon could help advance the clock, which is good to do if the plan is to spend more than a day or two in the new time zone.

A business traveler should try to schedule afternoon and evening meetings so that he or she is alert. If this is not possible, and the traveler wants to be at peak alertness in the morning, he or she should take a sleeping pill or attempt to advance the biological clock a few hours in the days before the trip. In developing a jet lag strategy, the most important fact is how long the traveler will stay in the new time zone. If the visit is only one or two days, it makes very little sense to attempt resetting your clock since you will only have to reset it again when you come home. Let's review San Francisco travelers arriving in Paris once again. They arrive at 8 A.M. Now their biological alerting is falling rapidly, since it is 11 P.M. the day before in San Francisco. If they have a large sleep debt they will feel terrible throughout the whole morning. In the afternoon, they may feel a little better, because their morning alerting starts to kick in around dinner time (9 A.M. San Francisco). But morning alerting is fairly weak, and by this time they have accumulated an even larger sleep debt, so they will still feel terrible. The ultimate irony is that by the time evening rolls around and they allow themselves to sleep, they cannot stay asleep for long because the evening alerting is just coming into play. Decide which time zone you would like to be in. If your stay is less than a week, you may want to stick to your home time zone rather than trying to adjust to the new zone.

What Is Your Sleep Debt Baggage?

Besides the duration of the stay, it is important to know the approximate size of the sleep debt you are carrying as you commence your travels. We have said that before you fly, as with driving, you should try to lower your sleep debt. However, if a conventional schedule must be maintained in the new time zone, you must understand that a low sleep debt will actually lead to more middle-of-the-night awakenings and fractured sleep. A somewhat larger sleep debt can help you stay asleep longer when you do want to sleep. So, if you are arriving in Paris in the evening, you may want to have some sleep debt to help you sleep that night. If you are arriving in the morning, you may want to have a slightly lower sleep debt so that you can make it through the day, or you may take a short nap in the late morning or midday.

CHAPTER 8 Sleep Deprivation: Consequences & Counter Strategies

Preparation

If you understand sleep debt and clock-dependent alerting, you can prepare knowledgeably. Most discussions of jet lag don't adequately address preparation before the flight. Many advisors recommend changing your watch to the new local time immediately upon boarding the flight, so that you start thinking on the new schedule. You could in theory change your watch a day or sooner, however be careful you do not miss your flight! Try to eat meals, go to bed, and get up an hour closer to your local destination time during the 24 hours before you leave. All of these techniques shift your thinking to the new time and your body to the new rhythm. All of the foregoing assumes a fairly long stay, certainly more than a day or two. It is also important to anticipate how you will be traveling from the airport to your final destination. Will you be driving or taking a cab? You must also think about delays and how this would alter your travel plans.

Resetting the Clock

There are three ways to reset the biological clock: exposure to bright light, ingestion of melatonin, and physical activity. The latter is less well understood. To accomplish their effects, these interventions must be scheduled at the proper time. There are, however, individual differences in how people react to each. If readers intend to follow some of the recommendations for travelers and shift workers in one or other of the many books that have been written, the recommendations should be tested a bit beforehand so one knows one's own level of responsiveness and what to expect.

If you are trying to shift your clock forward to an earlier time (called phase advancing) in preparation for flying east, spend some time in direct sunlight as soon as the sun comes up. To shift your clock back to later time (called phase delaying) in preparation for flying west, sleep late in the morning and get out in the sun as late in the day as possible. One should take melatonin late in the day if one wants to advance the clock, or in the morning if one wants to delay the clock. Do this first at some non-crucial time to assess your response to melatonin.

On the Plane

You really cannot control when your meals will be served on an airplane, but you can bring you own food to suite your needs. These little psychological tricks can make a difference. Whether you decide to nap or stay awake during the flight depends on whether you want to be awake or falling asleep when you arrive. For example, someone flying from San Francisco to New York with an evening arrival should avoid naps during the flight in order to be good and tired from traveling and ready for an earlier-than usual bedtime. On the trip home, the person might want to take a nap on the plane so he or she can stay up later and adjust to Pacific Standard Time again.

It should be noted that it is tempting to drink alcohol but the reader should keep in mind that alcohol can interfere with one's ability to adapt to time zone change and it is not an ideal sleeping agent.

Sleeping Pills

Some people use sleeping pills on planes to great advantage, particularly during red-eye flights or long international flights. Make sure that you have already used a specific medication several times and are familiar with its effects before taking it on a flight so you know that you will not be sedated upon arrival. Do not take a sleeping pill if you must change planes in the middle of the trip. Sleeping pills can cause a special type of amnesia (memory loss) called traveler's amnesia, in which a person may not recall events occurring during some period of time after taking a drug. This is ordinarily not a problem, because the person taking a **sleeping pill** intends to be asleep for 8 or more hours. It can be a problem when the drugs are taken to induce sleep while traveling, such as during an airplane flight, because the person may awake before the effect of the drug is gone, such as a 6-hour flight. This is how one can experience traveler's amnesia, where they have blacked out memory. Travelers who take sleeping pills to combat jet lag - particularly those who combine them with any type of alcohol - may experience a temporary but total lack of recall for events that occur while the drug is in effect. Remember not to mix alcohol with sleeping pills.

At the Destination

Some "experts" advise you to adapt to the local schedule once you arrive at your destination. This is too simplistic an approach. If you find it too grueling to do everything the locals are doing on their schedule, go ahead and take a nap-but try to time the nap to coincide with your dip in clock-dependent alerting. Plan on spending some of your first night awake. Your goal should be to try to do things on the local schedule, but do not torture yourself over it. Plan on getting adequate sleep over the 24-hour period, if possible, and use the principles that you have learned about clock-dependent alerting and sleep debt to help you cope until your clock has adjusted to local time. Keep in mind that many big cities are 24-hour cities where you might not have to adjust to the time zones.

Shift Work

Before World War II, relatively few people worked at night except for night watchmen, hotel clerks, bakers, criminals and a few others. After the attack on Pearl Harbor, extraordinarily high production goals dictated that factory workers churn out war material 24 hours every day, in three shifts of eight hours each. These shifts are still called the day shift (8 A.M. to 4 P.M.), the evening or "swing" shift (4 P.M. to 12 A.M), and the night or "graveyard" shift (12 A.M. to 8 A.M.). Workers sometimes are rotated so that no one works exclusively on the undesirable shifts.

Changing from one work shift to another is difficult in the same way as changing time zones. When workers change shifts, it is as if they had flown eight time zones away, such as going from Denver to Tokyo or San Francisco to London. When the biological clock is not alerting the brain, the sleep debt pushes it toward sleep. The biological clock is at its lowest ebb in the middle of the night, and people are more prone to distractions, lack of focus, poor memory, bad mood, and slow reaction times. Mistakes result. Nearly every major industrial accident in recent decades occurred after midnight in the early hours of the morning: among them are the Exxon Valdez (1989) grounding, the Chernobyl (1986) and Three-Mile Island (1979) nuclear accidents, and the Bhopal chemical plant disaster. Theoretically, people should be able to adapt to working at night and sleeping during the day, just as we can adapt to a new time zone after a few days. However, workers rarely if ever, completely adapt. Night workers typically revert to a daytime schedule on weekends and vacations when they want to see their children, spend time with their spouses, pursue outdoor activities, and lead as normal a life as possible. The only way they can participate in these activities is to break their nocturnal cycle, usually just when they are getting used to it. Studies show that workers commonly get two hours less sleep per night when their sleep time is during the day as opposed to during the night.

This brings up the major problem with **shift work**. Workers that work shift rotation never become fully adjusted to any single schedule. The brain is often fighting to go to sleep when work demands are being made and resisting sleep when bedtime arrives. One of the most common rotation schedules worldwide is one week per shift, followed by a counterclockwise change to the previous period (night shift to evening shift, evening to day, or day to night shift). This is the worst possible combination; a week is just long enough to become relatively acclimated to a schedule, and it is more difficult to make a counterclockwise than a clockwise change. It should also be pointed out that a counter-clockwise rotation is analogous to flying in an eastward direction. Just as workers are beginning to acclimate to a shift after one week, another eight-hour phase shift is imposed on their biological clocks. That phase shift is in the more difficult direction, demanding alertness just when clock-dependent alerting is lowest. For instance, going from the day shift to the evening shift is not a problem-starting at 4 P.M. and working until midnight only means staying up a little later than you normally would. But moving from the day shift back to the graveyard shift demands that workers start work just as clock-dependent alerting is descending to its weakest point in the day.

A large company in Utah changed from a one-week, counter-clockwise shift rotation to a three-week, clockwise rotation (day to evening, evening to night, and night to day). The three-week periods gave workers a week to make the adjustment to the new schedule and two weeks to maintain it. When it came time to rotate, it was to the later shift, which is easier to adapt to. More than 70 percent of the workers preferred the new schedule, and there were fewer complaints of sleep and various other health problems. *The company reported a 20 to 30 percent increase in productivity and significantly lower absentee rates!*

Healthcare Workers

Health care workers can be on duty for extended shifts ranging in duration from 12-24 hours, with day and night alternating as frequently as every few days. In any work situation, a tired and sleepy worker is less effective and, at some point, a higher risk for an error or accident. In 2002, a bill was passed implementing standard requirements for work hours for all resident physicians. Regulations prescribed no more than 80 hours of work per week, a maximum shift of 24 hours, at least 10 hours off between shifts, 1 day off per week and overnight on-call assignments no more than every third shift. The major catalyst for these national regulations was the notorious Libby Zion case in New York city.

Conclusion: Be Safe!

CHAPTER 8 Sleep Deprivation: Consequences & Counter Strategies

Friends do not let friends drive drowsy. The more we learn about the interplay between the alerting action of the biological clock and the sleep-inducing power of sleep debt and the more we apply the knowledge to many life situations, the more we can see what an exquisitely beautiful system we have for waking us up and putting us to sleep. Managing your sleep crises all comes down to understanding this subtle but powerful interaction between sleep homeostasis and clock-dependent alerting. You cannot necessarily adapt to any new schedule by just toughing it out.

In order to put the principles of the opponent process model to work during sleep crises, you have to evaluate the role of sleep debt and alerting in your own life-no standard formula works for everyone. The only hope for saving lives is for everyone to know when they are sleep deprived and to recognize that drowsiness is an extreme danger signal. If we know a friend or a fellow student has a risk for falling asleep at the wheel, we should not let them drive. In all hazardous situations, Drowsiness is RED alert!!

"Ama, ridi, sogna, e poi vai a dormire" -Love, laugh, dream, then go to sleep

CHAPTER 9 Dreaming and the problem of Consciousness and Useful Fiction

"If you can dream it, you can do it. Always remember that this whole thing was started with a dream and a mouse." - Walt Disney

Dreaming, Early Humans, & Supernatural

How did early humans understand their dreams? The earliest writings of mankind clearly include many opinions about the nature and origins of dream visions. A supernatural theme is often present in terms of dreams foretelling the future, often noting messages from supernatural sources, and higher meaning for the life of the dreamer. The origin of primitive religions may have arisen in part from a needed to explain dreams. Anthropologists have speculated that making the dream experience fit smoothly into the remainder of their existence posed a problem that primitive humans could only solve by introducing the concept of a spiritual self which could leave the body at night during the dream. This, in turn, may have played a role in the spiritual belief that each person has a soul which can leave the body temporarily during a dream and leaves the body permanently at death   How can a person experience a death of a loved one and then see them in a dream without there being a spiritual world.

In modern religions, dreams play an essential part and have specific meanings and interpretations. Here are more specific examples: Queen Maha Maya was the mother of the Buddha. She dreamt that she was chosen to be the mother of the "Purest-One." If you thought your parents put pressure on you, imagine what it would have been like to have been raised with that maternal expectation! In the Old Testament of the Bible, Joseph dreamed that he would someday have authority over his family and that they would bow their knee to him. In the New Testament, Joseph is informed of Mary's pregnancy through a dream. A cornerstone of the Muslim faith is the Prophet's Night Journey. In this account, the Prophet travels to Jerusalem and heaven in one night. According to Hindu thought, dreams are real and caused by the Supreme Brahman.

Dreams offer the wonderful privilege of citizenship into worlds with their own logic and seemingly no limitations. For early humans, there was no obvious reason to consider that one world was more significant than the other. The waking world had certain advantages of solidity

CHAPTER 9 Dreaming and the problem of Consciousness and Useful Fiction

and continuity, whereas the dream worlds offered the chance of communication and interaction, however fleeting, with distant friends, dead relatives, and even gods and demons. The dreams were accorded a validity equal to- and in some cases, greater than- that of waking life, though different in kind.

Today for the most part, dreams are much more peripheral to our waking lives. However strange, puzzling or disturbing a dream might be, we usually do not dwell upon it. Most dreams are quickly forgotten. It is the rare dream experience that we remember with the same accuracy and vividness that we remember our waking experiences. The major way in which the dream world is different from the real world is its discontinuity from dream to dream, night to night, and even from moment to moment within the dream. Dreaming is another place, or another world--a miracle of the sleeping mind that contemporary knowledge cannot begin to reduce to mundane neurophysiology.

S.B. Harary recounts out-of-body experiences which appear to be dream experiences in a book edited by D.S. Rogo, "Mind Beyond the Body." The following example is typical:

"Late one evening, while relaxing in bed, I gradually noticed the sound of a television in the next room along with the sound of water forcefully rushing in the bathroom down the hall. Intrigued and mystified by the sounds (my roommate was vacationing out of town and I had not left the television on or the water running), I decided to investigate. When I noticed my body still lying on the bed behind me after having felt myself get up, it became obvious to me that an out-of-body experience was occurring. My form seemed to be almost identical to that of my actual body, but the reality which I experienced appeared to have been 'rearranged.' During the experience, the atmosphere felt strangely 'charged' while the lighting in my apartment was not as I would have expected it to be. Furniture was moved about in odd positions. The portable television set, turned to its loudest volume, was resting on the floor immediately in front of my bedroom door. Down the hall, rusty-orange water gushed from brand new pipes into the bathtub. The front door of the apartment was partially opened. After I had turned off and moved the television set, and shut off the water faucet in the bathtub, I stood in the living room and considered reorienting the furniture. I decided that it would be too much trouble. Instead, I closed the front door and quietly concentrated upon my body which was lying back in the bedroom. Upon doing so, I immediately found myself lying in bed awake with full recollection of

the experience. The feeling was as if I had been directly transported from one point to another, not as if I were awakening from a dream."

Mind, Consciousness, and Mental Activity

The beginning of modern dream research began with the landmark discovery of REM sleep and reports of dreaming done by the Chicago researches, which included medical student William Dement. At the time of discovery an issue we struggled with was the definition of consciousness and unconsciousness. There have always been difficulties in defining and describing consciousness. We take our self-awareness for granted and we assume that everyone else has the same self-awareness. Mental activity is that which is *not* physical activity. It is the activity of an internal process that is called **consciousness**. Consciousness is that part of the human mind which is at the level of awareness. Thus, it can be distinguished from the **unconscious**, that part of the mind of which an individual is not aware. However, even though it may seem self-evident, we cannot prove that consciousness exists in anyone but ourselves. This is a topic that has engaged philosophers and psychologists for centuries.

For many years, it was assumed that sleep and the cessation of consciousness were synonymous. Though dreaming was obviously not unconscious, it was thought to be a very brief and fleeting phenomenon, occupying only a few seconds in an entire night of sleep. Since it is now known that dreaming goes on for substantial amounts of time, at least 20% of total sleep time, the notion that sleep is uniformly a state of unconsciousness is completely untenable. Human beings are unquestionably fully conscious during REM periods, and an altered form of consciousness appears to be present during non-REM sleep. An awareness of conscious experience during sleep implies continuity, reflection, and memory for whatever we were experiencing. The latter could even be abstract thoughts and abstract reasoning, feelings, or vague mental imagery. We are unambiguously conscious during our ongoing perceptual awareness of the external world while awake, and in our ongoing perceptual awareness of the dream world while we are in REM sleep.

Sleep Mentation

For purposes of an introduction to the subject of dreaming, it may be assumed that verbal reports of dreams are valid accounts of the dream experiences. Though dreams are often poorly

CHAPTER 9 Dreaming and the problem of Consciousness and Useful Fiction

remembered and incomplete, the dreamer usually knows whether his or her recall is good or not. Some experts feel that a process of "secondary revision" occurs when someone tells a dream. This revision is assumed to take the sometimes incoherent and fragmentary recall of a dream experience and fill in the gaps in order to transform the dream report into a coherent story that flows along in a logical manner. Even if there is such a process, dream reports commonly include sudden scene shifts and inexplicable intrusive and bizarre events. Except for dream content that might be embarrassing to the dreamer, there is usually no compelling reason for someone who is relating a dream or the content of any mental activity during sleep to be anything but as truthful and accurate as possible.

Though we are accustomed to having good recall of our waking conscious experience and for the most part take it completely for granted, the same cannot be said for our sleeping consciousness. Collectively, all thoughts, feelings, images, perceptions, hallucinations, and active dreams that take place during sleep are sleep mentation. Consequently, it is widely assumed by sleep researchers that some mental activity occurs continuously throughout sleep. However not all of this mental activity is the same as dreaming. Clearly, sleep mentation differs in many ways from waking mentation and consciousness, the most obvious being the absence of the absolute continuity that our journey through the waking day imposes. We also know that there is greater difficulty of remembering what we were thinking or dreaming during sleep. Given all these considerations, it is clear that we are never totally unconscious during sleep. It is also clear that we probably remember much less of the whole night of sleeping experience than we remember of our whole day of waking experience.

In the years since the discovery of REM sleep, many studies of sleep mentation have been conducted utilizing essentially the same methodology. Sleeping subjects are awakened at various times of the night, or in specific sleep stages, and are asked to recall and report what was happening, what was going through their mind, or what they were dreaming about. A skeptic might argue that investigators are studying verbal reports given in the waking state which may or may not be identical with the actual dream experience or sleep mentation. However, there is strong evidence that ongoing physiological events correlate with reported mental events. Although the correlations are not always high, we may nevertheless draw the conclusion that the report is a valid description of what was actually experienced during sleep.

Definitions

Sleep mentation can be abstract thoughts, words, or neologisms. There can be images which have clear form, or almost no form, such as kaleidoscopic flashing lights, and imagery may or may not be continuous. It is the opinion of most experts that vivid dreams are a series of adventures in a "real" world. By "real" world, we mean that dreamers experience themselves to be in an environment in which the dimensions of time and space essentially are the same as in the outer world, and in which objects are generally perceived in the same manner as in the waking state: up is up, and down is down. This vivid, complex experience is then contrasted with other types of sleep related mental activity which lack vivid perceptual awareness and which may be, as mentioned, abstract thoughts, unattached feelings, visual fragments, and other isolated sensations.

The most widely accepted definition of **dreaming** is a vivid, complex, hallucinatory experience, while asleep, generally accepted as real by the dreamer, and having a mostly logical progression in time. According to this definition, a static picture, however vivid, would not be a dream. An experience that is less than a vivid hallucinatory experience progressing in time falls under the general rubric "mental activity during sleep," and if it does not meet the criteria for dreaming, then it would be referred to as "nondreaming mental activity."

Many sleep researchers believe that dreaming (as defined above) always occurs in REM sleep and at no other time. Some investigators believe that dreaming can also occur during non-REM sleep, but they do not claim that dreaming is always present during non-REM sleep. A good generalization, if not precisely accurate, is that the mental activity during non-REM sleep consists of abstract thoughts, fragmentary images, and isolated feelings. It is now known that the processes which initiate REM sleep must begin in non-REM sleep, and that certain processes always associated with REM sleep occur at a lesser intensity during much of non-REM sleep. Since these processes are likely involved in creating the dream world, they could account for the occasional reports elicited from non-REM awakenings that conform to the above definition of dreaming.

Recall and Memory

Dreams are readily forgotten. Overall, it is clear that remembering a dream is usually more difficult than remembering a waking experience. There appear to be rare individuals that never recall their dreams. Many individuals remember dreams only occasionally, while others have vivid memories of several dream episodes almost every night. Though we assume that the verbal report is truthful, we cannot assume it is an absolutely accurate representation of the total dream experience.

When a sleeping person awakened from a very active REM period responds to the question, "Were you dreaming just then?" with "No I wasn't. There was nothing going on in my mind," most dream researchers assume that dreaming was going on, but was forgotten when the individual was awakened. This assumption is buttressed by the fact that some individuals who claim they rarely or never dream, in fact, remember dreams quite well when they are awakening from REM periods under experimental settings.

A fruitful area of psychophysiological investigation might be studies of how memory processes are altered during sleep. Evidence suggests that the transfer of memory from shortterm to long-term storage may diminish or cease during sleep. Some investigators feel that a dream will never be recalled unless there is an awakening during which the memory of the dream can be transferred to long-term memory storage. This hypothesis is currently untestable. Since brief awakenings occur commonly during REM periods, it is difficult to isolate instances in which it is possible to know with certainty that a dream was recalled from a REM period during which absolutely no awakening occurred.

The fundamental principle here is that the ability to remember the dream experience is the limiting factor in studying dreams and the process of dreaming. A subset of this is the dreamer's ability to communicate a description of any event. For this reason, relatively little is known about the dreams of very young infants and children.

Emerging concepts using new tools are confirming the role of dreams and memory. One school of thought is the synaptic homeostasis hypothesis. According to this hypothesis the universal purpose of sleep is to restore the brain to a state where we can adapt when we are awake. When we are awake, memories form as neurons that get activated together and strengthen their links. This hypothesis suggests that spontaneous firing during sleep may weaken the synapse, or contact points, between the neurons. Such a weakening, would return the synapse to baseline level of activity. This return to baseline, called synaptic homeostasis, could be the fundamental purpose of sleep by allowing more neuronal plasticity.

A common question asked is can we tell what a person is dreaming about during a sleep study. We cannot (yet). Recently, researchers have started to attempt to decode the contents of dreams using functional magnetic resonance imaging (fMRI) . In an experiment, they were able to predict the visual content of sleep onset activity using the fMRI. We are far away from being able to reviling the secrets of an individual's dreams, however, some scientist are trying.

Perception

Perception may be defined as the process of organizing and interpreting sensory information from the environment. The end product is the conscious awareness of our world and the continuous recognition of meaningful objects and events in it- a near infinity of material things, colors, sounds, kinesthetic sensations, awareness of body position, awareness of tastes and smells, and finally internal states of discomfort and pain. Our ongoing stream of consciousness and the ongoing perceptual processing when we are awake gives rise to an intertwined stream of thoughts, feelings, reminiscences, reactions, and behavioral responses. It is widely assumed by psychologists and neurobiologists that sensory information, perceptual processing, and elaborating responses with or without overt behavior are all essential for the brain to generate the experience of consciousness.

The most dramatic difference between consciousness in the real world and consciousness in the dream world is the complete absence of external perception. Yet, in the dream world, we see, hear, smell, and taste, and are aware of the position of our bodies. Where does the information come from? Also, in the dream world we respond to sensory information with thoughts and feelings, and we execute behavioral responses. We move around, walk, run, swim, and occasionally fly in the dream world. Yet at the behavioral end, our bodies and the muscles that would ordinarily execute the movements we make in the dream world are completely actively paralyzed by the brain itself. Perception of the real world is dramatically shut down at the onset of sleep, and shifted to the perception of an internal world when we enter a REM period.

Categories of Sleep Mentation

"Unconscious" Mental Activity

The concept of an active and motivating unconscious mind can be attributed almost entirely to the influential psychoanalyst, Sigmund Freud (1856-1939). He postulated that the manifest content of dreaming--the content of the conscious experience--was actively influenced and, to a large extent, determined by the activity of the unconscious mind.

One may wonder, however, whether or not unconscious the aspects of mental activity are active during dreaming including the lifetime storage of memories in our brains, the more elementary aspects of perceptual processing of which we are completely unaware of. There is no answer to this question at the present time. A corollary of this issue is whether or not the memories of dreaming are stored by the same mechanisms and in the same repositories as waking memories.

Sleep Onset Imagery

Under ordinary circumstances, we do not remember the precise moment of the onset of sleep. You can tell what time you went to bed but not what time you fell asleep. The retrograde amnesia surrounding the onset of sleep has been discussed. It is well known, however, that when awakenings occur shortly after sleep onset, mental activity in the seconds or minutes following the transition from wake to sleep is typically characterized by a kaleidoscope of images and experiences. For example, truck drivers say that when they are driving at night they are frequently startled by something dashing onto the road. It may even be seen in vivid detail as, for example, a deer. These is called **hypnagogic imagery**. This vivid imagery is associated with the onset of sleep. In the above example the driver experienced a micro-sleep with an associated hypnagogic image which, in the context of driving, was startling and arousing. The images associated with sleep onset are believed to derive, at least in part, from the abstract thoughts just before the transition into sleep. If someone has been thinking of a car, then one suddenly experiences seeing a car or being in a car. Another frequent phenomenon is a proprioceptive image, where individuals frequently have a very vivid "startle" in response to the feeling of falling. These "startles" are called **hypnic jerks** or **sleep starts**. They are generally unpleasant, and can be very disturbing when individuals experience them frequently. These phenomena fall under the category of parasomnias, which will be discussed later on in the texts. When the brain goes to sleep, and sensory imagery and perception is internally generated, the rain is freed from the constraints imposed by perception of the external world. Instead, it receives passively whatever sensory information the environment generates. When awake, the brain has difficulty ignoring anything, though attention may focus on a specific modality. In non-REM sleep, it is apparently possible for a single modality to be active, for example a powerful emotion without any antecedent, or an isolated proprioceptive experience such as falling, or a sound, or a visual image without sound. The word **hypnopompic** refers to the brief interval of waking up. Though less well studied, there is evidence that the momentary transition from sleep to wake can also be associated with hallucinations and sensory images upon awakening.

Inferences from Behavior

We can imagine that someday in the future there will be a computer or robot able to duplicate the behavior of human beings. At this point, human beings will be confronted with the question, "Can a machine be conscious?" This is a frequent question in scientific fiction. The tendency to attribute feelings and consciousness to animals and inanimate objects is called anthropomorphizing. The more closely the organism or apparatus resembles a human being, the more we assume it to be conscious. Later in this book, REM sleep without atonia is discussed, highlighting the behavior during REM sleep in a non-paralyzed animal seemed to be purposeful. If one can infer that conscious intent lies behind meaningful behavior in an animal, then one can infer conscious dreaming. We will also discuss a sleep disorder called **REM behavior disorder** in which the behavior is completely consonant with the conscious experience reported by the patient. A more convincing instance would be overt verbal behavior during REM sleep that is constant with the ongoing dream. This occasionally occurs, although the REM inhibitory process generally includes the muscles of speech.

The Reality of Dreams: Some Difficulties With Our Assumptions

Would a dreaming human brain in a bottle have any notion that it was *not* awake? How Real are Dreams? It is unlikely that the above question will be answered in our lifetimes. Nevertheless, the salient feature
of dreaming is that the dream experience is real while it lasts. It is these vivid, seemingly real experiences that excite us and draw our fascinated attention. The vivid reality of an unusual dream experience is part of what motivates us to tell our dreams to someone else. This raises the interesting question of whether or not many of our memories actually took place in dreams, despite assuming that they were memories of walking events. Some people attribute the experience of *deja vu* to the unknowing evocation of a dream memory by seeing something in a waking state. This evocation gives rise to the feeling that the object has been seen before. What

CHAPTER 9 Dreaming and the problem of Consciousness and Useful Fiction

we experience in a dream could change our behavior in the waking state. This allows for the possibility of utilizing a dream to practice the piano or practice a golf swing. In addition, a dream experience might change a waking attitude, particularly if it were not recognized as a dream experience. For instance, I was once very annoyed with a colleague who had referred to me as "a big windbag." I confronted him and he denied it. Could it have been a dream?

In today's society, we tend to confine the attribute of reality to the waking state and dismissing the dream world. For some, this means that the dream is trivial and unimportant, a shoddy facsimile of the real world; for others, the real dream world simply has a different set of rules and perhaps different purposes and goals in which one's aim may be to express certain aspects of the dreamer's inner life; and for a few others, unconcerned with rendering opinions, the dream is welcomed as the sole experience in which they can escape the offensive and incomprehensible bondage of time and space.

Despite their fascinating nature, possible clinical usefulness, and their ubiquity, dreams have only until recently been accepted by the scientific establishment as a topic worthy of investigation. Undoubtedly many reasons for this exist, but perhaps the most important is the status of the dream as a "private" event to which the dreamer is the sole witness.

Scientific and Philosophic Concerns About Dream Recall As mentioned above, the very nature of the dream experience renders it inaccessible to the close and detailed scrutiny by researchers that would allow us to conclude that the experience of dreaming, though not the specific content is identical for all humans. Even so, any discussion seeking to adorn itself with the mantle of scientific respectability should be obliged to formulate a clear statement of the nature and limitations of the subject matter; and yet, in nearly all "scientific" papers on dreams and more than a few books, we find that the universality of the nature of dreaming is simply taken for granted. The possibility that everyone does not know what is meant by a dream, or that we may be applying the label to more than one basic phenomenon is rarely, if ever, given serious consideration.

Dreaming is best to think of as a REM phenomenon. Many details of what was previsouly regarded as a comprehensive picture of sleep, were spuriously derived from observations confined to an overly small fragment of the sleep period. Following the discovery of rapid eye movements during sleep, observations covering the entire night have established that there are two distinct kinds of sleep with contrasting neurophysiological characteristics and mediated by different neuroanatomical systems. With the discovery of REM sleep, a historically important debate occurred. Specifically, could dreams occur outside of REM sleep or were dreams confined to REM sleep. Now we know that dreaming occurs in REM sleep and that it must be distinguished from fragmentary images that may occur in non-REM sleep. These fragmentary images are no longer considered dreams.

Nearly all descriptions of a dream include two points: a) a dream is mental and by implication "conscious" and accessible to recall, and b) a dream occurs during sleep. Of course, sleep may not be a necessary condition for the function of the mechanism which produces dreams. Identical mechanisms might also be responsible for such waking state phenomena as hallucinations, images, and visions. The typical "dream" is a description, given in the waking state, of an event that has already taken place. Accordingly, the presumed dreamer, now awake, has no way to localize the dream experience in time, except for his impression that it did occur in the remote or immediate past when he was asleep. Even the statement that one was asleep needs outside verification to be absolutely reliable. Fortunately, in contrast to the subjective experience of dreaming, the occurrence of sleep can be verified with brainwave recordings. Even if dreams do occur during sleep, the sole witness is the dreamer himself. There is, in the final analysis, no way to know absolutely if the dreamer's report is an accurate account of the actual dream experience.

Prior to the discovery of the association of REM sleep and dreaming in 1952, the leading experts in dreaming were psychoanalysts. The late Jules Masserman, who was a leading psychoanalyst, suggested the following in 1944:

"No dream as such has ever been analyzed--or will ever be analyzed--until we develop a technique of reproducing the dream sequence itself on a television screen while the patient is asleep. All we can do at present is to note carefully the patient's verbal and other behavior patterns while he is talking about his hypothetical 'dream' during some later analytic hour, remembering all the while that his hypnagogic imagery has inevitably been repressed and distorted in recollection, that it is described in words and symbols colored by his experiences not only before but since the 'dream,' and that in the very process of verbalization his description and associations are further dependent on his unconscious motivations in telling the dream at all, his transference situation, his current 'ego defenses,' his physiologic status and the

CHAPTER 9 Dreaming and the problem of Consciousness and Useful Fiction

many other complex and interpenetrating factors of the fleeting moment." (ie: people sometime lie)

To restate Masserman's position, any description of a dream experience, in psychoanalysis or out, may be highly distorted by faulty recall and the many factors which influence memory, or may even be a deliberate falsehood. Most clinicians who attempt to utilize dreams in therapy simply ignore these difficulties and accept dreams as all descriptions of experiences said to have occurred during sleep. Since any statement, accurate or not, is still a creation of the dreamer, and thus may reveal something about his personality, this definition is valid and useful for the practicing psychotherapist. Whatever arbitrary definition is chosen, a distinction between the report and the "real" dream must be maintained. Unfortunately, this principle is often ignored. One sees in the psychoanalytic dream literature an unqualified acceptance of a wide variety of questionable dream experiences, such as "blank" dreams described by psychoanalyst Bertram Lewin--dream experiences limited to a single non-visual perceptual mode--and "thought" dreams consisting only of a single word or idea. Obviously, these reports may actually reveal more about the process of remembering a dream than the nature of the dream itself. Prior to the discovery of REM sleep, to be an expert in dreams required the mandatory reading of Sigmund Freud's *"Interpretation of Dreams."* Freud consider imageless thoughts to be dreams. An interesting example of such an unusual "thought" dream is Sigmund Freud's famous "autodidasker" dream about which he says, "The first piece was the word 'autodidasker,' which I recall vividly." There were no sensory elements involved. As an imageless thought with its content limited to a single word, this would seem to be the smallest amount of experience that could qualify as a dream. Any further reduction would lead to the level of "dreamless sleep," the reports of which, as possible products of faulty recall, have no more likelihood of accuracy than any other report.

At this point we would seem to be impaled on the horns of a dilemma. To be consistent, if we accept *any* dream report, we must accept all dream reports, since we have no way of selecting only the accurate ones; in which case, we may be building up a somewhat illusory picture of the actual phenomenon. If we do not accept them, we shall be getting no picture at all.

Occasionally, the narration of a single dream "adventure" runs to several closely spaced typewritten pages when the verbal description is recorded and transcribed, and includes a variety of imagery and detail exactly as if it had happened in real life. Is it possible the true dream experience is always like the foregoing, and that the wide variation in the complexity of dream reports represents only a greater or lesser degree of decimation by faulty recall? Or, if the various reports are accurate portrayals (at least some of the time), should they all qualify equally for the status of a dream? Should the concept of a dream be limited to experiences containing sensory imagery or to experiences that possess movement or progression of events? What is to stop us from distinguishing a thousands types of dreams in the face of the bewildering variation in quality, complexity, vividness, and mode that exists in dream reports?

Useful Fiction

The penetrating analysis of any subjective phenomenon will lead inevitably to the brink of epistemological nihilism. This is true of the activities we attribute to the mind in the waking state as well as in the sleeping state. Yet, in our waking life we do not ordinarily concern ourselves with the validity of sensations, perceptions, and thoughts, or their ultimate reality. We accept all verbalizations concerning these concepts without qualification because long experience has shown over and over that they do correlate with observable events in the real world. Their function is to bridge the gap between sensory input and motor output; to help order the intervening processes that govern human behavior. According to the philosopher, Hans Vaihinger, dreams belong to the multitude of "useful fictions about consciousness" that "facilitate action" and serve to "make the interconnection between sensory and motor nerves richer and easier, more delicate and more serviceable."

A justification for the existence of the "useful fiction" of the mind is described in the following: A burning torch is brought close to a human observer. The observer says, "I feel heat." The sensation of heat correlates with the approach of the burning torch. From this example, we can envisage a two-fold process of enrichment. By repeatedly confirming that the burning torch and the report of feeling heat are inextricably linked, we reinforce our faith in the concept of the intervening sensation of heat. The sensation of heat may then be used to order other events in the physical world. Thus, although the ultimate reality of the conscious sensation cannot be verified, the concept is important and a useful fiction.

Let us see how all this applies to the problems confronting us in the dream world. The main stumbling block in the past has been that the sleeping person for all practical purposes, ceases to interact with his or her environment. In contrast to the waking state where a multiplicity of real events (external stimuli, motor patterns, for instance) can be observed to interconnect logically with sensations and thoughts, the sleeping state offers almost nothing to correlate with presumed mental events. Thus, in view of the complexity of dreams and the presumed simplicity of sleep, we were historically denied the possibility of looking at dream events in relation to a series of interconnecting real events. The dream had to be regarded as a psychic phenomenon completely divorced from the physical reality. Fortunately, the sleeping brain is much more than a pulsating blob of protoplasm. We now know that the complexity of the organism's functioning may similar to when it is awake. A myriad of observable physiological events may be correlated with dream reports if the effort is made. A more extensive and penetrating study of physiological activities and their variations throughout sleep, along with repeated testing of their possible relations to the psychic events making up dreams are needed. The nature and degree of these relationships, in turn, will allow inferences to be made about the role of memory in dreams and the possible multiplicity of basic dream types.

While not absolutely infallible, the combined study of physiological and psychological data throughout the entire night of sleep has enabled investigators to verify certain aspects of the dream experience, and perhaps more important, has stimulated a small but vigorous scientific interest in the process of dreaming and what is dreamed.

Is the Dream World Equal to the Real World?

Formerly, I Kwang Kâu, dreamt that I was a butterfly, a butterfly flying about, feeling that it was enjoying itself. I did not know that it was Kâu. Suddenly I awoke, and was myself again, the veritable Kâu. I did not know whether it had formerly been Kâu dreaming that he was a butterfly, or it was now a butterfly dreaming that it was Kâu. But between Kâu and a butterfly there must be a difference.

-- Chuang-tzu

What do we mean by "equal to" the real world? If we now look at a tree, we are aware of an infinity of perceptual information. Within the limits of the resolving power of our eyeball camera and the angle of aperture, we see all the details of the real object. We see the complex network of trunk and branches and countless leaves with their infinite subtle variations in shape and nuances of color. How many nerve impulses per second must traverse the optic nerves in perfect pattern and sequence to "create" this picture? Where is the film, the developer, and the completed picture? Before we become enthralled by the marvels of waking perception, let us ask our first question about the dream experience. If we see a tree in a dream, is it, or can it ever be, as complete, as detailed, as perfect as the image of a real tree?

In order to answer this question, a dreamer would have to hold the dream image (in this case, a large tree with summer foliage) in the focus of his attention sufficiently long enough to examine it in the minutest detail, and then to report back to us whether or not there was, in fact, thousands of branches and leaves which manifested a similar infinite subtlety of shape and color as a real tree.

In the waking state, there would be no problem in examining a drawing or painting of a tree and giving some report about its departures from a real tree. Indeed, it is difficult to think of a single example where there would be any problem in distinguishing between a real tree and a facsimile. On the other hand, an example where a casual glance might be misleading is real grass and AstroTurf, but a very close inspection would reveal the differences.

Our impression is that such an act of analysis, reflection and judgment seems beyond the capacity of the dreamer while dreaming. In Stanford's sleep research, however, we have compared vision and reality in the case of hypnotic hallucinations. We have instructed a deeply hypnotized subject to "see" a pair of shoes or some other object. We have then asked the subject to examine the shoes very carefully and to report exactly what they look like. It was almost always perfectly clear from the spontaneous report and interrogation that the hallucinated object was substantially different from its real counterpart. Often the hypnotized subject would express his puzzlement over the strangely incomplete and flawed image. It seems that dreaming is more detailed and complete than simple hypnotic hallucinations.

As indicated earlier, however, it is probable that such focused examination and analysis are beyond the capacity of the dreaming brain. Momentary reflections are not uncommon, but it seems rare that the dreamer pauses for any great length of time to carry out a meticulous examination of something.

There are some situations where we might be able to achieve great improvement in the report of the dream content. Lucid dreaming, (covered in more detail in Chapter 13) offers an

CHAPTER 9 Dreaming and the problem of Consciousness and Useful Fiction

opportunity. Also, we can study REM periods that occur late in the sleep period when the waking mind is better able to perform the recall and description process. We can also make the task more introspective with less emphasis on the items of content and more emphasis on the global nature of the experience.

If all this is true, then why are dreams experienced as reality? The usual answer has been that the sleeping brain lacks the capacity to make a judgment.

Therefore, a fleeting and disorganized, indeed even incomplete, world and experience is accepted as real. Now that we know that in REM sleep the brain is working just as hard as when it is awake, we cannot really support the impaired capacity explanation. Furthermore, as we shall see, many of the physiological events of REM sleep seem to replicate the events that would occur with an identical experience in the real world. This appears particularly true of eye movement patterns, perhaps of activity in various sensorimotor systems as well. Certainly, we are fully conscious and aware of ourselves. With regard to the dream world, we are oriented in time, place, and person. Thus, the dream world may be as complete and richly detailed as the real world. Why are dreams experienced as reality? The answer becomes a dream is real because it feels real when you are dreaming. The dream world is a real place with real people in it.

Once again, imagine if your brain were placed in a bottle and kept alive and healthy, and we had the power to exactly replicate in terms of brain activity some period of your recent waking life. We can postulate that you would exist in a world that would be exactly the same as the real world and you could not possibly know otherwise. If we assume that the experiences in the dream world occur because of brain activity, and that this brain activity is similar to that of the waking state, the miracle is how the brain replicates all of the sensory information to create the dream world we live in without any help from the sensory organs.

One could suggest that existing in the real world is much easier for the brain because all of the experience and organization of the sensory images are received passively. No creative effort is necessary. The elaboration of responses is the major task of the brain. This is apparently also true in the dream world, except that these responses are blocked at the level of the spinal cord by yet another active process. Given that the brain must perform the remarkable contortions of creating a world, living in it, responding to it, while carefully blocking all of the responses so the dreamer actually does not move, it is not surprising that the dreaming brain often seems to be more active than the waking brain.

Remembrance of Things Immediately Past in the Waking State

Let us assume that a dreamer could intentionally examine his surroundings and that such an intention could guide his dream motivations long enough to complete the task. Would a tree and all its leaves be real? Would he detect any perceptual flaws in the dream world? Many people seem to feel that the dream world is incomplete. However, once again they would also concede that we do not remember it very well.

Remembering anything at any time in meticulous detail is difficult. We might envisage the following experiment. Ten students would be invited to a professor's office, one at a time, ostensibly for a tutorial session. When they arrived, they would be admitted to complex but perfectly arranged anteroom. Once in the room a prearranged drama or skit is occurring. A hidden camera would record the student's movements. At the end of five minutes, the student would be escorted quickly into another room and asked to report what had happened in the anteroom. The inquiry would be conducted in the same manner that we conduct a REM period awakening. We would have a chance to observe how well a complex experience would be recalled in an analogous waking situation. How much of the activity would be forgotten? How much not noticed? How much inaccuracy would be revealed? What colors would be recalled? What things would be remembered in detail or only vaguely? From the data, we would develop some sort of score to express the degree of difference between the 5 minutes in the real world of the experimental room and the subject's recall of it. These differences might be impressively large, and the differences among subjects might also be impressively large. Such results would support the hypothesis that the dream world is equal to the real world but attenuated by imperfect sensory processes.

An analogous experiment was carried out by several dream researchers. They tested subjects in the usual way to elicit dream recall that could be used to predict the spatio-temporal pattern of vertical and horizontal rapid eye movements recorded just prior to the arousal. They found that the correlation of the predictions from dream recall and the recorded eye movement patterns was very poor, but significantly above chance when large numbers of instances were accumulated.

CHAPTER 9 Dreaming and the problem of Consciousness and Useful Fiction

They did exactly the same thing in the waking state. Correlation between prediction from recall and actual eye movement pattern was no better in wakefulness than in REM!

Imagine strolling casually down the Champs Elysees, and being suddenly plucked into a different setting: a raucous buzzer in the background, a dimly lit room in which a shadowy figure begins a relentless interrogation about what you had just been doing. One would surely forget a few things. Our task in describing the dream world is roughly as difficult, as the dreamer is forced into dream recall in a setting much as that just described. If we walk through a familiar room and are then asked to recall the experience, we will "remember" seeing our desk or bed not only because we actually saw them, but because we "know" from repeated experience exactly where our desk or bed are located.

Although we can probably account for less than perfect dream recall in terms of the special difficulty of the task, we must still ask if memory processes are different in sleep as compared to wakefulness. It is not surprising that memory processes appear to be impaired during sleep, particularly those associated with long term storage. The gate between short-term memory and long-term memory appears to be closed in sleep. Experimental data confirm this conclusion. If a bit of information is presented within a five-minute period before the onset of sleep, it cannot be retrieved after ten minutes of sleep have elapsed. If a similar period of wakefulness intervenes, retrieval is essentially unimpaired. This suggests that information introduced into short-term memory during sleep will simply decay over a finite time period, probably several minutes, after which retrieval will no longer be impossible. During wakefulness, by contrast, such information would be transferred to long-term memory, and potentially retrievable at a much later time. Until proven otherwise, perhaps the best hypothesis is that memory traces of dream experiences are not transferred into long-term storage except during wakefulness. Accordingly, as we progress through a dream, perhaps only the immediately foregoing five to ten minutes is retrievable from short-term memory. Dream experiences prior to that interval will have been forgotten. Finally, the rate of decay and accessibility may be extremely variable depending upon the content. Perhaps, for example, memory of color may decay more rapidly than memory of shape.

Individual Differences

Individual differences in dream recall have not received much attention. There are a few individuals who say they never remember dreams spontaneously. The same individuals show essentially normal amounts of REM sleep. However, even arousals from REM sleep may consistently fail to elicit dream recall in one or two such individuals. Is dreaming totally absent in these subjects? Can we have one human being who inhabits the dream world every night, and another seemingly similar human being who is always denied entrance? Some individuals appear to have unusually vivid and clear dream recall, approaching 100 percent in REM period awakenings.

Another question is whether or not such putative individual differences in dream recall correlate with individual differences in brain activity. They do appear to correlate with certain psychological characteristics. It is, therefore, quite possible that some as yet unidentified neurophysiological variable exists in sleep which would correlate well with individual differences in recall.

If the dream world is real, why don't we experience more astonishment and skepticism about some of its events as we might in the real world? There are three partial explanations that may be mentioned. The first is that if the bizarre or astonishing dream events actually took place in the waking state, our behavior might well be pretty much the same as in a dream. The second is that the sleeping brain cannot analyze events at the required level of logical function and is thus reduced to passive acceptance of whatever happens. A third possibility is that at a subconscious level, even though the dream world seems to fool us completely with its apparent reality, we nonetheless know that we are dreaming.

The circadian rhythm of wakefulness may oppose our accumulating an accurate introspective account of the dream world and its individual differences. It has been demonstrated that waking functions, including recall, peak in the daytime. They are at their lowest ebb during the night at around 3:00 A.M. to 6:00 A.M. Accordingly, dream recall should be more difficult than the recall of daytime waking experience.

Do Dreams Have a Purpose?

An important and subtle distinction that will be mentioned on more than one occasion is the distinction between "meaning" and "purpose." Almost anything can have meaning in that it arouses thoughts and feelings in an observer and may symbolize some definite concern or feeling of the observer. Thus, if one happened to notice an empty vase which ordinarily held flowers, it might resonate with an ongoing feeling of loneliness at the time, or lack of sufficient money. If one was looking at a sunrise, it might mean a spiritual renewal. However, it should be absolutely clear that the sunrise did not occur for the explicit and exclusive purpose of expressing the individual's yearning for spiritual renewal. On the other hand, when the event is not a recurring, cosmic, physical event, the sense of a meaning can easily mingle with an assumption of a purpose. A dream of a sunrise might be assumed to occur in order to express the dreamer's desire to begin new initiatives in his life and to become more aware of his yearning for change. This attribution of purpose to the occurrence of specific dreams remains hypothetical. Currently the purpose of dreaming remains controversial.

The one thing we can state with conviction about dreams is that they are often entertaining. Not without reason has the ubiquitous occurrence of dreaming during sleep been referred to as the "theater of the night." Dreaming has also been called the "theater of the absurd," and dreaming is definitely a theater in which the dreamer is also the star. Certainly, dreams offer many opportunities for us to have experiences that could not occur in the real world, and, if our dreams were a little easier to remember, the entertainment value for ourselves and others would be greatly increased. At some point, it is hoped that research on dreams will become a higher priority for human society, and some of the many unsettled questions that were posed in this part of the text will finally be answered.

Summary

A dream can be defined as a life-like experience during sleep that is accepted as real by the dreamer. Dreams create a "useful fiction." We can only assume that everyone dreams. The imperfection of dream recall is the major obstacle to the scientific study of dreams. Sleep mentation and fragmented imagery can occur in non-REM sleep. We do not believe that this non-REM activity is true dreaming. In dreams, momentary reflections on the surroundings are common, but a systematic detailed examination of objects in the dream world by the dreamer is very rare.

"For often, when one is asleep, there is something in consciousness which declares that what then presents itself is but a dream." - Aristotle

CHAPTER 10 The Psychophysiology of Dreaming

"The madman is the dreamer awake." - Sigmund Freud

Introduction

If the time ever comes when two dreamers can inhabit the same dream, then we would directly experience the dreaming consciousness of another person. This was dramatized in a science fiction modality in the recent motion picture *Inception*. Dr. Dement worked as a consultant and film crew came to the Sleep and Dreams class. It is impossible to know that anyone is conscious except yourself. Avoiding solipsist argument, if we were dealing with the perfect android, we could not tell with certainty whether it was or was not conscious. However, it is very easy to make inferences about our conscious mental processes from observing our behavior. If someone is looking at a red car and then says, "I see a red car," we are perfectly content to assume that he or she is experiencing a conscious perception of the red car. Behavior and verbalized consciousness in the waking state, in almost every instance, are precisely parallel. One assumes that the conscious experience is ineluctably linked and dependent upon the neurological brain activity that underlies both the perceptual component and the behavioral components. The "emergent property" of consciousness arises out of this neurological complexity. The term psychophysiology is thus intended to encompass the precisely parallel psychological and physiological events. Alternate terms might be subjective and objective, or mental and physical. In the years, immediately after the 1952 discovery of rapid eye movements during sleep, a great deal of research was carried out which established that the periods of rapid eye movements are also periods when humans dream. The overall approach in this research was to awaken sleeping laboratory volunteers either during periods of rapid eye movements or during periods when the eyes were not moving and to ask them if they had been dreaming. If volunteers said yes, they were asked to describe what they had been dreaming about. The results were dramatically clear. Awakenings that interrupted REM sleep periods were typically associated with vivid dream recall. When volunteers were awakened from non-REM sleep, vivid dream recall was very infrequent. Rarely in the difficult area of relating thoughts and images to bodily events has a relationship been so clear and robust.

Based on the fact that REM sleep is dreaming sleep, many all-night uninterrupted recordings of sleep have been carried out in humans to describe the total nightly amount of dreaming by measuring the total amount of REM sleep. Most people are very surprised to learn that they dream every night of their lives, and that their nightly adventures in the dream world usually occupy around two hours of their total sleep each night.

Mind and Body During Dreaming

Psychophysiology Parallelism

The relationship of mental and physical activities during dreaming is called the **psychophysiology of dreaming**. This area deals with the degree of correspondence during periods of REM sleep between the physical events taking place in the brain and body and the psychological events taking place in the dream. When we are in the dream world, can we assume that our brain is doing what it would be doing if we were having the identical experience in the real world? Since overt movements and behavior during REM sleep are prevented by the "shutting off" of spinal motor neurons, it follows that observable behavioral events are drastically curtailed. This, plus the general difficulty of precisely recalling dream experiences, makes the assessment of psychophysiological parallelism during REM sleep a challenging problem. In the following material, we list some of the overt behaviors that can be observed during REM sleep, and how observations of their manifestations have yielded understanding of the psychophysiology of dreaming.

REM Periods and Dream Recall

I (William Dement) as a medical student had the pleasure of working with Professor **Nathaniel Kleitman** and graduate student Eugene Aserinsky to observe rapid eye movements during sleep. One of the very first things we did was awaken subjects to elicit dream recall when eye movement potentials were present versus when eye movement potentials were not present. From a modern researcher's point of view, this study was not ideally designed. The awakenings were not counter-balanced. There was no precise terminology to qualify what constitutes a dream, so it was difficult to delineate between actual dream recall and unrelated responses. The investigators were aware of the condition of the awakening and therefore must have contributed considerable experimenter bias. Nonetheless, in 27 awakenings when eye movement potentials were present in the record, 25 awakenings produced dream recall. In 23 awakenings that were done when eye movements were not present in the record, only 4 awakenings produced dream recall. Since Kleitman, and his students did not yet have the concept of an eye movement period or of REM sleep as a state occupying an interval of time, some of their periods of no eye movement may have actually occurred within REM sleep. The number of eye movements per unit time varies from individual to individual and the number of eye movements within each successive brief epoch varies considerably. However, the difference in instances of dream recall from the two conditions was striking and justified their conclusion that rapid eye movements during sleep are associated with dreaming.

Although I was still only a medical student, Kleitman and I immediately undertook a second study. This involved designing a set of standard criteria and rating all reports as dreaming or not dreaming on this basis. Experimenter bias was not ruled out, primarily because at that time funds were not available for research. Neither the experimenter nor the subject received any funds, and it was not possible to hire someone to stay up at night with the experimenter just to do blind awakenings and interrogation. These refinements and many more have been carried out in the interim and the relationship between REM periods and dream recall has been repeatedly replicated.

In the mid 1950's, psychoanalytic interest in the dream was very pervasive and, accordingly, there was an intense interest in this phenomenon among psychiatrists who began to do research. Within a few years, the late Dr. Frederick Snyder was able to publish a table in which he summarized nine independent studies comparing REM awakenings against NREM awakenings. Rarely in the difficult area of relating recall of mentation with physiological events has a correlation been so robust. The basic conclusion is that REM sleep is associated with vivid dream recall and presumably consistently associated with vivid dream experiences. Non-REM sleep, by contrast, usually does not have dreaming in terms of a vivid experience but certainly is commonly associated with recall of mentation. Common mentation includes vague thoughts, static imagery, or a single sensation like a sound or a smell.

Factors Influencing Dream Recall

In the aforementioned studies, the percentage of dream reports from REM period awakenings was directly related to the time interval between the occurrence of REM sleep and the dream interrogation. When the awakening interrupted an ongoing REM period, the interval was as close to zero as possible. If the REM period ended and an interrogation was attempted a few minutes later, the percentage of dream recall was substantially reduced. We found that the percentage of REM recall decayed to the non-REM level in about seven minutes. Another element in eliciting maximum recall is to avoid distraction. The awakening stimulus is usually a question to focus the dreamer's attention immediately. Thus, the questions: "Were you dreaming just then?" or "What were you dreaming?" were used, and the dreamer in the usual laboratory situation would state whether he recalled the dream and describe the experience.

Length of Dream Reports and Length of REM Periods

Although there is considerable variability among individual reports, there is a consistent relation between measures that document the length of the reported experience and the length of the REM period prior to the arousal. In one early study (circa 1955), William Dement and Nathaniel Kleitman counted the number of words in dream reports from varying lengths of REM periods prior to the arousal and found a highly significant correlation. They also found that this correlation was highly significant for only about the first 10 to 15 minutes of the REM period. They found that there was no correlation when recall was elicited by awakenings preceded by REM periods 30 minutes or longer. They noticed that REM periods were interrupted every 5 minutes or so by a body movement and it appeared as if these disturbances progressively reduced the likelihood of recalling dream experiences that precede their occurrence. In addition, a REM period which contains such a body movement or interruption was more likely to be associated with a very marked shift in the action, almost as if they were two entirely separate dreams.

Time in Dreams

For some reason, many people believe that a great deal of dream activity can take place in a very short period of time. A dream that was described by the French writer, Andre Maury, in his book, *Sommeil et Reves*, published in 1861 was often cited as an example supporting this belief. In this famous dream, Maury found himself in Paris during the "Reign of Terror" following the French Revolution. After witnessing frightful scenes of murder, he was brought before the revolutionary tribunal and confronted by Robespierre, Marat, and other prominent figures of those terrible days.

Maury, in his dream, was questioned, condemned, and led to the place of execution surrounded by an immense mob. He climbed onto the scaffold where he was bound to the block by the executioner. The blade of the guillotine fell! Maury felt his head being separated from his body. He then awoke in extreme anxiety and found that the top of his bed had fallen down and had struck his neck in just the way the blade of the guillotine would have struck. Maury reasoned that the awakening stimulus (being struck by the top of the bed) must have initiated the dream, and that all of the dream imagery was compressed into the short interval between the initial perception of the stimulus and his awakening.

Sleep and dream research has shown, however, that dreams are almost certainly not instantaneous. A stimulus may modify an on-going dream but does not appear to initiate it. If the Maury account was a truthful report of his dream experience, it would have to be assumed that while he was dreaming of the French Revolution, by an amazing coincidence, the bed fell. It is not unlikely that a Frenchman of that generation would dream about the French Revolution with some reasonable frequency. Given that this is about the only dream ever recorded that supports the notion of instantaneous dream generation and that billions, maybe trillions, of dreams have been dreamed, researchers favor the hypothesis that this was a coincidental event. In one person's lifetime, there is an average of a quarter million dreams.

In 1956, William Dement and Nathaniel Kleitman reported a study which addressed the problem of the course of time in dreams in a more definitive manner. They used a special type of stimulus as a "marker" in the dream. This stimulus was created by ejecting several drops from a hypodermic syringe onto the faces of sleeping subjects after they had been in REM sleep for several minutes.

Subjects who were not immediately awakened by the spray were then allowed to sleep for several additional minutes (ranging from 1 to 8), before being awakened to report the dream. The investigators were able to accumulate 18 instances in which the cold-water stimulus was clearly "incorporated" into the dream at a specific point. The exact point of incorporation had to be absolutely clear. The amount of REM sleep between the presentation of the stimulus and the awakening to elicit recall was precisely measured from the polygraph tracings. Thus, on the one hand we had a precise portion of the dream story from the point of the stimulus incorporation to the awakening, and on the other, a precisely measured duration of REM sleep. By having individuals act out the dream story and measure the elapsed time, it was found that the two values were almost always very close. It was concluded that the rate of time during dreaming is approximately the same as the rate of time in the real world.

An example will clarify the experimental procedure and results: A subject who had dreamed he was acting in a play relayed the following narrative when he was awakened exactly 60 seconds after the cold water had been sprayed on his back.

"I was walking behind the leading lady when she suddenly collapsed and water was dripping on her. I ran over to her and felt the water dripping on my back and head. The roof was leaking. I was very puzzled about why she fell down and decided some plaster must have fallen on her. I looked up and sure enough, there was a hole in the roof. The rain was coming through. I dragged her over to the side of the stage and began pulling the curtains. They stuck and I yelled to the stagehands for help. We were all pulling the curtains when you woke me up."

The action that took place between the first appearance of water in the ongoing dream and being awakened would just about occupy 60 seconds in the real world. As it turned out, the spray of cold water was an excellent choice for the stimulus "marker." The appearance of water in the dream was usually abrupt and obvious, although often somewhat incongruous.

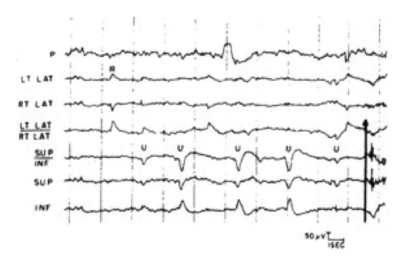

Figure10 - 1: These recordings were taken from REM periods just before the subjects were awakened for the purpose of giving a detailed account of the dream events immediately preceding the arousing buzzer. The direction of eye movements is indicated on the figure. "R" indicates eyes have moved to the right. "U" indicates eyes have moved upward. They illustrate the remarkable correspondence between eye movement sequences and events in the dream that we sometimes obtain.

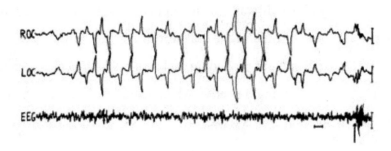

Figure 10-2 : The eye movements in the above figure were recorded from the outside corner of the left eye or LOC (left outer canthus) and from the outside corner of the right eye or ROC (right outer canthus). When the subject looks to the right, the ROC pen moves down and the LOC pen moves up; the opposite pen movement takes place when the subject looks to the left. The sequence of eye movements in this REM period sample is truly amazing. It consists of no less than 26 regularly spaced to-and-from horizontal movements with no vertical components whatsoever. Such a non-random and distinctly patterned eye movement sequence is a rare occurrence. When such a sequence does occur, it is an excellent opportunity to examine the relationship between dream content and direction of eye movements. In this example, a male subject was awakened (arrow) immediately following the horizontal eye movements and asked to report his dream. He stated that he had been watching a ping pong game between his friends. In the dream, he stood at the side of the ping pong table so that he had to look from side to side to watch the ball. He also reported that he had been watching a rather long volley just before he was awakened.

It has become a tradition in the Sleep and Dreams course to use water guns to awaken anyone who has fallen asleep in class (please see photo). They are then asked to describe what was the last thing they remembered to demonstrate retrograde amnesia.

Looking at the Dream: the Scanning Hypothesis

An earlier controversy was the development of the scanning hypothesis. It has been described earlier that the rapid eye movements associated with REM periods appear to be nearly identical to eye movements made when human beings scan their environments (see Figure 10-1 and 10-2.) The REM sleep eye movements are binocularly synchronous. The rotational velocity is identical to the velocity in wakefulness. REM sleep eye movements occur seemingly random in all directions. It is therefore not unreasonable, since the dreamer is inhabiting a complete dream world, that these movements represent the motor activity generated by the brain as the dreamer looks at the dream. The reader should recall that the extra-ocular muscles that rotate the eyeballs are not inhibited, unlike other voluntary muscles. In early studies at the University of Chicago, the experimenter (Dr. Dement) watched for clusters of eye movements that were exclusively or mostly in the horizontal plane and clusters of eye movements that were exclusively or mostly in the vertical plane. Subjects were awakened after clusters of movements that were either predominantly vertical or predominantly horizontal and also when there were very few eye movements. There was a very high correlation among these three types of awakenings and the imagery in the dream. Clusters of eye movements in the vertical plane were generally related to action that was appropriate. For example, one report was as follows:

I was walking across the field when I heard an airplane overhead. I looked up and realized that the airplane was dropping leaflets. I watched the leaflets flutter down to the ground and looked up to the airplane because they were changing color and I wondered who the pilot was. It was a little biplane that just circled around over my head as the leaflets fluttered to the ground. I looked down at them and up at the plane several times and then the buzzer sounded.

Clusters of horizontal movements were a little difficult to interpret because they were generally associated with any activity, but occasionally they were associated with specific activity such as looking around a yard or garden, or looking at various people sitting around a living room. When there were no eye movements for 20 or 30 seconds prior to an arousal, with no change in other indicators of REM sleep, the last dream images were often quiet where the dreamer reported "sitting and thinking" or staring at the horizon.

Close Up

The following series of experiments were conducted in a unique sleep laboratory - a rather spacious apartment overlooking the Hudson River in New York City. A special grant from the National Institute of Mental Health enabled me to defray half the cost of the apartment by converting it into a sleep laboratory with two bedrooms and several workrooms. The other half of the apartment was living quarters for myself and my family. In this way, I could work nights without leaving home. The work was carried out in the early 1960s.

A friend and fellow sleep-researcher, Dr. Howard Roffwarg, spent the night in a bathroom located between the two bedrooms, while I monitored the emerging EOG patterns on the polygraph. When I detected an eye movement pattern that was distinctive, yet fairly simple, I would arouse the subject by means of a buzzer. Then Roffwarg would obtain the dream narrative from the subject and attempt to predict the pattern of eye movements on the basis of the dream activity that the subject described. The method can be illustrated by the following dialogue between Roffwarg and a subject describing her dream:

"Right near the end of the dream I was walking up the back stairs of an old house. I was holding a cat in my arms."

"Were you looking at the cat?"

"No. I was being followed up the steps by the Spanish dancer, Escudero. I was annoyed at him and refused to look back at him or talk to him. I walked up as a dancer would, holding my head high, and I glanced up at every step I took."

"How many steps were there?"

"Five or six."

"Then what happened?"

"I reached the head of the stairs and walked straight over to a group of people about to begin a circle dance."

"Did you look around at the people?"

"I don't believe so. I looked straight ahead at the person across from me. Then I woke up."

"How long was it from the time you reached the top of the stairs to the end of the dream?"

"Just a few seconds."

On the basis of this interrogation, Roffwarg predicted, "There should be a series of five vertical upward movements as she holds her head high and walks up the steps. Then there should be a few seconds with only some very small horizontal movement just before the awakening."

The associated EOG tracing showed that the temporal sequence and direction of the eye movements were exactly as Roffwarg had predicted they would be. The actual sleep recording is reproduced in Figure 11-1.

The difficulty with such studies is that in most instances the eye movements are an extremely complex spatio-temporal mixture not suggesting much of anything. But in a subsequent analysis of the data from this study, we determined that Roffwarg was able to make predictions quite accurately in a fairly large number of trials. In our minds the matter was settled; we elaborated the "scanning hypothesis"-which has since been referred to somewhat skeptically by others as the "looking-at-pictures hypothesis." Roffwarg and I were convinced that we were able to account for the heightened sense of reality in dreams by hypothesizing that the brain is doing in the REM state essentially the same thing it does in the waking state - an internal sensory input is

somehow being elaborated. In other words, the dream world is "real" precisely because there is no detectable difference in brain activity.
-William C. Dement

One of the most common objections to the scanning hypothesis was based on the fact that newborn infants show REM periods with lots of eye movements. As far as we can tell, these eye movements are no different from those seen in adult humans. Since it seemed very unlikely that newborn babies were having visual dreams, many people regarded this as evidence that the eye movements were totally unrelated to dreaming. However, the possible lack of a relationship in infants is not crucial proof of a similar lack in adults.

A more critical test of the scanning hypothesis was conducted in persons who were blind from birth. Such persons are known to have dreams which are totally lacking in visual imagery and, hence, no scanning eye movements would be expected to be present in their REM periods. In a study of the famed blind pianist, George Shearing, sleep was recorded for one night in the same Chicago laboratory where eye movements were originally discovered. He had no eye movements on the EOG and reported only auditory dreams when awakened during REM periods. In this case REM periods were indicated by the typical EEG pattern and by muscle suppression in the EMG.

In another study of a group of blind patients, no eye movements were detected during REM periods in those who were blind from birth. Among those who became blind later in life, eye movements, saw-tooth waves in the EEG, and dreams with visual imagery were observed. Some investigators have attempted to draw conclusions from animal studies. For example, one researcher took movies of eye activity during the REM sleep of monkeys (he was fortunate enough to have one monkey that slept with its eyes half open) and thereby documented eye movements very similar, if not identical to, the scanning movements that occur during wakefulness. Similar results were obtained in the chimpanzee. In addition, there was some evidence that monkeys dream. One group of workers had trained monkeys to press a microswitch with the index finger when they saw a certain visual configuration. They noticed that these monkeys occasionally made the identical finger movements in REM sleep whereas untrained monkeys did not.

The type of rapid eye movements seen during sleep in the cats tends to undermine the scanning hypothesis. The feline eye movements consist mainly of bursts of small, quick, jerky movements all in the same plane (quite unlike waking movements). These unique movements, plus the fact that they are related to electrical activity discharged in the brain in bursts of spikes related to activity in the brain suggested to some that the eye movements of sleep are programmed according to rules quite different from those in effect during waking visual experience--rules that may not be related to vision at all.

It is interesting that in over five decades of debate about the scanning hypothesis, no one seriously attempted to confirm the original findings in human subjects. Finally, two laboratories independently attempted verification of the relationship between dreams and eye movement patterns. They both obtained negative results! However, it was apparent that investigators had not done their dream interrogations in a painstaking manner and, in addition, they had done relatively few REM period arousals.

One bothersome facet of previous studies was that some investigators apparently assumed one ought to be able to predict eye movements with almost 100 percent accuracy if the scanning hypothesis was valid. The tacit assumption was that 100 percent accuracy would be possible in the waking state at which time our memories for visual experience would presumably be excellent. Doubting this assumption, Stanford researchers decided to do two studies simultaneously: one in the waking state and one in sleep. They decorated their laboratory so that it would present a rich visual experience with many objects to look at.

Subjects were asked to sit with their heads in chin rests in order to avoid the complications of head movement. They were told that this was a study of visual recall and instructed to look around the room. The EOG was monitored continuously until a buzzer sounded and the subject was asked to relate what he had been looking at during the last fifteen seconds. Researchers interrogated the subjects and tried to predict the last eye movement. After a series of interrogations during wakefulness, the subjects slept in the laboratory, and a series of interrogations after REM arousals were conducted.

One very startling discovery emerged from this data. There was no significant difference between the results obtained during wakefulness and those obtained during REM sleep periods! Even in the waking state, eye movements could not be predicted with 100 percent accuracy. This demonstrated clearly that one would get negative results even in the waking state if the investigator used only a few trials. For nearly twenty years investigators had assumed that a high positive correlation would be easy to demonstrate in the waking state--and they were wrong. At this point the scanning hypothesis has not been refuted.

REM Behavior Disorder

In an earlier section, the ability to infer conscious experience and certain brain activity from observing overt behavior was discussed. The measurement of the temporal spatial pattern of eye movements is an example of this. Another example comes from a sleep disorder that was discovered in 1985 by Professors Carlos Schenk and Mark Mahowald at the Minnesota Regional Sleep Center. Professor Christian Guilleminault at Stanford also reported a case the same year. In this disorder, called REM behavior disorder (RBD), REM motor atonia -- the suppression of voluntary muscle activity -- is partially or almost completely impaired and the patient acts out the dream. RBD will be discussed in more detail in Chapter 19. The psychophysiological correspondence between behavior, verbalizations and dream reports has not been the focus of the clinical activity, but it is apparent that there is a high correspondence.

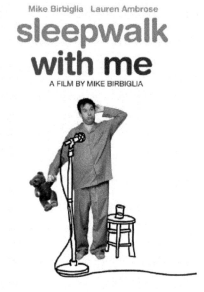

RBD was thought of as a unusual manifestation of sleep. However, the concept has now entered the mainstream of popular culture with the release of the movie, *Sleepwalk with Me,* in which Dr. William Dement has a cameo appearance.

Psychophysiological Parallelism and Lucid Dreaming

Another approach to this issue comes from the studies of a special kind of dreaming called lucid dreaming. These studies have provided powerful evidence for a correspondence between neurological activity and the dream experience. In addition to supporting both the scanning hypothesis and psychophysiological parallelism, the study has shown that lucid dreaming occurs during REM sleep. It was first noted by Aristotle, that we sometimes dream while knowing that we are dreaming, although it is doubtful that Aristotle used the term "lucid dreaming." In recent years, this phenomenon has achieved a much wider level of interest. Lucid dreaming is the subject of Chapter 13 and will be discussed here only in terms of its contribution to the evidence supporting psychophysiological parallelism in dreaming and REM sleep.

According to reports of "conscious and lucid" dreaming, the dreamer can possess a consciousness fully comparable in coherence, clarity, and cognitive complexity to that of the waking state while continuing to dream and remain in REM sleep. The experiments are based upon the fact that if subjects become aware they are dreaming, they can also remember to perform previously intended dream actions. Thus, if visual fixations in the dream sometimes show good correlations with polygraphically recorded eye movements, it seemed plausible that lucid dreamers could signal that they knew they were dreaming by means of intentional dream actions having observable physiological correlates.

Dr. Stephen LaBerge and his colleagues carried out the original experiments. They utilized five subjects trained in the method of lucid dream induction who were often able to have lucid dreams on demand. Standard recordings of brain waves, eye movements, and chin EMG were obtained as well as left and right wrist EMG recordings. The latter were utilized in the hope that a very strong clenching of the fists might show a burst of muscle potentials in an otherwise silent EMG. A variety of signals were specified which generally consisted of a combination of dreamed eye movements and a pattern of left and right dreamed fist clenches. The subjects demonstrated the signals during prerecording calibrations but did not practice further while awake.

Out of about 150 REM periods, 32 lucid dreams were reported, subsequent to spontaneous awakenings from REM sleep. The subjects reported signaling during 30 of these lucid dreams. After each recording, the reports mentioned the specific signals and the reports were submitted along with that night's entire polysomnogram to a blinded researcher who acted as the judge. The judge was asked to determine whether one (or none) of the polysomnographic epochs corresponded with the lucid dream signal. In 24 cases, the judge was able to select the appropriate 30-second epoch out of about 1,000 per polysomnogram on the basis of recorded and observed signals. The probability that the selections were correct by chance alone is astronomically small. All signals associated with lucid dream reports occurred during epochs of unambiguous REM sleep, scored according to the standard criteria.

The lucid dream signals were followed by an average of 1 minute (range 5 to 450 seconds) of uninterrupted REM sleep. The most reliable signal was a series of extreme horizontal movements: left, right, left, right. The most complicated signal consisted of a single upward eye movement followed by a series of left and right dream fist clenches in the order: LLL LRLL. This sequence is equivalent to the subjects' initials in Morse code. LLL equals three dots which is S and LRLL equals dot-dash-dot dot which stands for the letter L. The complexity of this signal argues against the possibility that the EMG discharges might be spontaneous and essentially random.

All cases of lucid dream signaling occurred during epochs recorded as REM sleep. The fact that lucid dreamers know they are asleep, can remember to perform previously agreed upon actions, and can signal to the waking world, makes possible an entirely new approach to dream research. These specially trained subjects can carry out all kinds of experimental tasks, functioning both as subjects and experimenters in the dream state. For the first time, sleepers can signal the exact time of a particular dream event, thus allowing for the convenient testing of otherwise untestable hypotheses. A researcher can ask a subject to perform any chosen action within the dream and the lucid dreamer can carry these directions out and signal when it happens. This signal also allows a clear mapping of mind/body relationships. LaBerge's studies at Stanford cover considerable ground showing the relationship between physiological changes in lucid dreamers' bodies and a variety of actions carried out by the dream bodies within the dreams.

LaBerge and his colleagues have confirmed previous studies of the rate of the passage of time in the dream world. In order to assess the rate of time, lucid dreamers were instructed to signal when they became lucid in their dreams and then to estimate an interval of ten seconds by counting to ten in the dream. The lucid dreamer then signals again to mark the end of the interval which can be directly measured on the polygraph record. They found that the average length of these ten second intervals was thirteen seconds which was also the average estimation of a ten second interval while subjects were awake. One example by a particularly capable subject was as follows:

By this point, fairly late in the morning, I was very determined to have the expected lucid dream. I felt especially motivated by all the filming crew being there, waiting for me to perform. So, when I found myself at the transition point between being awake and asleep I "made" it happen: my dream body began to float up in the air out of bed, a bed very much like the one that I knew my physical body was sleeping in. I waited until I was completely floating to be sure that I was really dreaming. But it seemed that I was being held back by something: my electrodes! However, I reasoned that these were only dream electrodes and I wasn't going to let my dream control me! At that point, I merely flew away not really caring about the "electrodes," which I presumed no longer existed. As I flew across the room, right through the wall, I signaled "left-right-left-right" to show that I was lucid. All of this so far took only a few seconds. I began estimating ten seconds, counting "one thousand and one, one thousand and two... "as I passed through the wall into the lounge area. Everything looked very dark and I felt that I wasn't very deep into my sleep until I saw a weak reflection of my face in a mirror. When I stared at it the room became very clear and lifelike.

Still counting, I decided that I'd like some action to report on later, so I grabbed a chair and playfully threw it into the air, watching it tumble and float. When I finished counting to ten, I signaled again. Next I was supposed to estimate ten seconds without counting, and this was when I got the idea that it would be interesting to fly to the polygraph room and actually watch a dream dramatization of my own signals being recorded. I needed to get there within ten seconds, so I flew right through the adjoining room, which was filled with boxes and chairs. For some reason, I let myself get caught up in avoiding stumbling on them, hoping to get to the polygraph before my next signal. In the distance, I heard a voice similar to mine doing the counting that I wasn't supposed to be doing! That puzzled me a little but I found it intriguing. Just in time, I arrived at the polygraph machine where several people were crowded around watching. I announced, "Hey, everyone, I'm doing it live!" as I signaled for the third time, seeing the polygraph pens flashing about wildly in my dream.

The same type of results has been obtained in breathing and breath-holding in specific patterns during lucid dreams which can be recorded on the polygraph. Lucid dreaming gives an even clearer picture of psychophysiological parallelism during REM sleep.

Summary

The salient feature of dreaming is that the experience is so vivid and detailed that the dreamer accepts it as real. It is likely that this heightened and unquestioned sense of reality is felt because it is a real experience inside the brain. We simply assume that if the brain, for whatever mysterious reason, is carrying out all the actions in the dream that it would carry out during the same activity in the waking state, it has no choice but to assume the activity is really taking place. Even when we occasionally know we are dreaming, the sense of reality is scarcely reduced. The dreamer is only prevented from rising out of his bed and acting out his dream by active motor atonia of REM. The miracle is that the brain is creating a surreal world and responding to it. How it does this is for future generations to unravel.

"Dreams: musical expressions of happiness and creativity." - *Kin Min Yuen, MD*

CHAPTER 11 The Content of Dreams

"Dreaming permits each and every one of us to be quietly and safely insane every night of our lives." - *William C. Dement, 1959*

Introduction

This chapter will address the question: What do people dream about? If we took a survey asking people what characterizes dreams in general, or their own dreams in particular, it is very likely that we would get many different answers. People tend to recall and recount the unusual. Thus, spontaneously reported dream experiences tend to be surprising, disjointed, bizarre, impossible, incoherent, absurd, nonsensical, and extravagant; a kind of temporary madness reflecting an alien, archaic world beyond the laws of time, space, logic, or morality. We might also feel that our personal dreams often contain expressions of unusual brilliance, wit, poetry, and intuition. Above all, dreams are fascinating and mystifying. We react to them almost like we do to special visiting lecturers; that is, we do not always understand what they are trying to tell us, but we depend upon them to entertain and inspire us.

Our friends or even casual acquaintances display an unparalleled generosity in sharing their dreams which they usually feel will be remarkably interesting to us. Have you ever heard anyone say, "Let me tell you about the dull dream I had last night?" If we should be so unfortunate as to lack both dreams and friends, there are always the movies or television where we can enjoy the latest unions of art and technology and the public dreams of cinematic artists like Ingrid Bergman, Frederico Fellini, Akira Kurasawa, Steven Spielberg, Tim Burton, and Christopher Nolan. Almost every popular TV series has devoted an episode or two to sleep and dreams. Our common knowledge of dreaming is anecdotal: a witch's brew of the occasional memories we have happened to seize or have acquired second-hand over the course of our lives. A comprehensive cataloging of what people dream about would be almost be like exploring the universe. Even the dreams of a single person, if that person has been very conscientious in recording them, would provide an overwhelming variety.

It is possible that the discovery and description of the ubiquity of REM sleep and the large amount of time spent dreaming each night changed our attitudes about their content. Previously, the occasionally recalled. very dramatic dream was thought to be a rare and startling occurrence and, as such, particularly meaningful and worthy of in-depth analysis.

Now we know that it is merely one of 10 to 20 or more dream stories that take place each and every night. A single REM period may contain more than one dream story. To completely describe what one person dreams about seems impossible, let alone the entire population. Moreover, how do we know whether dreams that are recalled are representative of the much larger domain of dreams experienced but not recalled? Some of these questions may be answered by REM period awakenings in the laboratory because the sleeper has no idea precisely when he will be awakened, and therefore cannot control what particular dream might be arbitrarily selected by the timing of the REM awakening. The analysis of thousands of laboratory-elicited dreams, in terms of a detailed content analysis, has rarely been undertaken due to its cumbersome nature and lack of funding.

Early Studies

Although interest and theories about dreams goes back to antiquity, attempts to carry out systematic observations on what people dream about are much more recent. One of the first of such efforts was carried out by women -- Mary Witten Caulkins, a young psychology instructor at Wellesley College in the 1890s. Introspectionism was in vogue at that time and the single paper Ms. Caulkins published on this subject in 1893, (as well as a paper by Sara Weed and Francis Hallam, a few years later) are the first reports emphasizing systematic content analysis of dreams. The methodology used by Ms. Caulkins was described in the following words: "Its method was very simple: to record each night immediately after awakening from a dream every remembered feature of it. For this purpose, paper, pencil, candles, and matches were placed close at hand." (What could go wrong?) Over the course of just a few months, Mary Caulkins and one assistant collected 375 dream descriptions which she then proceeded to examine and elucidate in a manner that can still serve as a model of scientific exposition.

Determinants of Dream Recall

As noted earlier, our knowledge of the dream world is derived almost entirely from the reports given by dreamers. Sleep researcher Dr. Stephen LaBerge has called the people who take dreams seriously "oneironauts," *oneiro* being the Greek root for dream. Thus, as we want astronauts to report accurately what they have seen and observed, we hope that oneironauts give us faithful reproductions of the major events and characteristics of their dream experiences.

There are two processes, however, which may distort any later description of a dream experience. The first is called **reconstruction**, in which the report is embellished with transformations, elaborations, and interpretations. While these maneuvers may promote a more coherent dream report, they may also constitute obstacles in empirical research on the nature of the dream as it was actually experienced. The other process is ***deduction***, in which the person giving the dream report logically fills in details of what must have happened. In other words, if the dreamer got in the car to drive to work and, in the dream experience, was suddenly at work, he or she might say, "I drove the car to work" which would not be an accurate description of the dream experience. The foregoing processes may also distort reports of experiences in the waking state, though probably to a lesser degree.

There are three general hypotheses derived partially from psychoanalysis via the dream report. These involve more fundamental mental mechanisms: the Repression Hypothesis, the Salience Hypothesis, and the Interference Hypothesis.

The Repression Hypothesis

This hypothesis states that defensiveness with respect to inner experience interacts with the content of dreams and this may affect dream recall and/or reporting. It could involve censorship at the waking level or censorship at the dreaming level. The repression hypothesis is not supported by content analysis of the dream. The content analysis approach has shown that most people dream about very obnoxious events quite regularly. One of the leading figures in the analysis of dream content, the late professor Calvin Hall, has stated, "There is no lack of dreams in my collection in which the most distasteful and hateful things happen. Fathers and mothers are murdered by the dreamer, the dreamer has sex with members of his family, he rapes, pillages, tortures, and destroys; he performs all kinds of obscenities and perversions; he often does these things without remorse and even with considerable glee." Calvin Hall categorized these dreams as "obnoxious."

It may be suggested that when such dreams occur, the repression process has momentarily failed. It is perfectly clear that most people have occasional dreams in which they do things they might never do in waking life. Such dreams may also be interpreted as things they would do in waking life if they knew they would not be punished, censured, or embarrassed. There is evidence that individuals who are rated as inhibited, repressed, timid, and pious in waking life tend to report fewer obnoxious dreams. A final possible interpretation of the occurrence of obnoxious dreams is that at a conscious level, the dreamer knows the dream is not real.

Salience Hypothesis

Salience refers to something that is striking or conspicuous or prominent. This hypothesis states that dream recall is positively influenced by neurophysiological arousal during REM, vividness and clarity of the imagery, and emotional impact of the experience. In other words, the conditions that are likely to heighten recall of an experience in any situation are also those that heighten dream recall. One test of this hypothesis is to compare individuals who are frequent dream recallers with those who are infrequent dream recallers. Such comparison shows that the dreams of frequent recallers are more vivid, more bizarre, and contain much more emotion. Furthermore, dreams that are reported spontaneously *the next day* by individuals are much more interesting, vivid, bizarre, and emotional.

Interference Hypothesis

This hypothesis states that dream recall will be inversely correlated with events that occur during dreaming, during awakening, and after awakening which interfere with the storage or retrieval of memories associated with the dream experience.

Any time subjects are awakened from a REM period and immediately distracted with an irrelevant question such as, "Would you like to know what time it is?" They are unlikely to recall a dream. Abrupt, but not startling, awakenings from REM periods are associated with a higher percentage of dream recall than awakenings which are gradual and prolonged. It is as if subjects focus on figuring out what is happening when they wake gradually and, thus, their attention is distracted from remembering the dream. A major interfering factor on dream recall is the spontaneous interruption of REM sleep. Most REM periods are interrupted every so often by a body movement. A study many years ago found that the mean duration of completely uninterrupted REM sleep in 20 young subjects was around five to ten minutes. This does not mean that REM sleep stopped. Full blown REM sleep is usually quickly reestablished after a body movement. A 25-minute REM period might thus include three to five closely approximated segments. Studies of the correlation between the length of the dream report and the length of the preceding REM period are highly significant only for the duration of REM sleep that follows a spontaneous interruption. One experiment that supports the interference hypothesis is the following: subjects were told they were participating in a study involving dark adaptation during sleep. They were awakened from REM periods and immediately asked to identify a series of objects above them in their dimly lit bedroom. After one minute, the "dark adaptation test" was terminated and the subject was casually asked, "Were you dreaming when I awakened you?" Under these circumstances, dream recall was virtually zero. This study demonstrates that dream recall can be impaired by a distraction or interference upon awaking.

Non-Reporters

Individuals who claim to recall dreams almost every morning while at home will recall dreams about 80-90% after laboratory awakenings from REM periods. In contrast, individuals who say they never dream at home sometimes fail to recall dreams when awakened from REM periods. Other home "non-recallers" recall their dreams at a lower percentage than home "frequent recallers" and, in addition, the recall is often sparse, and lacking richness of detail and events.

There are two theories that attempt to account for individuals who never or almost never recall dreaming in the laboratory or at home. The first postulates that they are simply terrified of revealing themselves. The second suggests that they sleep too deeply and are too groggy to focus, recall, and report. Of course, one source of deep sleep is sleep deprivation. Studies have shown that people with poor dream recall are likely to spend less time in bed and have a higher degree of voluntary chronic sleep deprivation. Vivid dreaming is reported more frequently by individuals who spend 8 - 9 hours in bed sleeping. An additional factor in this difference is the fact that the probability of awakening from a REM period is higher when sleep has lasted over 6 - 7 hours than when it has lasted only 4 - 5, simply because the percentage of REM sleep is much higher at the end of the night.

The Content Analysis of Dreams

Although a quantitative description of what people dream about is a daunting endeavor, some individuals have made the attempt. In the 1950-60's the leading methodologists in this area were Calvin Hall and his colleague, Robert L. Van De Castle of the Institute of

Dream Research. In their classic book, *The Content Analysis of Dreams*, they discuss the methodology of content analysis. Their somewhat unusual defensive posture is primarily because their method was developed in the heyday of psychoanalysis when the overt content of dreams was virtually ignored in the search for hidden meanings as windows into the putative "unconscious mind."

The development of a classification system requires some preliminary decisions: shall the dream content analysis system be one that can be used generally? Or shall it be formulated for the individual case? Obviously, general use is to be desired although, some argue, there is sufficient individual difference to discredit this approach. An additional benefit of general classification would be the detection of individuals who depart strikingly from general tendency. Should the system of classification be empirical or theoretical? Since dream theories change over time and one would like to accumulate a database of dream content, empirical information is obviously preferred for incorporation as data in a classification system. Finally, there is the issue of a unit of analysis. Comparison of frequencies of particular events in a dream, if they are to have any relative significance, must be for equal units of material. Units can be single words, phrases, sentences, lines, pages, or anything that would allow a regular measure of intensity.

The unit of analysis chosen by Hall and Van De Castle was the "dream report." A dream is an experience that occurs during sleep and a dream report is a description of that experience. The dream report should be limited to the description of the dream and should exclude statements which are comments upon or interpretations of the experience after the fact. How does the dream report differ from other types of verbal material? Much of what we speak or write in the waking state -- whether it be a request for help, an editorial, or an encyclopedia -- is done with the deliberate intention of influencing other people. On the other hand, making entries into a diary can be an impassioned process with no other intended aim than to describe the writer's experience. The former is instrumental -- intended to influence -- and the latter is representational. Dream reports should be regarded as primarily representational. Hall and VanDeCastle suggest the following scale-grouping and sequence: (1) setting and objects, (2) characters, (3) aggressive, friendly, and sexual interactions, (4) activities, (5) success, failure, misfortune, and good fortune, (6) emotion,

(7) modifiers, (8) temporal, negative, and oral, (9) castration. Despite these and other efforts no universal system currently exists.

What College Students Dream About

The following material will summarize norms that were obtained by a content analysis of 1,000 dreams collected from undergraduate students at Case Western Reserve University and Baldwin Wallace College between 1947 and 1950. The Case Western dreams were collected by Professor Calvin Hall from students in his psychology classes while the Baldwin Wallace dreams were collected by Professor Roland Cook from his psychology classes. Both instructors had assigned the recording of dreams as a project to be completed during the semester. Dreams composing the normative sample were selected in the following manner: five dreams were picked from each of 100 male and 100 female dream series. These dream series contained between 12 and 188 dreams. The selection procedure was a random one with the exception that any dream less than 50 words or more than 300 words in length was rejected. The ages of the dreamers ranged from 18 to 25. These norms should be appropriate for dream series collected in a similar fashion from other college students. They may not be appropriate for single dreams because when a person is asked to report one dream, he usually relates one that has an unusual or outstanding quality. In the sample, female dreams were on the average 8 percent longer than male dreams. Because of this difference in length of reports, it can be expected that the tables of Hall's results will generally show more dream elements for females.

Highlights:

- The setting in which dreams occur is more or less twice as frequently familiar to the dreamer than completely unfamiliar. Slightly more than half of men's dreams have an outdoor setting, while slightly more than half of women's dreams take place indoors.
- Nearly all dreams have several characters. Male dreams have significantly more male characters while female dreams have about half-and-half. The vast majority of characters in college student dreams are adults. About half are familiar and half are unfamiliar.

- Surprisingly, the number of dreams with aggressive social interactions is about the same for males and females. With each gender about half of the dreams have aggressive social interactions and half do not.
- Females had slightly more friendly social interactions in their dreams than male students. The overall percentage of friendly social interactions was about 40 percent of all dreams.
- Relatively few sexual interactions were reported. Male students reported a sexual interaction in 11 percent of their dreams, and females recorded a sexual interaction in 3 percent. The majority of these reports described sexual overtures and almost none involved sexual intercourse.

Recalling that the dreams were gathered in the 1950's, it seems likely that there was some inhibition in reporting sexual activities.

- The achievement outcome was classified as failure or success. Failure occurred in 13 percent of all dreams.

Success occurred in slightly less.

- The most common feeling or emotion experienced by the dreamer was apprehension followed by confusion. Females experienced somewhat more apprehension than male. Negative feelings are far more common than happiness.

Content analysis is extremely labor intensive. There are almost no comprehensive published studies. Unpublished information from dream series collected as recently as the 1980's and 1990's from the University of California at Berkeley, University of California at Santa Cruz, University of Richmond, and Crane Community College in Chicago are remarkably similar. Employing identical scoring as set forth by Hall and Van De Castle, Professor G. William Donhoff found relatively similar norms in college students of the 1990's. Not surprisingly, sexual content was somewhat more frequent. Donhoff reported that 13 percent of male students described dreams in which any kind of sexual interaction occurred and for females the figure was 10 percent. Formal content analysis, however, is the only way that we can actually formulate and quantify what we dream about.

Dreams from REM Sleep Awakenings

Most studies of dreams obtained from REM sleep awakenings (as opposed to morning-after reports) have had a specific purpose which precludes the type of systematic content analysis conducted by Calvin Hall. However, Dr. Frederick Snyder, one of the early REM sleep researchers, has published a paper entitled "The Commonality of Dreaming." The total sample includes 635 dream reports from REM period awakenings during 250 subject nights in the laboratory. The subjects were 18 male and 17 female middle-class college students.

The major difference from studies not involving a sleep laboratory is the presence of a researcher who awakens and questions the subject, obviating the processes that are involved in the transcription of remembered dreams -- it is less likely that the subject will reconstruct the dream. Since anyone who is awakening a subject has no idea what might be happening, one could make the claim that this is a random sampling of the dream content of these particular people. The report lengths varied from about 150 to more than 1,000 words. Snyder rejected very short reports but did not put an upper limit on length.

The physical setting in the dreams was usually familiar and could be described although often it was not a known place. The self was always present in the dreams and very frequently speech was mentioned and the content could be quoted.

An overall conclusion by Snyder about the results of his study was that dreams gathered from REM period awakenings were not impressively exciting. He assumed this reflected the fact that subjects could not choose which dreams to report or recall. They simply had to report what was happening when they were awakened during REM periods. The overall impression is that our dream lives are probably a little more mundane and commonplace than we think because the dreams we actually remember and the dreams that are recounted to us by our friends and family tend to be exciting and interesting.

Common Questions and Issues

The following are among the most frequently encountered questions and curiosities about dreaming. There is much more to be learned about dreams and dreaming. We may hope that the mind in sleep will become a higher priority in the quest for knowledge and that the world of dreams will be better understood and utilized.

Are Dreams in Color?

Many people believe that dreams are usually black and white, like old time movies. This is because the memory for color fades more rapidly than the memory for objects, spatial configurations, and items which have a name. The issue of color in dreams was specifically addressed by experimenters in 1960 when they awakened individuals from REM periods and asked a number of questions about the content. Color questions were carefully embedded in the general query so that subjects would not realize that the major interest was the presence or absence of color. They found that color was specifically recalled in 70% of their dreams while Snyder found color to be mentioned and recalled in 77% of "long dreams" and 61% of shorter dreams. It is likely that, as has been mentioned before, our recall of color in the waking state is aided by the fact that we know the color of many things: the sky is blue, the grass is green. However, when subjects are awakened immediately from REM sleep, they report that they really saw and experienced colors.

Do Babies Dream?

Since newborn babies have plentiful amounts of REM sleep, the question might be asked, do they dream? At the level of cognitive organization that must characterize a newborn human infant, this is a very difficult concept. The best assumption is that as the infant becomes conscious and able to respond to its environment, it probably experiences dreaming in much the same way it experiences the environment in the waking state. When the infant is able to talk it may begin to say things that suggest the recall of a dream. The occasional occurrence of talking in REM sleep also suggests a very early experience of dreaming in young children.

A very interesting question is when does a child learn to distinguish dream from reality. One three-year-old girl child reported, "I saw mommy riding in a roller coaster last night. I know it was a dream because mommy never rides in the roller coaster." It is commonly thought that nightmares at an early age are experienced by children as real and that phobias at this time may be the result of bad dreams which the small child thinks are real.

Do Animals Dream?

There is some evidence of activity and mental imagery appearing in animals. Since it has not been conclusively established that animals are conscious, and because dreaming is defined as conscious perception, we cannot be certain whether or not all animals dream. There is some evidence of activity and mental imagery occurring. Many people anthropomorphize the higher mammals such as chimpanzees, gorillas, dogs, and even cats, attributing thoughts and feelings to them. It is clear that REM periods in cats, dogs, and primates are characterized by heightened and highly organized brain activity. Also, their rapid eye movements are binocularly synchronous. If a perceptual awareness or consciousness is the psychological side of the physiological reality of highly organized neural activity, then it is reasonable to assume that at least the higher mammals may experience dreaming. One commonplace observation that pet-owners have made time and time again is that a dog will whimper, whine, or bark in its sleep; these vocalizations are now known to be associated with REM periods.

Compelling evidence that cats may be dreaming is based on the fact that the neural activity during REM periods is highly organized. This comes from observations by Michel Jouvet of Lyon, France, and Adrian Morrison in Philadelphia. They found that meaningful behavior occurs during periods of REM sleep without atonia. The researcher, through neurosurgical manipulation, was able to lesion the area of the cats' brain responsible of REM atonia. They were then able to observe the cat perform purposeful behavior, which was interpreted consistent with dreaming. This type of REM sleep occurs when the area in the brainstem that mediates the motor paralysis of REM sleep is put out of action. In human beings, the failure to show normal REM atonia is associated with very meaningful behavior, often accompanied by clear speech (see Chapter 19 which describes REM chronic behavior disorder). In REM sleep, the recordings from the cat's invasive electrodes demonstrate that characteristic ponto-geniculo-occipital waves (PGO waves) occur very frequently in cat. If dreaming occurs in animals, it seems that dreaming would be frequently interrupted by the periodic bursts of intense activity of the nonspecific PGO discharges. Some people believe that any mental and associated neural activity that occurs in the REM periods of animals would primarily represent instinctual behavior. Obviously, a wide variety of behaviors are related to instinctual drives in one way or another. Therefore, dream content could be fairly complex.

Can We Dream of Dying?

CHAPTER 11 The Content of Dreams

Individual dreams have been reported in which the dreamer sees himself, or herself, lying in a coffin, in a hospital bed, and so forth. However, content studies indicate that dreams in which the dreamer is actually killed, or dies, are extremely rare. The classic dream of falling-falling from an airplane, falling off a cliff, falling from some other high place is usually interrupted by an awakening accompanied by considerable anxiety or fright. There is a great feeling of relief that "it was only a dream." It may well be that the fear of dying in a fall from height induces such excitement and emotion that sleep simply cannot be maintained. In some cases, the problem is solved by the dreamer simply beginning to fly. In one instance, a dreamer reported a dream in which he fell off a cliff and was asked, "Well, did

you wake up before you hit the ground?"

The dreamer replied, "No."

The interrogator asked, "Well, what happened?"

The reply: "I bounced."

This brings up the controversy of whether or not one can experience in dreams something that has never been experienced. For example, can someone who has been blind from birth see in dreams? At the present time there is no satisfactory answer to this question. When examining the complex experience and minor details of dream scenes, it is very difficult to ascertain if the things being seen have never before been experienced.

Close Up

"In the late 1970's, the famous gorilla, Koko, was quartered in an enclosure behind our sleep laboratory in the basement of the old anatomy building. At the time, Dr. Penny Patterson was attempting to prove that Koko really communicated by sign language. One of my students, Steve Reich, had the idea that we could address the issue of nonhuman dreaming by "asking" Koko if she was dreaming in the same way we would awaken human subjects and ask them if they were dreaming. We would awaken Koko during a REM period and sign "what was happening?" or "what is going on?" If she signed something suggesting a dream, it would support the notion that she was indeed having a dream experience during the REM period.

With Dr. Patterson's guidance, Steve Reich set about learning sign language so he could "talk" to Koko. We thought he would simply spend the night in her "house," perhaps with

bars separating them, and watch for eye movements which we assumed would be readily visible in the same way they are in humans and many other animals. After Steve had made good progress in learning sign language, it was suggested that he should be introduced to Koko. This took place inside the enclosure in a kind of yard where Koko's house was located. After introducing Steve to Koko, Dr. Patterson and I turned our backs for a moment discussing various aspects of the study and possible problems and outcomes. When I turned back toward Steve and Koko, I saw that Steve's face was chalk white and he was completely immobile. It became obvious that Koko, squatting in front of him, was busily attempting to take his pants off. Dr. Patterson, noticing this, shooed Koko away and Steve zipped his pants and buckled his belt. The research project ended at that precise moment. The realization of his utter vulnerability, that the giant gorilla could have changed his life forever with a single squeeze of her powerful hands, was a research-ending shock for Steve. This represents just one of many unexpected research outcomes which we try always to anticipate, often without total success. With continued progress in developing sign language in gorillas, researchers may someday be able to "ask" Koko or some other gorilla what was going on just after they awaken from a REM period.
-William C. Dement, Some Must Watch While Some Must Sleep, 1972

Parapsychology and Dreaming

Many individuals believe that dreams can predict the future. They may have stories to tell about such dreams. There are those who take great stock in prophetic dreams, and others who merely laugh off the dream stories they hear. History is full of tales of prophetic dreams. Undoubtedly these dreams have had cultural influence in our society. However, there is no current scientific basis to believe that dreams reliably predict the future.

Telepathic Dreams at Stanford University

Turning to the determinants of dream content, we are confronted by one of the most fascinating questions of all: Is there a paranormal element that determines what we dream about? The existence of extrasensory perception, telepathic communication, or any event that transcends the physical laws of the known universe is definitely not proven; on the other hand, it has not been disproved. Certainly, folklore is replete with accounts of prophetic or telepathic dreams.

The most extensive studies of the phenomenon telepathic dreaming were conducted by Dr. Montague Ullman and his colleagues at the Maimonides Community Mental Health Center in Brooklyn. Results have varied from very poor to fairly good, and it should be noted that until recently, studies were not tightly controlled for bias and other confounding factors. In a typical study, "The procedures were designed to investigate the hypothesis that telepathic transfer of information from an A (or knowledgeable 'sender') to a sleeping S (subject or 'receiver') could be experimentally demonstrated." The sender was given sealed envelopes containing reproductions of famous paintings and instructed to open one of these during the night after the subject was asleep. The subject was awakened during REM periods and dream reports were obtained. Several independent judges were later asked to determine whether there was any correlation between the selected painting and the dream report.

The total number of correlations was ***not*** statistically significant, but there were several instances of unique correspondence between the painting and the dream. When the painting was Chagall's "The Drinker" (showing a man drinking from a bottle), the subject reported, "I don't know whether it's related to the dream that I had, but right now there's a commercial song that's going through my mind... about Ballantine Beer. The words are, 'Why is Ballantine Beer like an opening night, a race that finishes neck and neck?'

Close Up

We have conducted an experiment in telepathic dreams at Stanford, but before I recount our experience, just for the record, I should mention that some of my colleagues threatened to drum me out of our professional societies after they heard of the undertaking. They asked me why we were getting mixed up with such nonsense.

Anyway, back to our experiment. During the Winter Quarter of 1972, I had over 800 students enrolled in my course on sleep and dreams and I thought it would be fun and possibly informative if we conducted an experiment whereby the whole class simultaneously would try to "send a thought" to people who were sleeping in the sleep laboratory. We intended to test the premise that since single individuals might be able to transmit their thoughts into another individual's dreams once in a while, 800 people all sending the same message at the same time might be able to really blast through.

Six students from the class who felt they might have special "psychic" talents volunteered to be "receivers." These students prepared themselves by going to bed progressively more early so that on the day of the experiment they were able to arrive early at the sleep laboratory for the hook-up and be in bed at about 8 PM. Meanwhile, the class gathered at 9 PM at the Lucille Nixon Elementary School on the Stanford campus, about one and one-half miles from the sleep laboratory.

Our first problem was what to transmit. We finally selected several commonplace, unambiguous objects a horseshoe, a banana, and a key. We made slides of these objects as well as of the experimental subjects. We communicated by telephone with the sleep lab so we would know when our subjects began REM periods. The scenario went something like this: first, the laboratory technicians would inform us that a subject was having a REM period; then we would flash his picture on the screen to further identify him to those students who did not know him well; the class would decide on an object to concentrate on; and the picture of that object would be projected on the screen. The results were completely negative. None of the test images that the class "transmitted" were manifested in the dreams of the students sleeping in the laboratory.

In retrospect, there were many things wrong with this cumbersome and difficult experiment. In particular, one difficulty we did not anticipate was that we could not produce absolute synchronicity in 600 minds. It is actually quite hard to concentrate on a horseshoe for an entire minute, and the atmosphere of 600 students at a "happening" created additional distractions. Finally, showing a slide of the specific student to whom the "message" was being sent was an additional confound.

As tends to happen in this kind of study, we did get one very tantalizing though completely non-statistical result. During the third and fourth REM periods of one subject, the class was supposed to concentrate on the slide of a horseshoe. After "concentrating" on this image for one minute, we asked the lab technicians to wake the subject and see if he "got the message." There was no mention of shoes or horses, but the subject did give the rather unusual report that he had been dreaming of staring at himself in a mirror! Perhaps our class, or at least its female contingent, had actually concentrated more on the slide of the subject's good-looking face than on the less inspiring horseshoe.- William C. Dement

Creativity During Sleep

Persons who resent the amount of time they must "waste" in sleep have attempted to make use of the nocturnal hours by combining sleep with productive mental activity. The most prevalent of these attempts is sleep-learning. Extravagant claims have been made for various commercially marketed sleep-teaching devices, nearly all of which utilize acoustical repetition of the material to be learned. In the hands of legitimate scientific investigators, however learning during sleep has been completely unsuccessful.

No adequate demonstration has shown that using these techniques facilitates learning during sleep rather than during the succession of brief arousals that presenting the information induces throughout the night. Charles W. Simon and William H. Emmons conducted experiments in which the addition of continuous EEG monitoring enabled them to be absolutely certain material to be learned was presented only during sleep. These results showed that no learning whatsoever occurred during sleep. It should be noted, however, that in one test these experimenters used complex questions and answers and presented each pair only once; in another test, they used nonsense material. Some of the complex material might have been learned if it had been presented repeatedly. The nonsense material may simply have been disregarded because it had no relevance for the sleeping subjects. In view of these considerations, as well as recent demonstrations that low levels of both operant behavior and sensory discrimination can occur in humans during light sleep, the possibility that certain kinds of learning (perhaps low efficiency learning) might occur during sleep has not been conclusively eliminated. Moreover, additional processing of material presented during wakefulness may take place during sleep. For example, some investigators feel that memory consolidation occurs during sleep.

Artistic Creations

An accumulation of anecdotal evidence supports the possibility that artistic creation can occur during sleep. Most occurrences of such high-level mental performance have been attributed to dreams. Since we now know that dreaming is an intrinsic part of sleeping and occupies a specific and substantial portion of every night's sleep, we may assume that the opportunity for "creative dreaming" is potentially available to everyone.

Probably the most famous example of creative dreaming is the poem "Kubla Khan" by Samuel Taylor Coleridge, which Harvard scholar John Livingston Lowes has called "one of the most remarkable poems in the English language." In 1816, Coleridge published an account of its genesis. He had returned to a lonely farmhouse in Devonshire for reasons of ill health. Laudanum (a syrupy mixture containing opium) had been prescribed for a "slight indisposition," and its effects caused him to fall asleep in his chair while he was reading the following lines from "Purchas' Pilgrimage": "Here the Kubla Khan commanded a palace to be built, and a stately garden thereunto. And thus, ten miles of fertile ground were enclosed with a wall."

Coleridge said that he slept soundly for more than an hour, during which time he composed two or three hundred lines in a dream. Upon awakening, he took his pen, ink, and paper and instantly and eagerly wrote down the lines that have been preserved. He was unfortunately interrupted by a visitor who detained him for more than an hour. When he was able to return to the poem, some eight or ten scattered lines and a few sensory impressions were all that remained in his memory. "All the rest had passed away like the images on the surface of a stream into which a stone has been cast." In "The Road to Xanadu", a book containing a detailed analysis of "Kubla Khan" and "The Ancient Mariner", Professor Lowes has shown that Coleridge had already encountered many of the individual ideas or images expressed in his poems. But the creation of "Kubla Khan" certainly occurred during sleep. A fact that deserves more attention is that Coleridge was known to be a habitual user of laudanum. Little is known about the nature of sleep induced by narcotics, but evidence from sleep laboratories indicates that some measures of sleep remain within normal limits following habituation, and that the REM sleep is slightly enhanced. Coleridge also published a quatrain that came to him in sleep when he was definitely not under the influence of laudanum:

Here lies at length poor Col' and with screaming, Who died, as he had always lived, a dreaming: Shot dead, while sleeping, by the gout within, Alone, and all unknown, at E'nbro' in an Inn.

Another poem, "The Phoenix," was composed during sleep by the English essayist A. C. Benson. Benson wrote, "I dreamt the whole poem in a dream, and wrote it down in the middle of the night on a scrap of paper by my bedside. I have never had a similar experience before or since. I can really offer no explanation either of the idea of the poem or its interpretation. It came to me so apparently without any definite volition of my own that I don't profess to understand or be able to interpret the symbolism."

The Phoenix

By feathers green, across Casbeen,
The pilgrims track the Phoenix flown, By gems he strewed in waste and wood
And jeweled plumes at random thrown.
Till wandering far, by moon and star,
They stand beside the fruitful pyre,
Whence breaking bright with sanguine light,
The impulsive bird forgets his sire.
Those ashes shine like ruby wine,
Like bag of Tyrian murex spilt;
The claw, the jowl of the flying fowl Are with glorious anguish gilt.

In his book, *The Unconscious*, psychiatrist Morton Prince reproduced a long poem and an account of its creation while he was dreaming. Apparently, Prince awoke from the dream and wrote a description of it more or less unintentionally in poetic form. Robert Louis Stevenson, in his autobiography, *Across the Plains*, describes some of his dream life and credits his dreams for the plots of many of his stories, most notably *Doctor Jekyll and Mister Hyde*. The inspiration for the famous *"Devil's Trill Sonata"* composed by the Italian virtuoso violinist, Giuseppe Tartini, came to him in a dream in which he saw and heard the devil take up a violin and play the music that Tartini wrote down upon awakening.

The previous examples have been documented from autobiographical sources. There are literally hundreds of apparently apocryphal stories published without documentation. In the *Twilight Zone of Dreams*, Andre Sonnet credits dreaming for nearly every artistic and technological achievement accomplished by our race. He stated, for example, that the planetary model of the atom had come to Nobel Prize winner Niels Bohr in a dream. When

Professor Bohr was queried directly to obtain verification, he responded that he had never had a useful dream as far as he knew, and furthermore, it was Lord Rutherford who had conceived the model of orbiting electrons. It is clear that unsubstantiated claims of creative dreaming must be viewed with caution.

Problem Solving

As seen illustrated in Figure 11.1 (below)A dream was the inspiration for "the most brilliant piece of prediction to be found in the whole range of organic chemistry," the structure of the benzene ring. After many years of fruitless effort to solve the structural riddle of the benzene molecule ($C6H6$), Friedrich August Kekule, a German chemist, had a dream in which he saw six snakes biting each other's tails and whirling around in a circle. When he awoke, he interpreted the six snakes as a hexagon and immediately recognized the elusive structure of benzene. Another remarkable dream has been recorded by Hermann Hilprecht, a professor of Assyrian, in which a priest came to him and told him the true translation of the Stone of Nebuchadnezzar, which later proved to be correct. Also, originated in a dream was the frog heart experiment, the results of which established the foundation of the theory of chemical transmission of nerve impulses and earned a Nobel Prize for Otto Loewi.

Another Nobel Prize winner, Albert Szent-Gyorgyi, stated, "My work is not finished when I leave my workbench in the afternoon. I go on thinking about my problems all the time, and my brain must continue to think about them when I sleep because I wake up, sometimes in the middle of the night, with answers to questions that have been puzzling me." It is likely that artistic creation and problem solving occur in dreams more often than the documentation suggests. An intense waking preoccupation appears to be an important factor. This preoccupation depends, to some extent, upon the significance of the problem and the motivation of the individual seeking a solution. Thus, the problems being addressed by Kekule, Otto Loewi, and Hilprecht had occupied their energies for many years.

The creative and problem solving functions of dreams have only rarely been studied. The subject was approached once in 1892 by psychologist Thomas Child, who attempted to gather some statistics on this issue. In a questionnaire distributed to 151 male and 49 female college students, he asked: "During sleep have you ever pursued a logically connected train of thought upon some topic or problem in which you have reached some conclusion, and the steps and conclusion of which you remembered upon awakening?" Of 186 students who responded to this question, sixty-two (33%) answered in the affirmative. Some of the examples given were a chess game played in a dream, an algebra problem solved, a bookkeeping error found, and a translation of Virgil accomplished.

At Stanford University, we explored the phenomenon of problem solving in dreams through a series of problemsolving experiments involving 500 undergraduate students in three consecutive class meetings. Students were given a copy of a problem and an accompanying questionnaire and instructed not to look at the problem until fifteen minutes before going to bed that night. Before going to bed, the students were instructed to spend exactly fifteen minutes in an attempt to solve the problem. In the morning, they were instructed to write on the questionnaire any dream recalled from the previous night. If the problem had not been solved, students were asked to work on it for another fifteen minutes in the morning. The students' solutions to the problem were entered on the questionnaires, which were then returned to the instructor to be scored by several volunteers who were looking for solutions that could be attributed to dreams. The students were instructed not to discuss the problem amongst themselves until the next class meeting. Here are two of the problems and their solutions:

Problem 1: The letters O, T, T, F, F. form the beginning of an infinite sequence. Find a simple rule for determining any or all successive letters. According to your rule, what would be the next two letters of the sequence? Solution: The next two letters in the sequence are S, S. The letters represent the first letters used in spelling out the numerical sequence, "One, Two, Three, Four, Five, Six, Seven, etc."

Problem 2: Consider the letters H, I, J, K, L, M, N, O. The solution to this problem is one word. What is this word? Solution: The solution is the word "water" derived from the chemical formula H2O or H-to-O as given in the problem.

The total response represented 1,148 attempts at problem solving. Using a rather intricate scoring system, 87 dreams were judged to be related to the problem, 53 directly and 34 indirectly. If a solution was presented in the dream, the judges scored it as correct or incorrect, and whether or not the subject recognized it as such. The correct solution appeared only nine times all in the first experiment. On two of these occasions, however, the solution that appeared in the dream had already been obtained by the subject during the 15 minutes before bed. Of the 1,148 attempts, therefore, the problem may have been solved in a dream on only seven occasions. The following dream report contained one of these solutions:

"I was standing in an art gallery looking at the paintings on the wall. As I walked down the hall, I began to count the paintings one, two, three, four, five. But as I came to the sixth and seventh, the paintings had been ripped from their frames! I stared at the empty frames with a peculiar feeling that some mystery was about to be solved. Suddenly I realized that the sixth and seventh spaces were the solution to the problem!"

In the second experiment, there were 12 dreams classified as "mode of expression dreams" in which the answer, "water," was referred to either directly or indirectly. An example of the mode of expression dream was submitted by a 19-year-old male student. His dream recall was as follows:

"I had several dreams, all of which had water in them somewhere. In one dream I was hunting for sharks. In another I was riding waves at the ocean. In another, I was confronted by a barracuda while skin diving. In another dream, it was raining quite heavily. In another I was sailing into the wind." (This student perhaps did not recognize the word "water" as the correct solution because he had already solved the problem to his own satisfaction.)

While this experiment had several drawbacks in terms of design and controls, it was felt that it gave a valid indication of the possibility, albeit rarely evidenced, of problem solving during sleep.

The design of the experiment had several obvious shortcomings. Most of the anecdotal incidents of problem solving in dreams involved men who had been struggling with a particular problem for many years; our students had worked on the problem for only fifteen minutes. Even the most diligent and conscientious student had little incentive to obtain a solution, and this difficulty was probably more significant with the second and third experiment as the novelty decreased. In any experiment dealing with dreams, there is no assurance that the reported dreams are actually experienced by the subjects. Students could not really be prevented from studying the problem prematurely or discussing it among themselves. We are convinced however, that the dream solutions obtained in this experiment were valid examples of problem solving during sleep.

In the minds of many people, sleep is too commonplace to deserve careful consideration, and dreams are too "foolish" to suggest a logical course of action that might be carried out in the waking state. We may be losing enormous benefits afforded by the possibility that all of us are routinely presented solutions to our problems quite regularly in our dreams. Perhaps only the most perceptive dreamers possess the ability to recognize a solution that is presented in a disguised or symbolic fashion. Most of us, most of the time, are like the student who failed to recognize the word "water" as the solution to his problem even though he was deluged by water in his dreams! One can easily imagine Kekule shrugging as he awakened from the dream of the six circling snakes: "What nonsense! I must forget about snakes and concentrate on chemistry."

One can also imagine any number of ways that a dream might be utilized in this manner. For example, a woman may fancy herself in love with two men. She might experience marriage with each of them separately in a dream. She would then know which one to marry. There is the obvious conundrum of how one can be sure that dreams of the problem will take place, and if they do, how can one be sure to remember them? Lucid dreaming might be an answer. However, if one knows that one is dreaming, can the experience have the necessary power to change waking behavior? If the cancer dream mentioned below had been a lucid dream, it is difficult to imagine the same result with regard to my quitting smoking actually taking place.

A Dream that Probably Saved My Life

"I know, first hand, of another way in which dreams may occasionally have an extremely important problem solving function. Many years ago, (circa 1964), I was a very heavy cigarette smoker-more than two packs a day. Then, one night I had an exceptionally vivid and realistic dream in which I had inoperable cancer of the lung. It started with coughing up a little blood. I immediately had an X-ray and I remember as though it were yesterday looking at the ominous shadow in my chest X-ray and realizing that my entire right lung was infiltrated. The subsequent physical examinations in which a colleague detected the metastatic spread of the tumor to my axillary (arm pit) lymph nodes were equally vivid. Finally, and horribly, skull X-rays showed that the cancer had spread to my brain.

I felt the terrible anguish and dread of knowing my life was soon to end, that I would never see my children grow up, and that this would not have happened if I had quit cigarettes when I first learned of their carcinogenic potential. I will never forget the surprise, joy and exquisite relief of waking up. I felt I was reborn. Needless to say, the experience was sufficient to induce an immediate cessation of my cigarette habit. This dream had both anticipated the problem, and had solved it in a way that may be a dream's unique privilege. Only the dream can allow us to experience a future alternative as if it were real, and thereby to provide a supremely enlightened motivation to act upon this knowledge."
- William Dement, *Some Must Watch While Some Must Sleep*, 1972

Summary

We may presume that every human being dreams. The fact that a very small percentage of the population claims they never remember dreaming does not seriously undermine this assumption. Some people recall dreaming more than others. By and large, dreams mirror waking life in terms of settings and events, but from time to time-and the frequency is not really clear -- a dream is really unusual. This makes dreaming endlessly fascinating to the dreamer. Some individuals have recurrent nightmares and there is no question that severe trauma, such as in the Vietnam or the Iraq War, seems to permanently enhance this tendency.

The difficulties of the scientific study of dream content include the need to create a standard system of classification, deciding what its defining limits should be, and achieving a widespread acceptance of a standard unit of analysis.

In summary, dream research suggests that dreams are first and foremost an embodiment of thoughts through dramatization of life concerns and interests.

Dreams of the Famous

The following pages offer quotes and dream recollections from a broad range of well-known people. Read them at your leisure and for your own interest and pleasure. Maybe one day you will contribute to this list and if you know of any others please let us know.

Sylvia Plath recorded this dream in her journal: Marilyn Monroe came to her one night as a "kind of fairy godmother. I spoke, almost in tears, of how much she and Arthur Miller meant to us...." Marilyn gives Sylvia a manicure and they talk about hairdressers, Sylvia saying, "They always imposed a horrid cut on me." In the dream, Marilyn invites Sylvia to visit her during the Christmas holidays, promising a "new, flowering life."

Virginia Woolf, in a letter describing a recurring dream that overwhelmed her with terror, told of a dream of being alone, on the inside of a drainpipe, sliding towards its end. At the end of the drainpipe is madness: 'suddenly...I approach madness and that end of a drainpipe with a gibbering old man." She recorded another dream in her journal (in *A Sketch of the Past*): "I dreamt that I was looking in a glass when a horrible face --the face of an animal-- suddenly showed over my shoulder. I cannot be sure if this was a dream, or if it happened." (Both quotes taken from VW: *The Impact of Childhood Sexual Abuse on her Life and Work*, p. 105)

Thomas Carlyle: "Dreams! My dreams are always disagreeable --mere confusions-- losing my clothes and the like, nothing beautiful. The same dreams go on night after night for a long time. I am a worse man in my dreams than when awake--do cowardly acts, dream of being tried for a crime. I long ago came to the conclusion that my dreams are of no importance to me whatever." (quoted in William Allingham's ~ 7 Feb. 1868 (taken from *Oxford Book of Dreams*)

Fyodor Dostoevsky, *The Idiot*, 1869: "Sometimes one dreams strange dreams, impossible and grotesque dreams: on waking you remember them distinctly and you are amazed at a strange fact. To begin with, you remember that your reason never deserted you all through the dream, you even remember that you acted with great cunning and logic during all that long, long time when you were surrounded by murderers who tried to deceive you, hid their intentions, treated you amicably, while they had their weapon in readiness and were only waiting for some signal; you remember how cleverly you cheated them....But why does your reason at the same time reconcile itself with such obvious absurdities and impossibilities....One of your murderers turned into a woman before your very eyes, and from a woman into a cunning and hideous little dwarf, and you accepted it at once as an accomplished fact, almost without the slightest hesitation, and at the very moment when your reason, on the other side, was strained to the utmost, and showed extraordinary power, cunning, shrewdness, and logic? Why, too, when awake and having completely recovered your sense of reality, you feel almost every time, and sometimes with extraordinary vividness, that you have left some unsolved mystery behind with your dream? You smile at the absurdity of your dream, and at the same time you feel that in the intermingling of those absurdities some idea lies hidden, but an idea that is real something belonging to your true life, something that exists and has always existed in your heart; it is as though something new and prophetic, something you have been expecting, has been told you in your dream; your impression is very vivid: it may be joyful or agonizing, but what it is and what was said to you--all this you can neither understand nor remember."

Lewis Carroll, author of *Alice and Wonderland*. From *Through the Looking-Glass*, 1872: "I'm afraid he'll catch cold with lying on the damp grass,' said Alice, who was a very thoughtful little girl. 'He's dreaming now,' said Tweedledee: 'and what do you think he's dreaming about?' Alice said, 'Nobody can guess that.' 'Why, about you,' Tweedledee exclaimed, clapping his hands triumphantly. 'And if he left off dreaming about you, where do you suppose you'd be?' 'Where I am now, of course,' said Alice. 'Not you!' Tweedledee retorted contemptuously. 'You'd be nowhere. Why, you're only a sort of thing in his dream!' 'If that there King was to wake,' added Tweedledum, 'you'd go out -- bang! -- just like a candle!' 'I shouldn't!' Alice exclaimed indignantly. 'Besides, if I'm only a sort of thing in his dream, what are *you*, I should like to know?' 'Ditto,' said Tweedledum. 'Ditto, ditto!' cried Tweedledee. He shouted this so loud that Alice couldn't help saying 'Hush! You'll be waking him. I'm afraid, if you make so much noise.' 'Well, it's no use your talking about waking him,' said Tweedledum, 'when you're only one of the things in his dream. You know very well you're not real.' 'I *am* real!' said Alice, and began to cry."

Joseph Conrad, *Heart of Darkness*, 1902: "'It seems to me I am trying to tell you a dream -- making a vain attempt, because no relation of a dream can convey the dream sensation, that commingling of absurdity, surprise, and bewilderment in a tremor of struggling revolt, that notion of being captured by the incredible which is of the very essence of dreams...' He was silent for a while. '...No, it is impossible; it is impossible to convey the life-sensation of any given epoch of one's existence -- that which makes its truth, its meaning -- its subtle and penetrating essence. It is impossible. We live as we dream -- alone."

Katy Perry told *The Sun* newspaper that; 'I can do bitchy. Give me six hours of sleep instead of eight and I'll show you the bitch in me."

CHAPTER 11 The Content of Dreams

Lady GaGa tells *Rolling Stone* the Devil is trying to takeover and gives detail about her reccurring Illuminati dream: "I have this recurring dream sometimes where there's a phantom in my home and he takes me into a room, and there's a blond girl with ropes tied to all four of her limbs. And she's got my shoes on from the Grammys. Go figure -- psycho. And the ropes are pulling her apart." But it gets even stranger. "I never see her get pulled apart, but I just watch her whimper, and then the phantom says to me, 'If you want me to stop hurting her and if you want your family to be OK, you will cut your wrist.' And I think that he has his own, like, crazy wrist-cutting device. And he has this honey in, like, Tupperware, and it looks like sweet-and-sour sauce with a lot of MSG from New York. Just bizarre. And he wants me to pour the honey into the wound, and then put cream over it and a gauze.

Madonna had a recurring nightmare that a faceless, nameless man was trying to murder her (after her mother's death). Recently, while Madonna was in Paris staying at the Hotel Le Meurice, she had a recurring dream that she was a nun being hunted down by Nazis. She was so upset by the dream she couldn't eat. It was only later that she realized the hotel had been the headquarters for the Nazis during WWII. (p. 38, 113)

Another Madonna recurring dream: A friend says Madonna saw herself in her dreams as someone who was "so pure that she didn't defecate, or when she did, it was white." (p. 114) During the Blonde Ambition tour, Madonna dreamed Mikhail Gorbachev had attended her show and the first thing she could think of was "that Warren [Beatty] is going to be so jealous because I got to meet him first." (p.290) (From *Unauthorized Madonna* by Christopher Anderson)

Obama: Sleeps In Till 8! A self-described night owl, as a senator and (at the time) presidential candidate Obama did not get a ton of sleep whether at home in Chicago (he would stay up until one or two in the morning working on *The Audacity of Hope*) or on the campaign trail. His morning "sleeping in" time is 8 a.m., according to an AP questionnaire, but he probably wishes it were later: "I'm not naturally a morning person," he told CBS News. Becoming president has done nothing to help Obama narrow the sleep deficit. By the time midterm elections came around in 2010, Obama told Ryan Seacrest on his radio show that he gets a wakeup call every morning from the White House operator. "If I don't wake up the first time, they just keep on calling." During the debt-ceiling standoff in the summer of 2011, top Obama aide Valerie Jarrett told Reuters: "He's getting absolutely no sleep."

Eminem: Sleep experts often recommend black out shades to keep rooms dark and promote quality rest -- but rapper Eminem reportedly takes that to a whole new level, wrapping tin foil around the windows to keep the light out, according to *The Sun*. "He uses the technique as he's always jumping time-zones," an anonymous source told the UK paper.

Dorothy Parker dreamed she had the answer to the world's problems. In the morning, she found she'd scribbled a note to herself: "Hoggimous, higgimous, men are polygamous, Higgimous, hoggimous, women monogamous." (Garfield, *Creative Dreaming*, p.73)

Keith Richards (according to Mel Stewart): In the middle of the night Richards woke up and played the first sequence of "Satisfaction" on his guitar, straight from his dream. He usually leaves a tape recorder on all night by his bed; in the morning, he played it back, heard that first now-famous sequence. "After that," says Richards, "the rest of the night was just snoring." He didn't remember waking up and playing the song, but the tape was irrefutable evidence!

Elias Howe: In 1845 Elias Howe invented the sewing machine. He had a dream that helped him understand how the penetration of the needle would work in his invention.

George Harrison woke up humming some unknown song left over from a dream. While he was shaving, he looked in the mirror and realized he was humming the same tune, went to the guitar and started playing it; "All Things Must Pass," his first solo album.

Woody Allen and his dreams. "Dreams proliferate in Woody Allen's movies but he has surprisingly little use for real ones. The closest he comes to using them is to let his unconscious work on problems as he sleeps; when he gets into bed, he thinks of a problem in a script he's working on in the hope his mind will sort it out during the night." (From *Woody Allen: A Biography* by Eric Lax, p. 268)

Paul McCartney was said to have awakened from sleep and he immediately went to the piano and played "Yesterday", recalling it from a dream. Dr. Dement believes that Yesterday is the most profound and evocative song of all time.

The Beatles were in London in 1965 filming *Help!* and McCartney was staying in a small attic room of his family's house on Wimpole Street. One morning, in a dream he heard a classical string ensemble playing, and, as McCartney tells it:

CHAPTER 11 The Content of Dreams

"I woke up with a lovely tune in my head. I thought, 'That's great, I wonder what that is?' There was an upright piano next to me, to the right of the bed by the window. I got out of bed, sat at the piano, found G, found F sharp minor 7th -- and that leads you through then to B to E minor, and finally back to E. It all leads forward logically. I liked the melody a lot, but because I'd dreamed it, I couldn't believe I'd written it. I thought, 'No, I've never written anything like this before.' But I had the tune, which was the most magic thing!"

Albert Einstein: His theory of relativity was inspired by a dream whereby he was going down a mountainside ever faster and watching the appearance of the stars change as he approached the speed of light.

Nils Bohr: This Danish scientist dreamed about "atomic structure" and then came up with the theories we use today in chemistry about the atoms and their structure.

President Abraham Lincoln: Dreamed of his assassination and described the dream to his wife just a few days prior to his assassination.

Tommy John of the Yankees: "I always dream that I'm pitching a game and I'm late. I get to the park and I'm trying to put my clothes on and I can't get them on. I pull my socks up and they go down."

Dave Parker (Reds outfielder): "A ball I've hit badly on a certain pitch or a mistake I've made in the field will come back [to me in my dreams]. You kind of analyze yourself in the dream and you say, 'How in the hell did I miss that pitch?' or 'What was I thinking about when I dropped that ball?"

Joan Benoit Samuelson: "In one of my recurring dreams I'm in a department store, and I can't find my way out to get into the race. I either go up the 'down' escalator or down the 'up' escalator. I either wake up before the race starts, or the gun is going off as I'm running up to the start." She also has lots of missing-the-race-dreams: I'm trying to get to the start of the race, and I'm on time and running along to get there. Then I can only walk, and then I can only crawl, and then I'm bellying myself along the road. As I wake up, I'm hanging onto a limb at the end of a cliff."

Tim Mayotte had this dream just before going to Wimbledon for the eleventh time. He was playing John McEnroe in a match and he had a break point against him to win the set. Tim hit a beautiful backhand down the line, and the umpire called it "Out!" "What do you mean, 'Out?'" Tim hollered back. "That strawberry was in! Check the splotch!" No one was surprised that the tennis ball had turned into a strawberry just as Tim hit it, and Tim insisted that he, Connors, and the umpire all go check the mark left by the speeding berry.

"When you're dreaming with a broken heart
The waking up is the hardest part" - John Mayer

CHAPTER 12 Dream Anarchy Rules!

5-year-old son: "Dad, I want to be in a world where there are no rules."
Rafael Pelayo: "In your dreams..."

Introduction

As we progress in the 21st century, the prehistoric quest to discover why we dream will hopefully be answered. The source and meaning of dreams changed drastically during the 20th century. At the beginning of the 20th century psychoanalysts were the dream experts. With the discovery of REM sleep and its experimental manipulation the influence of psychoanalysts as sleep researchers decreased rapidly towards the end of the 20th century. Psychoanalytic theories did not stand up well to the science provided by the modern sleep lab. The debate over whether dreams are trivial or profound, triggered by external events or arising from our innermost yearnings and desires, garbled thoughts of the sleeping mind or symbolic messages from the supernatural, has been present since antiquity. However, lacking the constraints of modern science and scientific attitudes, human beings were able to endow dreams with a profound significance that were at the same time external because the dreams were thought to be supernatural. Dreams were thought to be derived from magical sources, from gods and demons and from the notion that the essence of the being, the soul, in fact, walked abroad and gathered experiences, portents, and messages of overriding importance ignored by the dreamer at great peril. One may remark that forgetting a dream must have very often saved someone from a difficult situation.

The composer Guiseppe Trarini (1692-1770) called one of his sonatas "The Devil's Trill" because when he was unable to complete it he dreamed he heard the devil play it. The trill was all he was able to remember of the dream

Legend has it that if an Iroquois Indian dreamed that he had stolen his neighbor's horse, when awake he was obliged to repay the theft by giving the neighbor one of his real horses. Is it a safe assumption that the dreamer creates the dream? If the dreams were from an external source, you would have to decide if the source was good or evil. Attempting to answer this question might cause you to lose some sleep! From a layman's point of view, the recall of a dream is usually accompanied by two feelings: 1) a sense of amazement that such a peculiar, startling, or unusual dream was actually dreamed, and 2) a curiosity about why this particular dream occurred on this particular night. This implies that the dream might have some special purpose or meaning. Why a dream is dreamed at any particular REM period, whether or not it has specific meaning that can be known by translation or interpretation, and whether or not it was dreamed to fulfill a specific knowable purpose are different questions, though they somewhat overlap.

The trickiest issue in relation to these questions is the difference between "meaning" and "purpose." This issue was also touched upon at the end of Chapter 9. We will try to clarify this issue with the following hypothetical story. Suppose a man is rummaging through a drawer looking for a pen and instead finds an old photograph that he has not seen for many years. It is a photograph of the man and his wife from twenty years earlier. He studies the picture. At that moment, what it means to him is the contrast between then and now and the significance of the intervening years, the work of marriage and the rewards of marriage. The photograph clearly is intensely meaningful and connects to an enormous number of memories and feelings. It would be totally absurd, however, to say that the photograph was in the drawer specifically for the purpose of being seen by the man at that moment because he needed to express thoughts and feelings about his marriage.

With regard to a dream experience, it is very difficult to separate meaning and purpose since the dreamer creates the dream (i.e., decides the photograph should be seen and then puts it in the drawer where it is certain to be seen). It is possible to conceive that the dream world occurs more or less at random with the exception of deliberate, volitional control of the dream as in lucid dreaming, a phenomenon to be discussed more fully in the next chapter. In this view, the dreamer experiences the random dream image and it nonetheless has meaning for him in the same manner as the chance encounter with the photograph mentioned above. The alternate view is that it is possible that the dreamer creates the dream, presents it to himself or herself, and the experience of the dream accomplishes a specific purpose -- such as discharging aggression or experiencing success.

External Sources Versus Internal Sources

CHAPTER 12 Dream Anarchy Rules!

The most consistent debate down through the ages regarding dreaming has concerned the true source of the dream: whether the content of the dream is generated from internal or external supernatural sources. At the outset, one should immediately state that the scientific study of dreams does not accept supernatural sources as an explanation. It is more than likely that many sources combine to form the dream story and its images at any given time. It is clear that the waking experiences of the dreamer and chance external events that occur during REM periods can occasionally influence the content of a dream. On the other hand, it has not been demonstrated that dream content can be consistently and predictably influenced by any kind of manipulation, however meaningful or powerful. Nonetheless, it is absolutely certain from experimental evidence that external stimuli can often or occasionally influence the dream content. If external stimuli intrude into the dream, the dream story that follows will usually be adjusted accordingly. Therefore, the source of dreams may be a combination of external environmental sources.

An example of this is a study published by Stanford undergraduates in which a train whistle, a very evocative sound of years past, was played to subjects during REM sleep. A typical effect was that the dreamer looked around and saw a train. He walked over to it, got in, and traveled somewhere. These effects seem to be more predictable with familiar and easily recognized sounds. In addition, when the sleeper transitions from non-REM sleep to REM sleep, we assume that a dream begins. Why does it begin in any particular manner? How is the first sentence chosen, so to speak? Is the raw material of dreams generated at random as REM sleep progresses or is it a dormant script waiting for REM sleep to give it life?

When people wish to know the meaning of their dreams, they are seeking something more profound than a chance noise being converted into a visual image. Many people may want to believe that the dream mirrors their most profound impulses, yearnings, fears, desires, and so on and it may help to resolve the latter. Proving that dreams have this function or purpose is difficult. It is even difficult to design an experiment that could settle the external versus internal debate. We are therefore in a situation where, to a large extent, the meaning that dreams have for us is the meaning that we confer upon them while awake. In view of the likelihood that both external and internal sources are operating, this chapter will consider both. In terms of external sources, we include sources inside the body that are presumably external to memory and drives that are organized in the brain itself, such as sound, light and touch. Thus, hunger, thirst, pain, shortness of breath, itching, and all sensations that arise in the body are considered external sources. For example, someone with sleep apnea may dream that they are drowning.

External Determinants of Dream Content

In the nineteenth century, it would be accurate to say that the prevailing view was that dreaming arose in response to bodily discomforts, sensations, external stimuli, and experiences that carried over from the previous day. Most of the written material in the popular and scholarly literature that describes the effect of the external determinants of dream content is anecdotal. Thus, even if an early investigator was assiduously compiling a dream diary and noting all the dream experiences that were recalled, the ability to relate the daytime events to the dreams of the subsequent night was unpredictable at best.

An upsurge of interest in this type of investigation occurred after the discovery of rapid eye movements. Investigators could carry out experimental manipulations in the daytime or in the evening prior to sleep, and then sample dreams from all the REM periods of the night for evidence of a response. One of the most dramatic of such studies was carried out in which subjects viewed movies of human birth, human circumcision, and a mother monkey eating her dead infant, before going to sleep in the laboratory. The dreams elicited from REM period awakenings were compared with dreams obtained on nights when the same subjects were shown movies whose content was emotionally neutral.

They reported a high level of effectiveness in influencing dream content. However, a great deal of, "after the fact", interpretation was involved. Researchers were forced to read into the dream story to find the content elements. Another investigator summarized these results by saying the real finding, given the dramatic nature of the movies, was the relative lack of effect on dream content. Most investigators feel that the external stimulus must in some way link up with a powerful internal determinant. They suggest that preconceptions about what will be a potent stimulus in terms of daytime experience can often be in error. Nonetheless, in the past, it seems difficult to imagine that viewing violence, blood, and mutilation do not have a powerful emotional impact which ought to show up consistently in dream content if such impact is a major dream content determinant. These days, violence, blood, and mutilation are routine video game experiences. Do you think video games have ever influenced your dreams?

Hunger and Thirst

The effect of stimuli arising from within the body has also been explored. There are many anecdotal accounts of explorers who were lost and starving and dreamed of sumptuous meals. A comprehensive study of effects of very prolonged starvation on the physical and psychological health of volunteers was carried out by Ansel Keys during World War II. Although this study would not seem ethical today, he could keep track of the dreams of his starving subjects. Somewhat surprisingly, he found no particular increase in dreams about food and eating. We must point out that this research was done before the discovery of REM, therefore, the dreams could not be verified.

William Dement and Edward Wolpert investigated the effect of thirst on dreams. Three subjects on five occasions completely restricted their intake of fluids for twenty-four hours or longer before sleeping in the laboratory. On each occasion the subjects reported that they were extremely thirsty when they went to bed, they had mouths so dry they were unable to salivate. Fifteen dream narratives were obtained under these conditions, and in no case, did the dream content involve an awareness of thirst or descriptions of actual drinking. Five of the dreams, however, contained elements that seemed related to the theme of thirst and drinking:

"I was in bed and was being experimented on. I was supposed to have a malabsorption syndrome."

"I started to heat a great big skillet of milk. I put almost a quart of milk in."

"Just as the bell went off, somebody raised a glass and said something about a toast. I don't think I had a glass."

"While watching TV, I saw a commercial. Two kids were asked what they wanted to drink and one kid started yelling, 'Coca-Cola, Orange, Pepsi,' everything."

"I was watching a TV program, and there was a cartoon on with animals like those in the Hamm's beer advertisement."

One subject for this thirst study recalled waking up spontaneously and immediately becoming aware of his painful thirst. On that occasion, he thought he heard raindrops falling on the window. When he looked outside, he saw the full moon and the stars. We assume that his desire for water was so great that he momentarily hallucinated the raindrops.

Dream Sequences: Can One Dream Influence the Next? Another approach, which was made possible by the discovery of REM sleep and the use of laboratory EEG techniques, was to examine multiple dreams of a subject on a single night. If we arouse a subject in every REM period, we are likely to get four to eight detailed dream reports. Will these dreams be similar or totally different? We might expect the dreams to be similar just because they occur on the same night. However, we occasionally obtain nightly samples where the dreams seem totally unrelated. A sequence from one night in the sleep lab started out with two hippopotami; then a taffy pull in the Russian embassy with the late Russian Premier Nikita Khrushchev as one of the pullers; next a motorcycle ride through a wheat field. In the last dream of the night, the dreamer was at his desk in Riverdale, New York, writing some sort of paper.

Is it possible that widely disparate dream episodes are related or linked together by some hidden thought or impulse in the mind of the dreamer? Even on the level of overt dream imagery, the degree and variety of possible relationships are almost infinite. In the most trivial case, five successive dreams might be said to be related to one another if each one contained the image of a tree or if there were people in each dream. At the other end of the scale, dreams might be related in terms of a complex thematic development or restatement that involves virtually the entire content of each successive dream.

The first study on this topic was done in 1956 by William Dement. By awakening subjects ten to fifteen minutes after the beginning of each successive REM period, they obtained thirty-eight nightly sequences of four to six dreams, each distributed among eight adult volunteers. Despite very careful scrutiny, they did not find the exact duplication of a single dream. Many people say they can wake up from a dream, go back to sleep, and continue the dream. In this study, no dreams in such a sequence were ever perfectly continuous. For the most part, each dream seemed to be a self-contained drama, relatively independent of the preceding or following dreams. Nevertheless, the manifest content of nearly every dream exhibited some obvious, though often trivial, relationship to one or more dreams occurring on the same night. In most cases, only contiguous dreams were obviously related. Some of the relationships seemed quite incidental yet intriguing, as in the following example:

". . . I went inside and started going up an escalator. I could see my wife up ahead of me four or five steps. The place was just mobbed. Then we were going down a hallway and I couldn't get to her. There were cakes of ice in the center of the hallway and people just milling in and out, everyone carrying suitcases and things. Then we started up this next escalator, and there was a girl standing beside me. She had a real shabby suitcase..."

"... He was collecting big hunks of watermelon, and I thought I'd get a job helping him, so I started picking them up, and some of them looked more like pieces of ice than they did like watermelon..."

Although the somewhat incongruous presence of ice illustrates a seemingly trivial relationship between the two dreams, this image might imply a deeper and more important relationship on the level of the underlying dream determinants (or maybe they were just cold that night). Thematic correspondence is more extensive in the following narratives elicited from two contiguous REM periods:

". . . I went in (a house on a hillside) and I had a feeling that I shouldn't be there or that it was somehow slightly naughty to be in there. Anyway, I was inside and I realized there was a gangster somewhere in the house. There was a third party in the room with us, and we were listening to something going on outside the room. Suddenly we had to escape and we all ran . . . there were three of us, my wife and I and some man, I can't remember who he was but he seemed to belong ...and we had to get away, so we jumped out the window. Then we got into the car, and I yelled at this guy for some reason, that I ought to drive. He didn't know how to drive our car, but there was something about him-like he was a movie hero or something-and he was taking over. He jumped behind the wheel, and he went roaring up the hill. Someone shot at us out the window as we ran off. "It started out with me telling somebody about a murderer. The

murderer was supposed to be in this house. I was telling two detectives a rather lengthy story about this gruesome murder. The idea was to lock them in this house with the murderer so they'd catch him. And wife, or some woman who was somehow related to me, was supposed to leave. So, she went outside and I locked them in. Just as I finished locking them in the house it occurred to me that this was a trick, and the murderer was this woman, and she was having me lock the detectives in the house so she could get me. Just as I went running down the porch stairs this horrible knowledge dawned on me. I ran out into the yard and was kind of looking at the house. It was an old house on a hill. The yard was kind of roundish. Suddenly she jumped out of the bushes and began running at me. She looked horrible. She was going to push me off the cliff-part of the hill was a cliff-or kill me somehow. Just before she got to me she changed into a tiger-a tigress. At that moment, I woke up crying out."

In each of these dream narratives, a house on a hill is the locale, and the dreamer leaves the house because of some danger. A gangster appears in the first dream and a murderer in the second. However, there seems to be a reversal of circumstances between the two dreams. In the first the danger is within the house, and a safe exit is made by the dreamer and his companions. But in the second the danger is on the outside, and the dreamer is unable to escape but must awaken in terror.

Another longer sequence revealed a very complicated scheme of relationships through four contiguous dreams. The dreamer seemed to be at the center of a kind of classical tragedy, in which those elements of strength which appear in the first dream are the very forces that vanquish him in the fourth. Although the dictated and transcribed dream narratives are far too long and detailed to be reproduced in their entirety, a summary of the elements will show an example of how the individual dreams might be interrelated.

Examples from Dr. Dement's dream logs when he was a student.

Dream narrative #1

"I was dreaming something about a woman. She was trying to do something about an inheritance, and I was trying to thwart her. I must have thwarted her pretty well, but she still had something she could do and in the dream, I am saying let me see your trump card. Let me just look at you. I went over and looked at her straight in the eye and I said, how could I

possibly be afraid of you? She just kind of turned away a little and just as you awakened me I was chasing her out and I shouted, you God damn bitch!"

Dream narrative #2

"I was watching a guy standing in the street and suddenly he raised his gun and shot this woman in the back. I was terrified and I ran into a little frame house. I was afraid this guy would come after me. I was feeling very frightened when you woke me up."

Dream narrative #3

"When the dream started out I was with Sarah Vaughan (the great jazz singer). She was trying to seduce me and she was kind of neurotic. She scorned any conventional approach. She wanted me to paint my arms one color and my body another and she wanted to strap me down. I finally got mad and left. I went into my own room and leaned on a chair. I leaned resting my right hand on the chair and turned on a TV set. Then I went into another room where dinner was being served. Suddenly, I looked at my hand. It seemed like we were in a family circle and I was in an entirely different family than I should have been, but when I looked at my hand it was bleeding. First I saw my little finger bleeding and then it looked like my whole damn hand was bleeding. Then I noticed a series of cuts clear up my arm. I was kind of stunned. Then somebody said that Sarah Vaughan had deliberately left a razor blade on the chair. Then there was a maid who came in and started to wash my hand."

Dream narrative #4

"It had something to do with the card game of bridge. We were sitting at the bridge table. Just three of us. These two women and me. There was something about how it took very long to play a hand. I was very puzzled about how the hand had been played. Then I looked at my cards and they were the wrong kind of cards. I felt very upset and humiliated. I didn't know what to do and then I woke up."

In the first dream the dreamer wins a very consummate victory over his female competitor in the competition for the inheritance. He scorns her "trump card" and chases her from the room. He is no longer certain of himself in the second dream. When a woman is shot in the back, (compare this cowardly act to the bravado of the first episode when the dreamer looked the woman straight in the eye) by a male associate who may represent the dreamer himself, he runs fearfully. In the third dream the dreamer is subjected to the humiliation of a passive role in love making with a celebrity. Immediately afterward he cuts his arm on a razor blade. Some would interpret this as a castration dream. Demonstrating how far the reversal of roles has gone in the reported dream series, the final dream finds the dreamer playing cards with two women and discovering that his cards are somehow "wrong". A far cry from the first dream in which he felt no fear of even his competitor's trump cards!

The continuing story of the foregoing four dreams might be summarized as follows:

A very competitive dreamer totally defeats a woman. Residual feelings of guilt develop and fear of retaliation may have been activated. The originally aggressive dreamer has been reduced to seeing a woman shot in the back and is very afraid. In the third dream, he is unassertive and then severely punished. In the fourth dream, he is completely defeated by women.

The fact that a subject who is being awakened during every REM period of the night to elicit dream recall remembers clearly the tremendous amount, variety, and richness of a single night of is simply amazing. As far as we know, the only available method of gathering this type of dream sequence material is this approach of awakening subjects during successive REM periods. However, this method has at least two important limitations. First, since one can never be certain in advance exactly how long an individual REM period will last, the awakening must occur shortly after the onset of REM sleep. An unknown amount of material is lost because the dream is prevented from reaching its natural termination. Secondly, the procedure of the awakening undoubtedly disturbs the dream pattern. Not only is the dream abruptly and unnaturally terminated, but a series of events, namely the awakening, the description of the dream, and the handling of the recording apparatus, might induce a spurious relationship of one dream to another.

External Sensory Stimulation During Sleep

Is it an Urban myth or a college dorm prank that if you put someone's hand in water while asleep that it makes them urinate? It is known that external sensory stimuli do not cause dreaming. However, if a stimulus happens to coincide with a REM period, can it influence the dream content? We already know that the answer is yes from studies with drops of water summarized in Chapter 10. Anecdotal literature is replete with examples: someone who dreamed of thunder awakened to hear the clatter of horses' hoofs on the pavement; someone who dreamed of a roaring conflagration awakened to find a candle flickering by his bed. One of the first studies of the relationship between stimuli and dream content which utilized the new technique of REM period awakenings was conducted in 1958 by William Dement and Nathaniel Kleitman. In this study, they inserted three different, relatively non-specific stimuli into REM periods. The first stimulus was a 1,000-cps pure tone which sounded for five seconds at a level slightly below the awakening threshold of REM sleep. The second was a flashing 100-watt lamp placed where it would shine directly into the sleeper's face. The final stimulus was a few drops of cold water ejected from a hypodermic syringe. From our discussion of the course of time in dreams in Chapter 10, the reader will remember that the water drops were especially effective. The stimulus was presented after the characteristic change in the EEG and rapid-eye movements had signaled the start of a REM period. If the stimulus did not awaken the subject, he was allowed to sleep for another few minutes before being awakened and asked to report his dream recall. The dream reports were subsequently examined to determine whether the stimulus had been incorporated into the dream. Incidence of stimulus incorporation varied from 42 percent for the cold-water drops to 23 percent for the light flashes and 9 percent for the pure tone. It should be noted that the water drops, as common sense would tell us, was most easily recognized in the dream reports. Nonetheless, there appeared to be a kind of hierarchy of incorporation. In addition, although a stimulus was presented fifteen times during periods of NREM sleep, no REM periods were initiated and no dreams were recalled on these occasions. Since the time of this early study, several investigators with various objectives in mind have conducted studies using external stimulation. One investigator used spoken names that were either emotionally significant or neutral to the subjects. The names were presented below the threshold of arousal during REM sleep. The study found an incorporation rate of about 54 percent but no differences in incorporation between emotional and neutral stimuli. It was noted that although perception of external stimuli can occur during REM sleep, their origin is perceived as a part of the dream. Other investigators used recordings of the subject's own voice and of other voices as their stimuli. When the subject's own voice was played, the principal figure of the dream was more active, assertive, independent, and helpful. When another's voice was played, the main figure was unequivocally passive.

Hoping to provide a conclusive demonstration of the effects of stimuli on dream content, several freshmen in the Stanford University Sleep and Dreams class of 1970-71 conducted an exhaustive study. (The results of this study were published in the Stanford Quarterly Review, Winter 1972). This study involved elaborate procedures and statistical analysis, and independent judges were used to rate the amount of incorporation of each stimulus. The students chose as their stimuli taped recordings of twelve very familiar and evocative sounds such as a rooster crowing, a steam locomotive, a bugle playing reveille, a dog barking, traffic noise, and a speech by Martin Luther King, Jr. The subjects were monitored per the usual procedures, and the sound tape was played starting at approximately ten seconds after the onset of a REM period.

The students found that the sound clearly influenced dream content in 56 percent of the recorded dreams; the locomotive sound was the most effective (82 percent) and traffic noise the least (28 percent). A strong incorporation of the steam locomotive is illustrated by the following report:

"I dreamed I was riding in a train. I was driving the engine, and the train was in Branner dormitory, and right close to the engine there was this pit. It was about two or three stories long, and it was still open, and the train kind of chugged down into it, and it was real scary. I was dreaming the whole time. When I was going into the pit, it was amazing because there were some people at the top of the pit watching me go down."

Anxiety

A good example of the effect of anxiety on dream content is the effect of the sleep laboratory and its equipment and wires on a naive subject. Such a subject, seeing the rather impressive equipment, and having wires attached to his or her head, will feel uneasy and will worry a little about electricity and the possibility of getting shocked. The dreams, if subjects are awakened from REM periods on their first night, will often clearly show the laboratory situation. One study found that more than a third of all dreams recalled from REM period awakenings depicted the laboratory situation. An example is as follows:

"I dreamed I was lying here and something went wrong so that any second I was going to be electrocuted. I wanted to tear the wires off, but suddenly realized that my hands were tied. I was very relieved when you woke me up."

The frequency of laboratory depictions dropped precipitously by the second night presumably because the subjects had learned and were confident that the equipment and the electrodes attached to their heads were not at all dangerous.

On those occasions when dreams occurring on later nights do reflect the laboratory situation, they are usually much less fearful:

"[I dreamed] you came in and told me there was a big party going on next door. We decided to call it a night and go to the party. After the electrodes were off, I put on a tuxedo and went over. A whole bunch of people were dancing and I saw this girl standing in the corner."

It is clear that external events, previous experiences, stimuli delivered during REM periods, and a variety of other conditions can influence the content of dreams though never with absolute consistency. Beyond this, little is known. The sleep laboratory offers a powerful tool to investigate a development of dream content and the rules, assuming there are rules, of dream formation. However, at the present time dream research is certainly not on society's front burner. There currently is only minimal federal funding that supports dream research. The likelihood that federal support for this type of work will flourish depends on public demand.

Internal Determinants of Dreams

It is important to look back to early days to understand why there is so little funding for sleep research. One of the people who had the most influence on the thinking of his time and beyond was Sigmund Freud (1856-1939). Indeed, Freud's work radically shifted the emphasis of dream theory away from physiology and external events and toward psychology, or more properly, a new psychology emphasizing the role of the pre-conscious and unconscious levels of the mind. The first edition of his most influential book, *The Interpretation of Dreams*, was published in 1899. It contains all the essentials of the psychoanalytic method of interpreting dreams to arrive at their latent (hidden) content. These principles form the foundation of the psychological discipline known as psychoanalysis which is both a clinical method, and a theory of personality, behavior, and abnormal psychology.

The Psychoanalytic Theory of Dreaming

In the introduction of *The Interpretation of Dreams*, Sigmund Freud states that he would prove "that there is a psychological technique which makes it possible to interpret dreams, and that, if that procedure is employed, every dream reveals itself as a psychical structure which has a meaning and which can be inserted at an assignable point in the mental activities of waking life." He found that the method of free association, which was useful in dealing therapeutically with neurotic symptoms and their meaning, could also be applied to interpreting dreams. In this method, patients were instructed to suspend all judgmental and inhibitory editing of their thoughts and simply to communicate without hesitation everything that came into their minds and or consciousness as they began focusing on a dream image. This process would involve **free-associating** to consecutive scenes or images of the dream. The psychoanalyst then interprets the associations based on his knowledge of psychoanalysis and of the patient.

It is hard to underestimate the importance of Sigmund Freud. He was a worldwide rock star before there were real rock stars. He was as well known in his generation as Steve Jobs is to yours. According to Freud, when the work of interpretation is completed, it will be obvious that a dream is the fulfillment of a wish. Having arrived at this conclusion, Freud was immediately confronted with the problem of unpleasant dreams, anxiety dreams, or even nightmares in which dreadful harm threatened the dreamer. These dreams express outcomes that could hardly be something the dreamer wished for. To account for such dreams, Freud had to invoke mechanisms of distortion and disguise which would serve to hide the actual meaning of the dream from the dreamer's conscious awareness. He modified his initial wish fulfillment hypothesis in the following manner; "A dream is a disguised fulfillment of a suppressed or repressed wish." In Freud's day, many human impulses were, in fact, not acceptable by society. Overt expression of sexual impulses, aggressive impulses, and more primitive oral and anal impulses were not openly discussed and of course never acted upon, except by social deviants.

The overt conscious experience of the dream was called by Freud the **manifest content**, and the hidden wishes, motives, and driving force of the dream he called the **latent content**. The goal of dream interpretation is to arrive at the latent content. All of the devices by which this hypothetical latent content is transformed into the manifest content is called the "dream work." Transformation of a latent thought or wish into a manifest image could involve several processes which are analogous to translating foreign text into English, or to understanding a rebus puzzle where concepts or words are translated into pictures. The latent content, or the wish, was generally thought to be triggered or activated by some event that occurred on the previous day.

Condensation refers to the connection between the manifest dream content and all the latent dream thoughts that appear to be related to it. Freud felt that condensation was typically carried out on a large scale. Dreams in general appeared to be brief, meager, and laconic in comparison with the range and wealth of dream thoughts. In short, he opined that many wishes could be condensed into a single image. He noted that writing a description of a dream rarely filled more than half a page, while the analysis of the dream and all the dream thoughts underlining it might occupy six, eight or a dozen times as much space.

Displacement means that a latent dream thought is simply displaced upon a manifest image without any obvious relation. Thoughts underlying it might occupy six, eight, or a dozen times as much space.

Symbolism is a process with which everyone is familiar, though in many cases the choice of a symbol can be highly idiosyncratic. In psychoanalytic dream theory, symbols generally have a similar form to that which is symbolized. However, sometimes quite different objects appear to symbolize the same dream thought, or many dream thoughts are represented by the same symbol. It is also possible for these processes -- condensation, displacement, and symbolism -- to operate simultaneously to create a single dream image. **Secondary revision** is a final process by which the dream work is hypothesized to add interpolations and additions to the manifest dream content so that it becomes more coherent.

Freud noted that undistorted, wishful dreams often occurred in children. Many disguised wishes that appear during dream interpretation seem relatively trivial and should not be able to provoke censorship and distortion that takes place. This was explained by the hypothesis that a conscious wish can only become a dream instigator if it succeeds in awakening a related forbidden unconscious wish thereby obtaining reinforcement from it.

Much psychoanalytic theory of dream interpretation has been oversimplified. The unconscious wishes almost defy conscious characterization. In the Freudian world view, an unconscious wish is related to the most primitive needs and totally selfish behaviors that are necessary for the survival of the organism. The urge to devour and incorporate is the "oral wish." The perception that no other person exists and that the mother is part of the self is the "megalomania" of infancy. The totally selfish solution of an impulse to simply kill or crush any obstacle related to this. Later in life, the drive is to reproduce. However, Freud felt that sexuality was part of the infantile impulses not yet directed in terms of mature genital behavior. These impulses would derive their power from the most primitive, unconscious psychic energy. In the course of achieving consciousness, they would be endowed with more specific content. For example, an impulse to buy an ice cream cone might be energized by the unconscious wish to incorporate the mother or the mother's breast into the self.

The interpretation of dreams of patients by the method of **free association** yielded a great deal of information about thoughts, wishes, and fantasies not ordinarily expressed in Victorian society. A later unconscious wish that becomes important as the infant matures is the presumed attraction of the infant to the mother and the resentment of the intrusive father. This gives rise to the Oedipal wish to kill, destroy, or eliminate the father. Primitive sexual energy is labeled libido. The wish to take the mother from the father was assumed to provoke fear of retribution in the form of castration. This fear in males is repressed and transformed into a host of neurotic anxieties related to male-male competition.

Even though these concepts seem ridiculous to us now it was once considered the scientific state of the art of dreams. These concepts supported an interest in sleep research and the discovery of REM sleep. One wonders what Freud might have thought of our modern understanding of sleep, realizing neither he nor his patients remembered even the tiniest fraction of their entire dreams lives. The psychoanalytic method and theories of dream interpretation have neither been proven nor disproven because they have not lent themselves easily to scientific method. The self-interpretation of a dream by means of free association carries the possibility of self-deception. The major elements leading to a conviction that these notions were correct involve the realization that many human impulses were indeed ignored or repressed in society at the turn of the previous century (1900). These impulses were found to lie behind many actions and behaviors if the latter were analyzed with total honesty.

Although Sigmund Freud was one of the most dominant figures of the first half of the twentieth century, his influence has faded without his theories ever being definitively tested. Freud's disciples who practiced psychoanalysis and dream interpretation were more like religious believers than scientists. The development of a scientific test of psychoanalytic theory was never really addressed. The relative success of this approach in treating neurotic patients seemed to validate the theory. Today, the purely psychological approaches have been largely replaced by molecular, neurochemical approaches. In a more general sense, some psychoanalytic concepts simply defy a definitive proof, particularly the concept of the unconscious and of psychic energy.

The concept of repression, on the other hand, has been verified many times in the recovery of traumatic memories accompanied by strong emotion. Similarly, Freud's initial observation that when a repressed traumatic memory was recovered and discharged, the neurotic symptom tended to disappear, has been confirmed many times. This psychoanalytic therapeutic approach to neurotic and hysterical symptoms played an important role in the development of dream interpretation.

Close Up

When I was a medical student serving a clerkship on the psychiatry service (circa 1953), I was assigned a depressed patient whose complaint involved a repetitive dream that evoked great terror. In this dream, the patient would be walking in the forest or other similar situations and would become aware that a giant bear was stalking him. The patient was a male. Sooner or later, accompanied by overwhelming terror, the bear would appear, trap the patient, and attempt to bite off his penis. Imagine the excitement of a young medical student, steeped in the Freudian tradition, at hearing such an obvious dream of castration. The patient was a man whose age was sixty-five. I could not initially understand this dream since the patient had never lived in the wilderness and had little contact with bears of any kind. However, in the course of associating to the dream image, there was a sudden recovery of a very traumatic experience he had undergone as an early adolescent.

In the early 1900s, in most communities, masturbation was severely forbidden. It was thought to be harmful, and certainly sinful, and totally unacceptable. To prevent young boys from masturbating, there was the occasional use of a bizarre device which was a metal ring with small pins or spikes. This device was attached around the penis, just tight enough so that any increasing size could be severely painful. The memory that the patient uncovered in my presence was an episode in which he had an erection and because of the tight ring developed priapism, i.e., his erection could not subside. In the course of several hours, it became extremely painful and the possibility of gangrene became an urgent issue. The parents could not detach the ring. In a state of panic, they called the fire department which was the only emergency service available. With the aid of their tools, the firemen managed to detach the ring. This experience had been completely forgotten by the patient.

The terror, embarrassment, and stigma of this event returned in full force during the recall. Even from a more liberated vantage point, I could not help but feel the force of this trauma for the young boy. The identity of the bear was also revealed. He recalled that the fireman who saved him was a burly man, with a dark, full beard. Most impressively, the recurrent nightmares vanished and the patient was cured. This spectacular success in my two-week clerkship may have played a part in the A+ grade that I received in psychiatry.- William Dement,

Jungian Theory

Though somewhat younger, Carl Gustav Jung was a contemporary of Sigmund Freud. Jung was initially one of a small group of "disciples" who worked with Freud and learned the methods of psychoanalysis as a therapy and a theoretical approach to the psyche together with the theory of dream interpretation. His major difference from Freud was that he was never able to agree that every dream was a "facade" behind which psychological meaning lay hidden.

Jung also differs from Freud in his use of associations. Freud, felt he could uncover a dreamer's complexes through a chain of associations, often straying far from the original word or image. Although Jung believed that associations could reveal a patient's complexes, he insisted that both dreamer and analyst stick very closely to the dream script because it would give a more direct, immediate statement of the unconscious state of the dreamer.

Unlike Freud, Jung did not work with free associations, but with amplification. The dreamer and the analyst contribute associations and analogies and stay focused on the dream material itself. For Jung, the dream provides the easiest access to contents of the unconscious. He treats dreams as facts, as real experiences and feelings. He places great value in the power and creativity of the unconscious mind but does not feel the latent content is more important than the manifest content. He sees dreams simply as a different language, the language of the unconscious; a language that needs to be deciphered with a different vocabulary (i.e., symbols.)

CHAPTER 12 Dream Anarchy Rules!

Jung believed that dreams arise from both the personal unconscious (linked individually to the dreamer's personal life, history, feelings, etc.) and from the **collective unconscious.** The collective unconscious underlies and links all mankind and is expressed through myth, fairytale, and the like (as the Internet in the 21st century). A major difference between Jung and Freud is in Jung's concept of the collective unconscious; the collective unconscious deals with larger, more universal conflicts. Jung said the personal unconscious must always be dealt with -- made conscious --before access to the collective unconscious is possible.

Jung felt that most dreams unfold in a dramatic way and dream elements include setting, protagonists, problem, and solution. He distinguishes between "little" and "big" dreams, little ones have to do with the individual dreamer's life and big ones tap into a larger, universal world that is expressed through archetypes. (Such collective unconscious dreams often take the form of single, vivid, almost abstract images and they stay with the dreamer for much longer than dreams relating only to the personal unconscious.)

Jung had a creative attitude toward dreams: he asks the question, "What is this dream telling me? What is it for?" while Freud asked, "What is the dream repressing? What is it covering up?" For Jung, the dream compensates for the conscious attitude of the dreamer; for Freud, the dream occurs to preserve sleep, and to protect the dreamer from his repressed wishes.

For Jung, the dream is the main instrument of the therapeutic method. He believed that no dream interpretation is accurate unless the dreamer feels it to be so. The analyst cannot force an interpretation that the dreamer rejects. Jung felt there is no danger in an analyst misinterpreting the dream because soon enough the unconscious will, through dreaming, correct the mistake. "Consciousness can be trained like a parrot, but not the unconscious." If a physician and patient are mistaken in their interpretation, their unconscious will correct them. Jung felt it was imperative for analysts to assume ignorance in the face of a dream; never to think they know what a dream means; that it's impossible to know what a single dream means. He urged analysts to place the dream in its context: which means to find out from the dreamer exactly what is going on in his conscious life, and what associations the dreamer has to the dream elements. For instance, if the dreamer dreams of a place, what feelings, images, thoughts, memories, etc. does that place evoke for the dreamer? In interpreting dreams, Jung subscribed to no universal or standard symbols; the individuality of the dreamer always affects the meaning of the dream. For Jung, the same motif will have very different meanings if it is dreamed by a child or by a man of fifty.

Hobson-McCarley Theory

Most psychologists and behavioral scientists today feel that conclusions about dream interpretations from psychoanalysts were somewhat inaccurate. In sharp contrast, in 1977, a new way of viewing dreams was developed. This new way was put forward by Allan Hobson and Robert McCarley of Harvard University. They considered peculiar activity of the visual system most easily seen in a cat. In addition to brain activation and muscle atonia, feline REM sleep is characterized by bursts of activity called **p**onto**g**eniculo-**o**ccipital waves (PGO) waves originating in the brain stem (the pons) and being transmitted through the visual system. These bursts of activity are dramatically evident in lateral geniculate nuclei (the thalamic visual relay station) and the occipital (visual) cortex. These bursts also affect the oculomotor system and are accompanied by nystagmoid jerks of the eyeballs. The eye movements that seem to be triggered by PGO waves do not look like eye movements that would be executed if the cat was looking at something. Simultaneous with PGO and eye movement bursts, there is also twitching of the whiskers and twitching of the muscles. Hobson and McCarley have reasoned that this activity is non-specific. That is, it does not have any specific patterns that might give rise to perceptual experiences -- rather the PGO activity is much more like banging a drum with a big stick, or hammering on the wall. Accordingly, it probably causes greatly increased non-specific brain activation. Hobson and McCarley stated that this non-specific activity cannot be ignored by our brains, that it must be accounted for by the cerebral cortex, or the higher brain centers. They hypothesized that dream content is generated more or less at random as the brain attempts to account for this pseudo information coming in over the visual system. It is also likely that similar information comes in from some brain stem source over the auditory system though this is less certain. From all this, Hobson and McCarley conclude that dreams have no purpose, and that what we dream about is essentially random. They felt that the only meaning dreams might have is derived from thinking about them during wakefulness in relation to other meaningful aspects of the dreamer's life. The Hobson-McCarley theory is an antithesis of the theories of Freud and Jung. It may well be that the actual truth is a combination of purposeful and random processes. Long dreams do appear to interrupt themselves several times. Hobson and McCarley have made much of bizarre shifts in the dream content, and have hypothesized that they are caused by the human analog of a very intense burst of PGO activity.

Nevertheless, it is impossible to deny that every so often a dream has great meaning in depicting what is going in one's life. It may also be so long and coherent that it cannot be random images. In addition, we must ask, if an occasional dream appears to be a very long and coherent story, when was the story written? Was it written before the REM period which contained it? Was it put together a second before each successive image? This would mean that when a dream begins, no one including the dreamer knows where it will end. Even this possibility seems to be denied by an occasional complete, coherent dream story which shows no evidence of meaningless meandering as it progresses to its dramatic denouement. The following dream illustrates this point:

This dream took place in 1968, when we were working with a compound called parachlorophenylalanine (PCPA) in cats. This compound depletes brain serotonin which has the effect of appearing to release the phasic activity of REM periods into the waking state. We regarded this as an animal model of the psychotic state and were working with unbelievable intensity to study the phenomenon and to understand all its ramifications and implications. Most of our observations were 24 hours a day and seven days a week, so that the amount of labor was awesome. About one year into this project, I developed serious doubt that the chemical we had been purchasing from the manufacturer was in fact PCPA. I had the tremendous anxiety that we had wasted one whole year of very demanding work. It was Friday when I developed serious concern. I wanted to get samples of the compound analyzed by mass spectroscopy, but could not get it done until the following Monday. Thus, I had to wait over the weekend with intense anxiety and the hope that I was wrong. Friday afternoon I went to a reception at the home of a friend, Professor Nathan Oliveira, one of Stanford's leading artists, and he was telling me about a recent trip to Sweden where he had a show of his paintings. He mentioned that someone had asked him if he was Jewish. He is, in fact, Portuguese, and he said in jest, 'there are no Jews in Portugal.' That night I had the following lengthy vivid dream.

It took its roots from the movie, "Around the World in 80 Days," and from events of the day before. In the dream, I was on the trail of the lost tribe of Portuguese Jews. I would pursue them over mountain and plain and arrive at the warm ashes of a campfire only to realize that the lost tribe of Portuguese Jews was still ahead of me. With considerable intensity, I pursued them across the United States, across the Atlantic Ocean, across Europe and Asia with considerable hardship, always arriving a little too late. Finally, I arrived in the Russian port of Vladivostok, realizing they had already sailed for California. I requisitioned a boat and went out to sea. There were raging storms, and toward the end of the voyage the boat sank. Buffeted by waves, I struggled through the surf and was finally thrown up on the beach totally exhausted and bereft of all possessions. As I crawled up the sand totally defeated, my head bumped something. It was a signpost. I looked up and the sign said, "Ha, ha- We were here all the time! Signed, the Portuguese Jews. "I woke with a feeling of overwhelming loss. Certainly, the dream was mildly humorous. But it depicted that I had gone all the way around the world, i.e., a whole year, and had accomplished absolutely nothing. And it reflected the intensity of the work and the tremendous loss and waste at the end. The point is that the dream was a coherent and meaningful whole. I could not possibly believe that it occurred as a random process. On Monday, the compound was confirmed to be PCPA and the year was not wasted. I was incredibly relived and cannot express how happy I was.

Does Dreaming Have a Function?

Despite thousands of years of speculation and all our recent research, why we dream remains unclear. Some argue that dreaming is important in the consolidation of memory. We will review some of the experimental data that supports this controversy. It may well be that REM has different functions in different animals. It is also possible that the function of REM sleep may be different from the function of dream recall, if dreaming has a function at all. We can speculate that being able to recall our dreams may be a byproduct of some other essential neurological function of REM sleep. An analogy would be our ability to sing which may be a byproduct of our need to exchange oxygen and carbon dioxide for us to breathe. However, we could also speculate that from our ability to sing we developed speech. Dreaming may play a role in some of our higher cortical functions such as creativity. Whether a specific individual sings may not be important but singing has influenced our culture, as have our dreams. An approach suggested by Dr. Robert Stickgold of Harvard among others, is to consider dreaming within a larger neurocognitive framework of "off-line" memory processing during sleep. The basis of this approach is that dreaming reflects the activity of the brain and that this activity necessarily includes the reactivation of memories and emotions from earlier experiences. If any neural activity in the brain leaves the activated networks altered, dreaming must modify the networks storing memories and emotions. Dreaming then becomes the conscious experiences of these activated networks in the process of being modified. In support of this concept there is research that suggests dreaming represents the conscious experience of brain mechanisms that perform "off-line emotional and memory reprocessing during sleep".

In addition, dream research suggests that memory functions are reflected in the content of dreams. In humans and rodents, patterns of neuronal firing that occur when learning a task while awake are reactivated during post-training sleep. In the hippocampus of rats, simultaneous recordings from large numbers of neurons have shown that specific patterns and neuronal firing sequences observed as the awake rats sought out food on a circular track were replayed during subsequent sleep. In humans, positron emission tomography (PET) studies have shown that brain regions activated during learning of a task were selectively reactivated during the next night's sleep, which supports the concept of sleep being important during memory consolidation. Further supporting a role of sleep and dreams in memory processing are studies of cognitive performance in humans immediately after awakenings from non-REM and REM sleep. Brain imaging studies have shown that patterns of human brain activation differ across the states of waking, REM sleep, and non-REM sleep. Regional cerebral blood flow is generally decreased during slow-wave sleep. In REM sleep, there is a reactivation of some cerebral regions along with further deactivation of others. The imaging pattern suggests a shift in global brain function in REM sleep away from conscious executive control with decreased activity in the prefrontal cortex activity. At the same time, there is increased activity in hallucinatory and emotional processing via increased activity in sensory association cortices and the amygdala, anterior cingulate, and medial orbitofrontal cortex. Of interest, cognitive testing performed immediately after awakenings from REM and non-REM sleep suggests that the sleeping brain is biased toward less common associations and more flexible cognitive processing. Studies have found a tendency toward the activation of less common associations after awakenings from REM sleep. This, in theory, would play a role in creativity.

Arguments against REM having a role in memory consolidation have been made. Notably the neuroscientist, Dr. Jerome Siegel of UCLA has questioned the validity and effect size of research attributing a special role of dreams in memory processes. Total REM sleep can be suppressed with several medications including antidepressants yet they do not appear to degrade memory function. Brain lesions that cause complete loss of REM sleep in humans do not seem to abolish memory. Dolphins and other cetaceans have excellent memory and cognitive capabilities, but do not have REM sleep.

Arguing that REM does not have a function in human memory because dolphins do not have REM may not be entirely valid since by the same argument we would conclude that human legs have no function! However, the functions of REM are not yet completely understood. It may turn out that the brain is robust enough to have several mechanisms available to enhance memory and that REM sleep is just one of them. This would explain why REM appears to be associated with some memory processing in some studies but that the absence of REM does not always interfere with memory. Part of the difficulty in unraveling this mystery is that what some researchers are calling dreaming and REM sleep may not be the exact same thing. The superficial brain wave sleep recordings of REM may be altered by various medications but some other deeper physiological process not yet identified may be still be active despite the medication effects. If an oceanographer tried to understand the ocean by only looking at the waves on top of the water and never looking underneath the waves, little progress would be made. In sleep we have historically relied on superficial brain waves measured over the scalp to explain how sleep and dreaming work. Newer techniques including FMRI and PET scans are allowing us to look "underneath the waves". The concept that REM enhances learning is very attractive. Yet we need more research before we can be sure why we dream.

Summary

There is a great distinction between meanings and purposes of dreams. These are hard to separate because the dream is created by the dreamer. It is, therefore, hard to judge if dreams are random or if there is an unconscious purpose. The major technique by which psychoanalysts attempt to arrive at the hidden meaning of a dream is called free association. In this technique, a patient will think about an image from the dream or the whole dream, and will then, without inhibition or censorship, say whatever comes into his or her mind. This usually occurs without much difficulty, and the thought usually seems important and clarifying. If the psychoanalyst has worked with the patient for a long time, he or she knows the patient and can see often where ideas and associations are leading, or that they relate to certain aspects of the patient's personality or life. The fact that psychoanalysts come to know the patient extremely well through many hours or sessions, introduces an obvious bias in arriving to an interpretation. Post-Freudian sleep research is leading to new theories. The discovery of REM sleep permanently changed the direction of dream research. Furthermore, the correlations between eye movement activity and dream content dramatically enhanced the conclusion that in many aspects the dream world is a real world. Neuroscientists Hobson and McCarley learned from studying REM brain activation in the cat that PGO sharp waves occur that cause bursts of apparently non-specific activity throughout the brain. These scientists concluded that dreams are random events resulting from the brain's attempt to make sense of the non-specific signals coming in through the visual system and other sensory systems. Accordingly, they feel that dreams are without deep meaning or purpose.

However, it may well be possible, that dreams routinely combine content that is meaningful and content that is meaningless. REM sleep may have multiple function, some of which are psychological and physiological. We can all look forward to the future dream research unrevealing the ancient quest towards the meaning of our dreams.

Nathan Olivera

"Leave a note on your bed" Sleep on the Floor by The Lumineers

CHAPTER 13 Lucid Dreaming: a special case

"You know you're in love when you can't fall asleep because reality is finally better than your dreams." - Dr. Seuss

Introduction to Lucid Dreaming

We all know how powerful and terrifying nightmares can be, but would they affect us as much if we knew we were dreaming? There is no question that when dreamers realize they are dreaming, nightmares may be stripped of their power to terrify. Most accounts of this type of experience describe waking up almost simultaneously with the realization that one is dreaming. If the dream continues with the dreamer entirely aware that he or she is dreaming, this is a **lucid dream**. This unique experience is truly an altered state of consciousness. Although historical references abound with accounts by individuals who had lucid dreams, the term, lucid dreaming, may have first been coined in 1913. The term achieved a much wider acceptance among the general public in 1968 when Celia Green, an English parapsychologist, published a book entitled *Lucid Dreams*.

We know that in certain pathological states, hallucinations can occur when we are awake. Thus, strange sights or events may intrude into our reality. We will ordinarily be surprised, upset, and skeptical. In the dream world, we will typically accept bizarre events without surprise. This may be because, whatever the process, the dreamer writes the script in some way. Nonetheless, the dreamer has no conscious idea what is going to happen next. In lucid dreaming we remain in the dream world, but events are less of a surprise. In some cases, lucid dreamers appear to be able to control the content of their dreams which greatly extends the possible implications of this phenomenon.

Given that we do not understand how the brain organizes and processes sensory information to perceive the real world coherently, it is truly phenomenal that we perceive a nonexistent world in REM sleep. However, it is more amazing that we can alter the world at our will in lucid dreaming. Thus, we can decide to create the Taj Mahal in our dream and explore its chambers, or we can create the world of our own living room. We can even decide to seek forbidden pleasures and raptures in the privacy of our individual dream worlds.

CHAPTER 13 Lucid Dreaming: a special case

Historical Notes

Per Dr. Stephen LaBerge (*Lucid Dreaming*, 1985), the earliest lucid dream report in Western history is preserved in a letter written in 415 AD by Saint Augustine. Later accounts are sprinkled throughout history until finally the Marquis d'Hervey de Saint-Denys published a remarkable book, *Dreams and How to Guide Them*, in 1867. This book documented more than 20 years of personal dream research in which Saint-Denys described the sequential development of his ability to control his dreams. The sequence is the typical sequence described by many lucid dreamers. First, the individual increases his dream recall. Next, he becomes aware that he is dreaming, learns to awaken at will, and finally learns to direct the course of the dream. How any person can master this process poses a challenging mystery.

The scientific community owes the term "lucid dream" to Frederick Willens Van Eden, a Dutch psychiatrist and well-known author. Van Eden did serious research into lucid dreaming and used the term in a paper presented to the Society for Psychical Research in 1913, describing 352 dreams in which he knew he was dreaming.

In her remarkable book, *Studies in Dreams*, published in 1921, Mary Arnold-Forster describes her unusual ability to control her dreams. She describes many dreams of flying, dreams of going to specific places in the world that she had always wanted to explore, and even dreams that she herself totally constructed. Despite all this, she did not specifically and concretely address the issue of whether she knew she was dreaming. The following is an account of one of her remarkable dreams of flying:

"In my dream I was present at a party given in the rooms of the Royal Society in Burlington House. Lord Kelvin, Lord Rayleigh, Sir William Ramsay, my brother-in-law, Sir Arthur Rucker, and many others whom I knew, were there. They were standing together in a little group, and my brother-in-law asked me to explain to them my method of flying. I could not explain how it was done, only that it seemed to me much easier to fly than to walk. At his suggestion, I made some experimental flights -- circling around the ceiling, rising and falling, and showing them also the gliding or floating movement near the ground.

They all discussed it critically as though they were rather 'on the defensive' about the proceeding, looking upon it, I think, as a new and doubtful experiment, rather savoring of a conjuring trick. Then Lord Kelvin came forward and, speaking with that gracious manner that his friends so well remember, said that he felt the power of human flight to be less surprising, less baffling than the others seemed to think it. 'The law of gravitation had probably been in this case temporarily suspended. -- Clearly this law does not for the moment affect you when you fly,' he said to me. The others who were present agreed to this, and said that this was probably the solution of the puzzle. An assistant was standing behind the group of men, and to show them that flying is not really difficult, I took his hand, and begging him to have confidence in me and to trust to my guidance, I succeeded in making him fly a few inches from the ground. Since then, when I fly, if people notice the flight at all, which is very seldom, Lord Kelvin's explanation always seems to satisfy them. His reply also gave me the second formula that I can make use of in a dream in case of need, and, like the original formula, it is always successful.

I have sometimes fancied in the middle of a flight that I am losing my power to fly; I have begun then to drift downwards in the air, and have failed to rise again, easily. At such moments, the 'word of power' comes into my mind, and I repeat to myself, 'You know that the law of gravitation has no power over you here. If the law is suspended, you can fly at will. Have confidence in yourself, and you need not fear.' Confidence is the one essential for successful flight, and confidence being thus restored, I find that I can fly again with ease." -Arnold-Forster, 1921

Mary Arnold-Forster also discusses her "super dreams" in which some difficult problem was solved or in which she had abilities such as speaking languages or doing mathematics that she did not have in the waking state. Overall, this is quite a remarkable book.

In the 1970s, Anne Faraday's popular books, *Dream Power* and *The Dream Game*, had a great impact on public awareness of dream consciousness. These were followed by the 1974 book written by Patricia Garfield, *Creative Dreaming*, which contains a wonderful collection of tools for lucid dream work, as well as a great deal of fascinating information. Finally, it remained for Dr. Stephen LaBerge to bring to the study of lucid dreaming the additional power of physiological sleep recordings and to establish for the first time the ability to look at physiological changes in connection with changes in dream content.

Who is the Lucid Dreamer?
A lucid dream implies a lucid dreamer. Obvious as that may seem, there are subtleties here. First, who exactly is the lucid dreamer? Is the lucid dreamer identical to the person we seem to

be in the dream? Or to the person who is asleep and dreaming? The question of the identity of the dreamer is in a certain sense mysterious; to solve the mystery, we first need a list of suspects.

The most obvious suspect would seem to be the sleeper. It is, after all, the brain of the sleeping person that is doing the dreaming. But sleepers have a perfect alibi: they weren't there at the time of the dream, or any other time either -- they were in bed asleep! Sleepers belong to the world of external rather than internal reality -- because we can see and objectively test that they are asleep. But dreamers belong to the world of internal reality -- we cannot see who, how, or what they are dreaming. So, we must turn our attention to the denizens of the dream world. In a dream, there is usually a character present whom the sleeper takes to be himself. It is through the dream eyes of this dream body that we normally witness the events of the dream. The dream body is ordinarily who we think we are while dreaming, and this seems the obvious suspect. But actually, we only dream we are that person. This dream character is merely a representation of ourselves. I call the character the "dream actor" or "dream ego." The point of view of the dream ego is that of a willing or unwilling participant -- apparently contained within a multidimensional world (the dream), much as you probably experience your existence right now.

That the dream actor is not the dreamer is shown by the fact that there are some dreams in which we apparently play no part at all. In these dreams, we seem in varying degrees to witness, from the outside, the events of the dream. Sometimes we dream, for example, that we are watching a play. We seem to be in the audience while the action unfolds on stage. In this case, we are at least represented as being present, though passively observing.

- LaBerge, 1985

Characteristics of Lucid Dreaming

Lucid dreaming presents a number of exceptions to the ordinary properties of the dream world. The non-lucid dreamer tends to consistently suspend reflective judgment of events. For instance, if a purple kangaroo hops into the living room, one does not muse, "this is impossible," and begin to try to understand how it could possibly occur. The non-lucid dreamer tends not to engage in the deep thought one might bring to bear while studying or pondering a problem particularly for long periods of time. The non-lucid dreamer tends not to feel a need to remember the events of a dream. Furthermore, there is a sense that the dream impels the dreamer that the dreamer cannot use his will to decide where to go and what to do. Thus, in marked contrast to non-lucid dreaming, lucid dreaming is characterized, indeed almost defined, by reflective awareness, being conscious that a dream is occurring, understanding the dream, making the dream happen, and so forth.

Upon realizing that he or she is dreaming, the dreamer can gain much more control of the dream events. The memory of lucid dreaming appears to be essentially equal to that of the waking state. Lucid dreams involve more complex thought as a rule than ordinary dreaming. According to LaBerge, the lucid dreamer is simultaneously an involved actor in the dream and an observer of the dream. Thus, the apparent vividness of a lucid dream may be to some extent an artifact of the relative ease of detailed recall of the experience. Nonetheless, many lucid dreamers state that the dream has some quality of seeming to be enhanced beyond the usual dream experience.

Almost all lucid dreams seem to begin within a dream period, and by inference a REM period. Thus, the dreamer is dreaming when a transition to lucidity occurs. Sometimes the reverse appears to occur, that is the dreamer forgets that he is dreaming and ceases to dream lucidly. An awakening usually terminates the lucid dream. When the lucid dreamer experiences a fading of the dream, he expects to awaken. Sometimes there is a false awakening in which the dreamer thinks he has awakened and may not recognize that the dream has continued. This would seem to be an end of lucidity. This is in reminiscence of the movie, *Inception*.

The foregoing "out-of-body" experience allows us to see how the dreaming brain can produce experiences that can strongly implicate a supernatural world. In fact, many of our magical and supernatural beliefs may have arisen from the dreaming aspect of our existence.

When Does Lucid Dreaming Occur?

It is not clear what actually causes a dream to become lucid. In a group of 330 student respondents to a questionnaire at Stanford University, 85 percent answered "yes" to the question, "Have you ever had the experience of dreaming and knowing that it was a dream while you were dreaming?" Sixty percent of these students claimed to have had lucid dreams more often than "once or twice." Other individuals and other surveys have set lower numbers of lucid dreamers at around 10 percent, and several surveys have found only one or two lucid dreams in hundreds of dream reports. Analogous to sleep apnea which was thought not to exist three decades ago by almost everyone, a similar situation may pertain to lucid dreaming. For a non-lucid dreamer, such experiences probably seem very rare, but an active awareness of their possibility will lead to surprisingly frequent encounters with individuals who report them. No differences in personality variables have been found and it has been assumed that virtually everyone who can recall dreams with relative ease has the potential for lucid dreaming.

When lucid dreamers keep track of the clock time at which they awaken from lucid dreams, we see a marked tendency for such events to occur later in the night. If we remember that three-fourths of REM sleep occurs during the second half of the night, this finding should not seem unusual. In addition, the lighter sleep at the end of the night favors awakening with the potential of remembering the content of the dream.

Can Lucid Dreaming Be Learned?

There is a market of tools that makes it possible to increase the frequency of lucid dreams. Whether everyone can "learn" to have lucid dreams is uncertain. However, several workers have attempted to develop techniques by which this phenomenon may be experienced more readily. The most noteworthy has been developed by Stephen LaBerge, a founder of the Lucidity Institute and an outstanding investigator and advocate in this area. His device is called the "Nova Dreamer". It is a mask that covers the sleeper's eyes and contains two red light bulbs placed so that one is over each eyeball. Ceramic strain gauges in the mask sense the occurrence of rapid eye movements and REM sleep, causing the dream light bulbs to begin to flash. This sensory input may either be incorporated into the dream or make the dreamer become aware of the dream. Realizing the lights are a signal, the dreamer will begin to dream lucidly. According to the developers of the dream light, when used in combination with mental concentration, the device can increase five-fold or more a person's chances of having a lucid dream. The developers and investigators in this area stop short of the claim that everyone can learn to dream lucidly. Much more lucid dreaming research needs to be done in relation to optimal age, optimal use of various techniques to induce lucid dreaming, interaction with the physiological state, states of sleep deprivation, and so forth.

Possible Functions of Lucid Dreaming

The possibility of fulfilling wishes by dreaming about them is not here set forth in the Freudian sense. Rather, it would be the direct gratification of something for which the dreamer yearned, "May your fondest dreams come true." Lucid dreaming offers the opportunity to make one's wishes come true in dreams. An obvious possibility is having a sexual dream combined with the ability to throw aside all inhibitions because the event is known not to be real. In the age of AIDS, this could be a major advantage of the lucid dream. Adventure and exploration are possible; this might simply mean one's dreams could be more exciting than usual. The occurrence of lucid dreaming also offers the possibility of rehearsing various aspects of the waking world. Could one practice the piano? Could one practice ice-skating? Could one practice throwing the football?

In general, could one improve waking physical skills? To prove that this is possible, more research will be necessary. Another aspect of practicing in a dream could involve reducing performance anxiety. Almost all performance anxiety is reduced by repetitive experience. However, there is a paradox in that if the dreamer knows the experience is a dream, then is it bereft of its usefulness in reducing anxiety? Again, more research is needed.

One would expect that it might be possible to explore many interesting possibilities through lucid dreaming. Would I like to be a cowboy? Let's try it during a night of dreaming. Problem solving can occur in dreams, but whether or not lucid dreaming would have an advantage in this regard has not been claimed. However, once again proof requires more work. Finally, investigators have speculated that the dream might have a healing quality. For instance, if one dreams of health, one might be able to overcome illness. There are many aspects of brain and body interaction that remain beyond our knowledge. Nonetheless, there are intimations that sleep has untapped potential for healing, and dreaming may represent the peak of this possibility.

There is mention of nightmares being a symptom of post-traumatic stress disorder (PTSD). This obviously means that dreams of the trauma and that these dreams are of a disturbing nature. Lucid dreams could be an important tool for the recovery of PTSD victims and it is unfortunate that this stigma could be the preventing factor surrounding this type of therapy. If a person suffering from PTSD has a distressing dream about their trauma, it could be very beneficial to re-experience the trauma while having more control and less fear. This gives the opportunity for exploration of other possible outcomes (as seen in attending the exam to view the reactions while only half dressed) or the exploration of feelings in general. A person in therapy for PTSD could be instructed in their therapy session to use this lucidity to their advantage. If the dreamer becomes lucid, they could be instructed to change the setting or situation of their trauma (as seen in running back home to put clothes on) and use this shift to initiate exploration. This is only some of the possible connections of lucid dream therapy to PTSD. There are a variety of advantages to lucid dreaming. It is easy to apply these to the notion of post-traumatic stress therapy. The dream ego is less afraid of threatening dream figures or situations. For this reason, there is less resistance to confront these figures or situations. Using appropriate techniques for manipulating lucid dreaming, the dream ego can get in touch with places, times, situations, or persons that are important to the dreamer.

Quick Guide to Lucid Dreaming by Elisa Desiree Lupin-Jimenez

1. Improve **dream recall** by recording anything you can remember from your dreams right after you wake up in a designated dream journal. As you practice remembering your dreams each morning, you will find that you will be able to recall many more dreams in greater detail. After all, what's the point of having a lucid dream if you can't remember it?

2. Recognize **dream signs** (e.g. strangely colored objects, unusual movement, odd behavior) and tally them up in your dream journal. Notice which types of dream signs appear more often in your dreams.

3. Perform **reality tests** by asking yourself several times a day if you are dreaming then using a constant "totem" (like the spinning top from the movie *Inception*, or my own version of a totem, which involves checking the time twice) to test reality. Pay close attention to discontinuities in your totem. The goal with these reality tests is to develop the habit of questioning whether you are in a dream or not so that when you dream, you will remember to test reality.

4. Use the **MILD** (Mnemonic Induction of Lucid Dreams) technique, developed by Dr. LaBerge. As you are preparing to fall asleep, focus on remembering to recognize when you are in a dream. You can repeat this phrase out loud as an example of focusing on your goal: "Next time that I dream, I will remember that I am dreaming." You should tell yourself that you will

lucid dream when you fall asleep; the more you believe it, the more likely it is that you will lucid dream.

Summary

The phenomenon of lucid dreaming is a special example of extending the forms of existence by a combination of an apparently natural talent and a learned ability. The mystery of lucid dreaming is that reality is, in a sense, granted by the dreamer in a paradoxical situation of knowing that he is dreaming. Yet, at the same time experiencing the dream as a form of reality. It has often been said that dreaming frees man from the inexorable bonds of time and space. However, in ordinary dreaming this freedom is only partial since the dream controls the dreamer. In lucid dreaming, these bounds are totally discarded. The dreamer controls the dream and is free to explore the unbounded universe of his or her imagination.

In the dream world, the dreamer cannot tell what will happen next. However, in lucid dreaming, dreamers feel that what happens is less of a surprise. They can sometimes even control events. Memory of lucid dreaming appears to be essentially equal to that of the waking state and involves more complex thought as a rule than ordinary dreaming. Possible functions of lucid dreaming are fulfilling wishes, exploring alternative choices, improving physical skills, and solving problems.

Almost all lucid dreams seem to begin within a dream period, and therefore a REM period. There is a tendency for lucid dreams to occur during the second half of the night when three-quarters of REM sleep occurs. It is also possible to increase the frequency of the occurrence of lucid dreams with practice.

"Caution: Cape does not enable user to fly." — Warning label on Batman Costume

CHAPTER 14 Sleep Disorders & Walla Walla

"The town so nice, they named it twice."

What Are Sleep Disorders?

Sleep disorders are a broad range of health conditions. The reader may question why sleep disorders are included in an introductory text in Sleep and Dreams, the simplest reason is that the students who have taken this course in the past have gone on to recognize sleep disorders among themselves and family members. One way this course helps you is by helping others. Sleep disorders are easy to recognize once you suspect them. The best news is that sleep disorders improve once correctly addressed. We will discuss a variety of symptoms. It should be recognized, however, that sleep disorders could be present even when there are no symptoms at all. If you sleep alone, how would you know if you snore?

Since sleep medicine is a relatively new field, many physicians have not prioritized sleep disorders as a part of their routine care. There is increasing awareness of the impact of sleep on health. When evaluating a person's sleep there are four dimensions that should be considered: 1.) the amount of sleep 2.) the quality of sleep 3.) the timing of sleep 4.) the sleeper's state of mind.

The Symptoms of Sleep Disorders

The six major symptoms that suggest the presence of a specific sleep disorder are:

1) Unrefreshed sleep

2) Excessive daytime fatigue, and/or sleepiness

3) Difficulty falling asleep or staying asleep

4) Unusual or violent behavior during sleep

5) Snoring

6) Motor restlessness in the evening

A single night of insomnia, while it may be unpleasant, should cause no concern if sleep on subsequent nights seems completely normal. Persistence of the above symptoms related to sleep should always be evaluated.

In contrast to heart disease, arthritis, and many other illnesses which are much more common in seniors, sleep disorders can occur at any age. It is particularly concerning that relatively little information has been accumulated about the sleep disorders of college, high school students, and young children. Even more alarming that there is very little student awareness about sleep in general. Moreover, as several studies have shown, there is also very little awareness about sleep disorders among doctors and other health professionals. The National Commission on Sleep Disorders Research, created by the U.S. Congress, surveyed primary care physicians, the gate keepers of medical practice, and found that they know next to nothing about normal sleep, its regulation, the effect of sleep deprivation on daytime function, and about sleep disorders. No specific studies have been carried out to evaluate the sleep knowledge of doctors who practice in health centers that serve America's youths, but there is no reason at all to believe they are any more aware than their primary care colleagues. As a result, the pervasive nature of sleep disorders affects all age groups has not sufficiently penetrated our health care priorities. In view of this situation, it is in their best interest for everyone in America to become more knowledgeable about sleep disorders. Key goals in this area are as follows.

(1) People should be aware of sleep disorder symptoms. A wide variety of sleep problems give rise to a small number of obvious symptoms of which anyone can be made aware. By far the most important and most common is fatigue.

(2) People are should know that sleep disorders are common and effective treatment is available.

(3) People who learn about sleep disorders will become a knowledge resource for their families, their friends, and, to some extent, for their community. People who are educated about sleep disorders can usually recognize a serious problem not only in themselves but in a family member, friend, or colleague.

What Is Abnormal Sleep?

Sleep problems can develop slowly over time. This insidious progress makes drawing a sharp line to know when the disorder began difficult. Among many sleep disorders, the line between normal and abnormal is often a question of the severity. For example, an occasional nightmare is certainly not a sleep disorder, and would be considered perfectly normal by sleep specialists. On the other hand, several nightmares every night is definitely unusual and deserves attention. We all know that great excitement or stress can lead to difficulty sleeping, particularly when occurring in the evening hours close to bedtime. However, the symptoms from a single night of poor sleep does not create a syndrome or diagnosis of a sleep disorder.

Being chronically sleep deprived is also a condition where defining a sharp boundary between normal and abnormal is almost difficult. The American Academy of Sleep Medicine (AASM) has defined and coded a disorder called **Behaviorally Induced Insufficient Sleep Syndrome** (they could have easily called it college student sleep disorder). Afflicted persons suffer from severe, chronic fatigue and/or very troublesome, unintended sleep episodes.

A more commonplace phenomenon with a wide range of severity is snoring. Breathing should be silent. More than half of all older adults do not breathe quietly during sleep. Snoring may be the marker of obstructive sleep apnea syndrome which can result in death.

History of Sleep Disorders Medicine

Introduction

Although sleep medicine is a relatively recent subspecialty the practice of diagnosing and treating sleep disorders has become a recognized medical specialty with the creation of The American Board of Medical Specialties as a unique medical field in 2005. It now has the same level of academic credentialing and recognition as internal medicine, pediatrics, surgery, and all the other medical specialties. Stanford University has the largest accredited training program for sleep medicine fellows in the country. For many years, narcolepsy was the only specific sleep disorder that had been described. The term restless legs syndrome (RLS) was first introduced by Karl A. Ekbom, a Swedish neurologist and surgeon, in 1945, although the earliest description of restless legs associated with sleep disabilities possibly came from Sir Thomas Willis, an English physician, in 1672. The term obstructive sleep apnea was first used in 1965.

Only after the development of the Sleep Disorders Clinic at Stanford in 1970 did the discovery and acceptance of sleep disorders accelerate. Prior to the establishment of the Stanford Sleep Disorders Clinic, the management of patients never included the objective measurement of their sleep. At present, almost 100 sleep disorders have been described. Most of these discoveries occurred while examining sleep in individuals who felt that something was wrong with their sleep.

The Discovery of Sleep Apnea

One of the most important, events in the history of sleep disorder medicine occurred in Europe when

Figure 14.1: Nathaniel Kleitman Ph.D (1895-1999), the first person in the world to devote his entire professional career to the study of sleep and he lived to be over 100.

obstructive sleep apnea was described in 1965. It was described simultaneously in France and Germany by neurologists interested in sleep. This important discovery was initially widely ignored. A landmark meeting in Bologna, Italy, helped bring attention and worldwide interest in sleep apnea. At this meeting recognition of sleep apnea having a role in cardiovascular disease was acknowledged. They also clearly identified snoring and excessive daytime sleepiness as important diagnostic indicators. One of the young presenters Dr. Christian Guilleminault helped establish the Stanford Sleep Disorders Clinic.

Clinical Sleep Medicine Initiated at Stanford University

Until 1970, the practice of medicine ended when the patient fell asleep. The first clinical service created specifically to diagnose and treat patients with sleep disorders was initiated at Stanford University in 1970. The Stanford Sleep Disorders Clinic introduced the novel approach of examining the sleeping patient by utilizing all-night polygraphic sleep recordings as a clinical diagnostic test. The all-night diagnostic sleep test eventually came to be called "polysomnography" and has become the clinical diagnostic standard.

Clinical Significance of Excessive Daytime Sleepiness

By 1975, excessive daytime sleepiness had been clearly recognized as an ominous complaint for sleep disorders patients, and as a dangerous phenomenon unto itself. Given the importance of impaired daytime alertness as a signal of abnormal sleep, it is puzzling that the basic researchers of the time who were devoting their careers to the investigation of sleep had little or no interest in the impact of sleep on daytime function.

Figure 14.2: A 1970 student volunteer being hooked up for a sleep test

The neglect of daytime sleepiness is even more difficult to understand since it is now widely recognized that sleepiness and the tendency to fall asleep during the performance of hazardous tasks is one of the most important problems in our society.

How Many Sleep Disorders Are There?

For the reader to appreciate the breadth of sleep medicine, the full listing from the **International Classification of Sleep Disorders** is included here.

International Classification of Sleep Disorders

Every medical discipline must have a system to classify its content and is known as a nosology. The first nosology for sleep medicine was published in 1979.
This nosology system has gone through several changes that reflect the dynamic changes in our field. The current system, which undoubtedly will change in the future, is the **International Classification of Sleep Disorders** (ICSD.) An outline of the ICSD 3 is described below and will serve as an overview of a spectrum of clinical sleep medicine:

<u>Insomnia</u>
- Chronic Insomnia Disorder
- Short-Term Insomnia Disorder

- Other Insomnia Disorder

ISOLATED SYMPTOMS AND NORMAL VARIANTS
- Excessive Time in Bed
- Short Sleeper

<u>Sleep Related Breathing Disorders</u>

OBSTRUCTIVE SLEEP APNEA DISORDERS
- Obstructive Sleep Apnea, Adult
- Obstructive Sleep Apnea, Pediatric

CENTRAL SLEEP APNEA SYNDROMES
- Central Sleep Apnea with Cheyne-Stokes Breathing
- Central Apnea Due to a Medical Disorder without Cheyne-Stokes Breathing
- Central Sleep Apnea Due to High Altitude Periodic Breathing
- Central Sleep Apnea Due to a Medication or Substance
- Primary Central Sleep Apnea
- Primary Central Sleep Apnea of Infancy
- Primary Central Sleep Apnea of Prematurity
- Treatment-Emergent Central Sleep Apnea

SLEEP RELATED HYPOVENTILATION DISORDERS
- Obesity Hypoventilation Syndrome.
- Congenital Central Alveolar Hypoventilation Syndrome
- Late-Onset Central Hypoventilation with Hypothalamic Dysfunction
- Idiopathic Central Alveolar Hypoventilation'
- Sleep Related Hypoventilation Due to a Medication or Substance
- Sleep Related Hypoventilation Due to a Medical Disorder

SLEEP RELATED HYPOXEMIA DISORDER
- Sleep Related Hypoxemia

CHAPTER 14 Sleep Disorders & Walla Walla

ISOLATED SYMPTOMS AND NORMAL VARIANTS

- Snoring
- Catathrenia

Central Disorders of Hypersomnolence

- Narcolepsy Type 1
- Narcolepsy Type 2
- Idiopathic Hypersomnia
- Kleine-Levin Syndrome.
- Hypersomnia Due to a Medical Disorder
- Hypersomnia Due to a Medication or Substance
- Hypersomnia Associated with a Psychiatric Disorder
- Insufficient Sleep Syndrome

ISOLATED SYMPTOMS AND NORMAL VARIANTS

- Long Sleeper

Circadian Rhythm Sleep-Wake Disorders

- Delayed Sleep-Wake Phase Disorder
- Advanced Sleep-Wake Phase Disorder
- Irregular Sleep-Wake Rhythm Disorder
- Non-24-Hour Sleep-Wake Rhythm Disorder
- Shift Work Disorder
- Jet Lag Disorder
- Circadian Sleep-Wake Disorder Not Otherwise Specified (NOS)

Parasomnias

NREM-RELATED PARASOMNIAS

- Disorders of Arousal (From NREM Sleep)
- Confusion Arousals

275

- Sleepwalking
- Sleep Terrors
- Sleep Related Eating Disorder

REM-RELATED PARASOMNIAS

- REM Sleep Behavior Disorder
- Recurrent Isolated Sleep Paralysis
- Nightmare Disorder

OTHER PARASOMNIAS

- Exploding Head Syndrome
- Sleep Related Hallucinations
- Sleep Enuresis
- Parasomnia Due to a Medical Disorder
- Parasomnia Due to a Medication or Substance
- Parasomnia, Unspecified

ISOLATED SYMPTOMS AND NORMALVARIANTS

- Sleep Talking

Sleep Related Movement Disorders

- Restless Legs Syndrome
- Periodic Limb Movement Disorder
- Sleep Related Leg Cramps
- Sleep Related Rhythmic Movement Disorder
- Benign Sleep Myoclonus of Infancy
- Propriospinal Myoclonus at Sleep Onset
- Sleep Related Movement Disorder Due to a Medical Disorder
- Sleep Related Movement Disorder Due to a Medication or Substance
- Sleep Related Movement Disorder, Unspecified

CHAPTER 14 Sleep Disorders & Walla Walla

-

ISOLATED SYMPTOMS AND NORMAL VARIANTS

- Excessive Fragmentary Myoclonus
- Hypnagogic Foot Tremor and Alternating Leg Muscle Activation
- <u>Sleep Starts (Hypnic Jerks)</u>

<u>Other Sleep Disorder</u>

Appendix A:

Sleep Related Medical and Neurological Disorders

- Fatal Familial Insomnia
- Sleep Related Epilepsy
- Sleep Related Headaches
- Sleep Related Laryngospasm
- Sleep Related Gastroesophageal Reflux
- Sleep Related Myocardial Ischemia

Sleep and Mental Health

Complaints about sleep are common among patients with psychiatric conditions. Sometimes these complaints are due to the psychiatry condition or are co-morbid or a secondary sleep disorder. In either case, psychiatrists will often have to address sleep concerns. It is not a coincidence that the original Stanford Sleep Disorder Clinic was administered under the Psychiatry Department. Psychiatrists are the few physicians that routinely ask about a patient's sleep. The association of the so-called "nervous breakdown" with disturbed sleep has been known since antiquity. The agitation, anxiety, emotional pain, and fear associated with a rapid onset of schizophrenic psychosis or a severe mood disorder is certainly sufficient to produce considerable sleep disturbance.

Depression is almost always associated with complaints of insomnia. Sleep laboratory studies have confirmed this strong association although the sleep disturbance is often not as severe as the complaint would suggest. An episode of major depression is diagnosed when people report that they are overwhelmed with negative emotions such as feeling sad, gloomy, hopeless, and/or helpless. Their self-esteem withers and enjoying anything is impossible. There may also be anxiety, tension, and agitation. Guilt, shame, and feelings of unworthiness over past sins and failures often dominate waking thoughts and patients feel trapped and desperate. There are also physical symptoms, such as loss of appetite, psychomotor agitation, a whole host of somatic complaints, and in almost every instance of severe depression, complaints of sleep disturbance are present. It is not uncommon that the sleep disturbance is the primary complaint. The association is so consistent that many physicians automatically, but incorrectly, assume that anybody complaining of insomnia is depressed.

The **bipolar** (manic-depressive) mood disorder is characterized by the occurrence of one or more manic episodes, usually accompanied by one or more major depressive episodes. It appears to be quite different from unipolar depression. The alternations in mood can occur very rapidly, occasionally every day, or a single alternation may occur over months and there may be a normal mood between an episode of depression and an episode of mania. Sleep mechanisms and biological rhythm mechanisms have been causally implicated in this illness. The mania can occur with extreme rapidity and, as noted, is typically directly preceded by a marked reduction in sleep tendency. In contrast, the switch to the depressed phase often occurs following a major sleep period, as if the sleep itself caused the switch.

Not surprisingly, the manic phase is typically accompanied by reduced amounts of sleep and sometimes almost total insomnia for a period of several days or more. Prior to the development of lithium treatment, very severe mania could lead to death, presumably from profound exhaustion and sleeplessness. When the switch to the depressive phase occurs, it is typically accompanied by lethargy and greatly increased amounts of sleep. This is in marked contrast to simple depression where sleep is typically reduced in amount and disturbed by multiple awakenings. Continuous polysomnography recordings have not been done in large numbers of bipolar patients, but in a few such studies, the amount of daily sleep during the manic phase appears uniformly reduced and in a few instances, has approached zero. One important issue is whether a sleep deprivation effect is operating which would ultimately oppose the insomnia occasioned by the manic episode and produce more normal sleep or even trigger a switch to the depressed phase.

Schizophrenia is a severe mental disorder in which the victim may be out of touch with reality. It is often called a thought disorder and has a typical cluster of symptoms that include delusions, hallucinations, and breakdown in orderly mental processes such that thoughts are irrational, reverberative, and disorganized. Emotions are also abnormal and inappropriate either in being blunted or being irrationally excited or angry. The illness of schizophrenia usually has its onset in adolescence sometimes developing rapidly or sometimes slowly. An acute schizophrenic onset is always accompanied by profound agitation and sleeplessness. Sleep researchers with recordings every night have studied several acutely ill schizophrenic patients. They showed a period of extraordinarily severe insomnia which lasted a week or more.

One of the more serious psychiatric conditions that is clearly associated with disturbed sleep over a long period of time is the **post-traumatic stress disorder (PTSD)**. It received greatly increased attention in the wake of the Vietnam War and now the Iraq and Afghanistan wars. In general, PTSD is characterized by the re-experiencing of a traumatic event in the form of repetitive dreams, recurrent intrusive daytime recollections, and dissociative flashback episodes. Some clinicians have suggested that the sleep disturbance associated with PTSD including the recurrent dream anxiety episode may be a diagnostic "hallmark." Decades after the precipitating event, patients continue to report disturbed sleep and the aforementioned anxiety dreams.

Seasonal affective disorder (SAD) is a mood disorder occurring in the winter months characterized by diminished energy, hypersomnia, overeating, and other depressive symptoms. Some investigators feel it should also be classified as a disorder of the biological clock. It is now generally accepted that bright light administered early in the morning has a significant ameliorative effect on SAD.

Insomnia as a Predictor of Mental Illness

The National Institute of Mental Health studied sleep disturbances in one of its epidemiological populations. Such a population is a large and representative group of individuals which Institute investigators are following over time. The psychiatrists who are investigating sleep reported that 40 percent of those with insomnia and 46 percent with hypersomnia suffered from a mental disorder. In contrast, only 16.4 percent of those with no sleep complaints had a psychiatric problem. They also found that individuals who had insomnia at the baseline interview that was still present in a follow-up interview a year later were 70 times more likely to develop a major depressive episode than those whose sleep disturbance had been resolved by the second interview.

Troubled Minds, Troubled Sleep

It seems reasonable that a healthy human being should have healthy sleep. Sleep disturbances in association with severe physical illness are easily understood in terms of pain and discomfort. The sleep disturbances in association with mental illness can be understood in terms of stress and anxiety. Many investigators, however, feel that a more specific role of sleep mechanisms in mental illness is likely. There is a need for more research to fully understand the connections between sleep and mental illness.

Clinical Vignette

A young woman flew out from the East Coast to see me without an appointment. She came to my office and told the receptionist she would wait until I showed up. When I arrived, I saw immediately that she looked awful and was obviously desperate. She had been forced to leave a Peace Corps assignment overseas because her insomnia was so bad. I tested her nighttime sleep with polysomnography and her daytime sleep tendency using the MSLT over the course of two full 24-hour periods. Right away I noticed that the MSLT showed her to be less sleepy in the daytime than the average person off the street. During the nighttime testing began, she fell asleep fairly quickly and slept well the whole night. The same thing happened the second night. Yet after both nights, the patient perceived having slept very little. When it came time to discuss the results of the tests, I sat down with her and went over the EEG recordings from both nights. In my most comforting and reassuring, yet authoritative manner I said, "I have been doing these sleep recordings almost all of my adult life, and your tests show absolutely perfect, normal sleep. These are just beautiful sleep spindles, and this is wonderful slow-wave sleep." I reassured her with great firmness that she didn't need to worry about the quality of her sleep. The patient flew back East and had no further problems with insomnia.

Interestingly, if people who suffer from sleep state misconception are given sleeping pills they often say that they slept more than usual, even though the all-night recording has shown that their sleep time didn't actually increase at all. In these cases, the sleep medication doesn't induce sleep, but rather fosters a perception of having been asleep.-WCD, *The Promise of Sleep*

Dr. Dement's Walla Walla Project

Walla Walla, Washington is my hometown. Situated in the southeastern corner of the state, it is a somewhat isolated community of about 25,000 people living inside the city limits and another 30,000 in the nearby surrounding area.

In 1992, my mother was still living. I visited her as often as possible, and maintained relationships with a few old friends and several local physicians. When it became crystal clear to me that the vast majority of Americans were not receiving the benefits of sleep medicine, I knew that Walla Walla was no exception. It occurred to me in a flash of inspiration that the very best place to launch a crusade to change the way society deals with sleep and bring sleep medicine into the healthcare mainstream could be in my own hometown.

In an instance of extremely good fortune, the first primary care physician I approached in the Walla Walla Clinic was immediately enthusiastic about my intentions. I have since learned that only a small minority of primary care physicians are interested in tackling this area or any new and different health care issue for that matter. The perfectly valid reason for this is that primary care physicians are already overworked and underpaid.

I was also fortunate to be joined in what has come to be known as the "Walla Walla Project" by my friend and colleague in sleep medicine, the late Dr. German Nino-Murcia. Dr. Nino-Murcia became keenly interested in this project and at the same time had a flexible schedule that enabled him to accompany me on my visits to Walla Walla. At the start of the project, we meticulously reviewed the records of nearly 1,000 consecutive patients at the Walla Walla Clinic. Numerous complaints of fatigue were noted in these records, but a true sleep history was never recorded. Of the charts reviewed, only two patients had been referred (to Seattle) for sleep evaluations; one with a presumptive diagnosis of obstructive sleep apnea (OSA) and one with severe insomnia related to depression. Because the prevalence of obstructive sleep apnea with unambiguous symptoms is well over ten percent of the general population, at least several dozen patients out of the 1000 should have received a diagnosis of obstructive sleep apnea, not to mention other sleep disorders. This lack of recognition on the part of Walla Walla physicians was not surprising; the National Commission on Sleep Disorders Research had already highlighted the serious lack of primary care involvement in recognizing sleep disorders. After the baseline chart review, Dr. Nino-Murcia and I conducted a two-day, on site, training session for all interested health professionals. The instruction included basic information on the physiology and pathologies of sleep, diagnostic procedures, and treatment strategies. A respiratory therapist was trained in polysomnographic testing.

Initially, polysomnographic data were recorded on computer discs in Walla Walla and mailed to Dr. Nino-Murcia for interpretation. Several of the interested Walla Walla primary care physicians then began diagnosing and treating sleep disorders patients. Weekly telephone conferences were held with Dr. Nino-Murcia in which all the cases were discussed in detail. Then, as the Walla Walla physicians became more comfortable with polysomnogram interpretation and treatment modalities, only the more complicated and difficult cases were discussed. Although the majority of cases involved obstructive sleep apnea, other sleep disorders were diagnosed, including Restless Legs Syndrome, Narcolepsy, Delayed Sleep Phase Syndrome and Psychophysiologic Insomnia.

Since it was the first time anything like the Walla Walla Project had ever been attempted, it began slowly. However, from today's perspective, the results of this demonstration project have been extraordinarily gratifying. During the first two years, more than 200 patients had polysomnography testing. Of those studied, 88 percent were diagnosed with OSA and most were treated with continuous positive airway pressure (See Chapter 18).

The Walla Walla physicians were initially amazed by the large number of their patients who had serious sleep disorders. All of these patients had previously been seen at the Clinic on multiple occasions. Yet, their sleep disorders were not recognized until after the Walla Walla Project was well underway. Fewer than five percent of the polysomnographic tests conducted in Walla Walla were normal.

By 1995, the Walla Walla physicians participating in the project had acquired the skills and experience to manage any sleep disorder entirely on their own. At least three physicians have learned to score and interpret polysomnograms. Currently, multiple patients undergo polysomnographic evaluation every night. The weekly sleep disorders conference has become a Walla Walla mainstay. Regular attendees include internists and family physicians, neurologists, pulmonologists, a psychiatrist, ENT surgeons, respiratory therapists, and a cardiologist. This multidisciplinary conference is quite useful in clinical decision-making in difficult cases and also serves as an educational forum.

One outstanding primary care physician, Dr. Richard Simon, is now a diplomat of the American Board of Sleep Medicine. Along with several colleagues, Dr. Simon has founded a full-service sleep disorders center that has passed all the accreditation criteria of the American Academy of Sleep Medicine.

Even more exciting, sleep has entered the mainstream of Walla Walla society as a fully qualified member of the basic triumvirate of health – good nutrition, physical fitness, and healthy sleep. A sleep curriculum is being introduced into Walla Walla's two middle schools and high schools. Material covering the nature of sleep, sleep deprivation, biological rhythms, and the essentials of healthy sleep is being taught in Walla Walla's three colleges. I also have future plans to someday include Walla Walla's largest institution, the State Penitentiary. We know that many prisoners throughout America are suffering from undiagnosed and untreated or misdiagnosed and mistreated sleep disorders. How can someone who cannot stay awake be rehabilitated?

There is one thing I must make absolutely sure that everyone understands. Primary care physicians are absolutely not responsible for the neglect of sleep disorders in America today. They are as much the victims of the lack of medical school teaching as anyone else. If we wish to identify a villain at this point in time, I am sorry to say that the blame should probably be laid at the door of the Congress of the United States. The final report of its own National Commission on Sleep Disorders Research could not have been more clear and decisive. The Commission characterized the pervasive lack of awareness about sleep disorders as a national emergency, and its most urgent recommendation to the Congress was the immediate implementation of an effective national awareness campaign to be carried out by the federal government. Appropriating the small number of funds ($1.3 million) that would allow an effective national awareness campaign to be launched was identified as the number one priority for Congressional action.

The Commission's leading advocate in the Congress, Oregon's Senator Mark Hatfield, stated it clearly at a press conference on Capitol Hill in January 1993 (I flew to Washington D.C. to participate), "America is a vast reservoir of ignorance about sleep disorders." As one of his last actions before retiring, Senator Hatfield introduced legislation creating the National Center on Sleep Disorders Research within the National Institutes of Health. However, the funds that would support a national awareness campaign were never made available even to the National Sleep Center. Were Senator Hatfield still in the Congress, things might be different.

In Walla Walla, as of the last inquiry approximately 9000 seriously ill citizens have been diagnosed and treated, first at the Walla Walla clinic then at the Kathryn Severyns Dement Sleep Disorders Center located in St. Mary's Hospital. The results of the Walla Walla project have been a major revelation. In an alarming opposite of what one would expect to encounter in the general population, 85 percent of the polysomnography tests in Walla Walla have revealed very severe illness! This means that even the data so far in Walla Walla may remain the tip of the iceberg. It also means that several thousand human beings will receive clinical salvation from the Walla Walla Project, and the potential for salvation elsewhere is staggering. Because of the lack of public awareness and medical school teaching, untold numbers of undiagnosed and untreated or misdiagnosed and mistreated sleep disorders victims in primary care populations elsewhere are getting progressively worse. I fervently hope that the community-wide success we have seen in the sleepy little town of Walla Walla will go a long way in overcoming the resistance and denial that is still rampant throughout most of the world.

The key to success in Walla Walla has been the presence of an excellent resource, the fully-accredited Kathryn Severyns Dement Sleep Disorders Center directed by a highly qualified, board-certified sleep disorders specialist, Dr. Richard Simon. The small size of the community makes it very easy to contact and educate local physicians. There is a two-way synergy in this relationship. The sleep center must have referrals and local doctors can do enormous good for their patients without the need to become sleep experts. There are many things primary care physicians can do with little experience, such as identifying and treating Restless Legs Syndrome. However, they cannot easily get into the business of establishing a clinical sleep laboratory, training polysomnographic technologists, and acquiring the expertise to manage difficult cases.

It seems to me that every small community in America can replicate the Walla Walla Project. A special insomnia program is being started in Walla Walla. We are looking at the patients in cardiology practices. High school students are wide-awake in their classes. I swear that when I walk through the one large shopping mall, people seem more energetic, friendlier, happier than they were several years ago. On every visit, I try to guess which of the people I encounter on the

CHAPTER 14 Sleep Disorders & Walla Walla

street or in stores is still alive because the Walla Walla Project was successful. Moreover, every physician in or near Walla Walla refers patients to the Sleep Disorders Center.

We still have miles to go before we sleep. In other communities, the rivers of sleep disorders continue to flow past the unseeing eyes of primary care physicians. In Walla Walla, we have clearly shown that these patients can be easily identified and managed within the constraints of office-based medical practices. Although many patients have a problem with excessive daytime sleepiness, we have learned that these patients instead come to primary care physicians with complaints of fatigue or lack of energy, not sleepiness. Moreover, we have learned that most victims never complain to their doctors. The loudest and clearest message from Walla Walla is that doctors who truly care about their patients must proactively ask about their sleep and about daytime fatigue. "How are you sleeping?" must be as routine as "How are you feeling?"

Primary care physicians are in an ideal position to diagnose sleep disorders early. Concern about healthy sleep must always be one of the highest priorities throughout medical practice and certainly in primary care.

Our Walla Walla experience suggests another exciting possibility. Primary care physicians have the unique opportunity to make the entire community aware that sleep is an important component of good health and quality of life. People everywhere do not know that sleep deprivation creates a sleep debt that must be paid off. People are also not aware that obtaining extra sleep is the only way to pay off their sleep debt. The consequences of inadequate sleep - especially during dangerous, soporific activities such as driving - can be devastating.

During adolescence, the need for sleep does not decrease and may actually increase; yet most adolescents drastically reduce their nightly amount of sleep. Education is compromised as students sleep in class; lives are endangered as students drive home from football games and parties at night. Family physicians, internists and pediatricians are ideally suited to bring this information to the families and schools in their communities. By instructing patients in the simple rules of sleep hygiene and educating them about the consequences of sleep deprivation, many problems can be simply and inexpensively remedied. This is truly preventive medicine.

Our studies have established beyond the shadow of a doubt that sleep disorders permeate primary care patient populations. Sleep disorders cause or exacerbate other illnesses such as atherosclerotic heart disease, heart failure, asthma, headaches, obesity, depression and hypertension. In addition, we have learned that almost all primary care patients also experience transient sleep disorders which they never mention. Again, community physicians must take the initiative.

What has been achieved and what is now going on in Walla Walla is for me one of the most rewarding and exciting accomplishments of my entire life. Everything I am working for, and everything I believe sleep medicine can do for my fellow human beings, is being done in Walla Walla. Hundreds of people have literally been rescued from premature death in this sleepy little town, and the lives of many more have been greatly enhanced by restored health and energy. If it can happen there, it can happen anywhere. By William C. Dement

Conclusions:

Sleep medicine is a vibrant and growing field of medicine. People that complain about their sleep can now routinely see a physician with specialized training and board certification in this field. Much of these patients will improve when their condition is addressed properly. The biggest barrier to improve their conditions is their knowing that they will get better and their physician's reluctance to help them make sleep a priority in their health.

Fortunately, educational efforts and acknowledgment in sleep disorders can make a big impact in a community as described in the Walla Walla project.

"No person has the right to rain on your dreams." -M.L.K.

CHAPTER 15 Insomnia

"I can't sleep 'cause my bed's on fire." - Psycho Killer

Clinical Vignette #1

In August 1991, an extended family was vacationing at a California beach - two sisters, Helen S. and Rose C., their husbands, and their five children. The summer idyll was interrupted by the news that the Iraqi armies of Saddam Hussein had swept into the neighboring emirate of Kuwait. Two days later Helen's husband, a reserve officer in the US Air Force, was called to active duty. Overnight, he boarded a transport plane bound for the Middle East. No one knew what lay ahead and Helen worried that her husband could be in combat within a matter of days. In the midst of all of this, Rose's husband suffered a sudden heart attack while jogging on the beach. In critical condition, he was evacuated to a nearby large city hospital via helicopter, and Rose went with him, leaving her three children in the care of her very anxious sister. Alone with five children and worried sick about both her husband and her brother-in-law, Helen was so consumed by stress and anxiety that she couldn't sleep. Who could blame her? After three nights of lying awake imagining worst-case scenarios in the Persian Gulf and in the city hospital, Helen was so exhausted she could not take it anymore. Wisely, she made an appointment at a local clinic and asked the doctor for something to help her sleep. Despite her extremely stressful circumstances, the doctor would not prescribe medication, saying, "Sleeping pills are addicting." Instead, he suggested a glass of warm milk, a hot bath, and soothing music.

After another sleepless night, Helen was simply too exhausted to care for the children in her charge. She called her mother, who lived about 200 miles away, and asked if she could bring the kids and stay for a few days. Packing up all their belongings, she set out for her mother's house. Although still terribly worried, Helen started to feel some relief at the prospect of her mother's help. But with the relief came the first dangerous hint of drowsiness. No longer masked by all the stress and anxiety, the sleep debt that had been accumulating over the previous four nights began to assert itself.

In the mid-afternoon, as Helen was driving through a small town, the unthinkable happened. Without slowing the car in the slightest, she drove through a red light. Helen still does not remember approaching the town or seeing the red light, but she does remember the piercing, nightmarish squeal of brakes as a pickup truck smashed into the side of her car. She and her two children in the front seat sustained minor injuries. Her sister's three children in the back seat were all killed instantly. Witness Testimony: National Commission on Sleep Disorders Research, 1990

Definition and Characteristics

This chapter will review insomnia, which is a term loosely used by the public as a complaint of being unable to sleep. Other than pain, insomnia is probably one of humanity's oldest medical complaints. Imagine you could sleep whenever you wanted to, but did not HAVE to sleep. How would your life be different? Lots of things would change if we could avoid sleeping. For example, think of the money you would save when you travelled if you did not NEED to sleep. Travelers often spend more money on hotel rooms than on airfare. If we did not need to sleep, the university would not need dorms. If we hypothetically could develop a medication that could give you the refreshing feeling of having slept 8 hours would you take it? For many of us the need for sleep is an inconvenience. Yet we have to sleep for reasons that remain unclear. In this chapter, we will discuss what happens when we cannot sleep and develop insomnia.

The simplest definition of insomnia is trouble either falling asleep or staying asleep to the point that a person perceives it bothers them the next day. If staying awake does not bother a person the next day, it is not insomnia. There has to be some degree of **subjective** daytime impairment. The above definition of insomnia is a description of a complaint or a symptom. Over time, the insomnia symptoms can become an insomnia syndrome.

In medical terminology, a **symptom** is something a patient subjectively complains about that usually cannot be measured. It is what the patient feels. A **sign** is something that is observable or measureable by someone other than the patient. For example, a person may complain of having a fever, which would be a symptom. If that person's temperature measurement confirms the fever, it is a sign (temperature, heart and respiratory rates are collectively called vital signs). A syndrome is a collection of associated symptoms and signs that typically occur together. An analogy is pain, which can be both a symptom and a chronic syndrome. Insomnia is similar to pain in that the symptoms can change over time and lead to a syndrome. If somebody has had insomnia for longer than 3 months, then that person has probably developed a chronic insomnia syndrome.

CHAPTER 15 Insomnia

What is it like to live with insomnia? Many patients seeing a sleep physician for the first time may have had insomnia for many years, often spanning decades. If you ask a chronic insomniac why they waited so long they may say they felt it was their own fault they had insomnia. They may blame themselves or think they are trapped in situations that force them to not sleep. They will often complain they have "tried everything" and that "nothing works." It is not difficult to recognize when a person has progressed from just having insomnia as a symptom, to it becoming a chronic syndrome. Simply listen to them describe their sleep and perhaps ask a few questions. Insomniacs talk a certain way about sleep. If a person ever tells you, they are worried or scared that they would not be able to fall asleep at night or they are worried that if they wake up at night and that they will not be able to sleep they have probably developed chronic insomnia. They talk about **trying** to sleep while everyone else typically says they go to sleep. It is a very insomniac way of talking to say they "try to sleep." Healthy people do not "try" to get hungry and do not "try" to breath. It should be effortless. We simply eat and breathe. The only way to try to sleep is by staying awake and that is the crux of the problem. People with chronic insomnia can progressively get worse because their poor sleep experiences lead them to think and behave in a negative way about sleep. The thought of sleeping wakes them up. Insomniacs are trapped in a paradox. The paradox of insomnia is the paradox of sleep. We have to sleep for reasons that are not entirely clear, however, we are vulnerable to being attacked while sleeping. From a certain perspective, sleeping is illogical. If you were a predator and wanted to capture your prey, logically the easiest time to do it is while your prey was sleeping. If that is true, then by further logical extension you would think that by the laws of God, Mother Nature, natural selection, biology, or however you wish to think about it, that some animals would not sleep or that the animals that slept the least would have an advantage over all the other animals. However, that is not the case. In every biological niche that we know of virtually all known animals sleep. Even single-cell organisms have a basic rest/activity cycle. Typically, carnivores, such as lions, sleep much more than herbivores, like giraffes. Humans are the top predators and our sleep needs are between those two animals. The need for sleep seems to defy simple logic.

For insomniacs, the more logical and analytical they are in their attempts to sleep the more likely they are to get worse. For example, it is logical for an insomniac to think that if they cannot sleep then at least they should stay in bed and "rest". However, we know that sleeping and resting are two different physiological events. The more time a person stays in bed resting the more habituated they will be to lying in bed awake and the less they will sleep. Often they spend more time awake than asleep in their bedrooms. This will add to the vicious cycle of worsening insomnia. The correct approach is to spend less time in bed. The less time the person stays in bed the more powerful the homeostatic drive will become. They are then more likely to actually fall asleep and they will eventually sleep deeper. This is counterintuitive to insomniacs. If the right approach to sleep were logical, we would probably have fewer insomniacs in Silicon Valley. Chronic insomniacs are trapped by their internal logic, which leads to misconceptions about sleep.

Insomniacs become hyper-vigilant when they try to sleep. The consequences of their insomnia have become so stressful that the thought of sleeping wakes them up. Sleeping has become such a hassle for them that they often do not look forward to the evening and going to sleep. They may say they "dread" having to go to bed, they may feel their sleep is out of their control. When we feel out of control, we do not feel safe. When we do not feel safe, we will avoid sleeping or sleep as little as possible. The ensuing sleep is often insufficient and unrefreshing. Insomniacs want to sleep so badly that they will crave sleep or romanticize the times in the past they remember sleeping better. Often after suffering through years of poor sleep, insomniacs resign themselves to thinking that nothing will help. When an insomniac tries something new to help himself or herself sleep, they inadvertently make themselves worse because they stay awake to monitor if whatever they did is helping them sleep! The following vignette will illustrate this last point:

Clinical Vignette #2

Several years ago, a patient, Judith, was taking zolpidem (Ambien) in a 10mg dose with moderate success. The manufacturer began marketing a new formulation of zolpidem with an increased dose of 12.5mg. On a Monday, Judith called me asking if she could try the new the medication. The day she called the samples of the new medications had just arrived in my office. I told her to stop by, she would be the first person I give the medication to, and she can let me know how it worked. She came by the next day and picked up the sample of the 12.5mg zolpidem. The following morning, she called me upset and yelling. "Did you give me a placebo?" she asked. I was startled by this and reassured her that I gave the samples I just received. She then said, "Are you sure, because 10mg always worked and 12.5mg did nothing at all!" I started to wonder in my mind if I had somehow received bad samples. She then went on to tell me she was calling me from the pharmacy. She was so convinced that I had given her a placebo that she actually went there and asked the pharmacist to show her what the new 12.5mg tablet looked like. She then said, "Lucky for you, what you gave me looks like what they have here." While on the phone with her, I thought about the hypervigilance of insomniacs. I told her to keep taking the medication for the next couple of nights and give me a call. At the end of the week, she called and said she was sleeping better and continued taking the 12.5mg does after that. The novelty of simply changing the dose of the medication had heightened her hypervigilance. Once Judith became suspicious that I was giving her a placebo she got angry which made her insomnia worse. A change that should have improved her sleep briefly exacerbated her own insomnia.

As mentioned, an insomniac will often admit they dread having to go to bed at night. They may feel guilty or tortured by their lack of sleep. During the day, they may try to block out or forget the terrible night they had, but they are reminded every night of the problem. The insomnia symptoms can wax and wane. Some nights can be horrible and some nights are not so bad. On rare nights, they may feel they have gotten a good night's sleep. Nevertheless, those good nights can be frustrating because they also are reminded of how good sleep can be. How they will sleep on any given night becomes almost unpredictable to them. If they take sleeping pills, this adds another layer of complexity and concern. They may go to bed anxiously wondering when and how much sleep medication they will require. They wonder at night if they should just take a single pill as they were instructed or take 2 pills instead since the previous night one pill "did not work." Then they might wonder if they take 2 pills will they prematurely run out of medication. Are there any refills left at the pharmacy? Should they alternate the pill with something else at night? They may go through a litany of thoughts and decisions on how they will approach the night, yet still feel they did not sleep well. Typically, insomniacs, when describing what bothers them the most about their insomnia, they say that it is not the lack of sleep *per se*. They are no longer expecting a good night of sleep. What actually bothers them the most is how the poor sleep affects them the next day. They may think, "Tomorrow depends on how well I sleep tonight." If you think your day is predicted by how well you sleep the previous night, then you must monitor your sleep by either staying awake or sleeping as lightly as possible. This will ultimately worsen and reinforce the insomnia pattern.

Since chronic insomniacs have become hyper-vigilant to the thought of sleeping, they will often fall asleep when they are not "trying" to fall asleep. For example, they may fall asleep easily at the movies or at a play. Typically, insomniacs sleep better away from their bedrooms. Most of us sleep better in our bedrooms, insomniacs do not. They have associated their bedrooms with places of frustration and lack of sleeping. This is possibly why insomniacs will sleep better away at a hotel. They may inaccurately attribute their improved sleep on the hotel's mattress brand or model. It should not be surprising the hotel industry sells many mattresses to its guests! Sometimes it is difficult for someone who sleeps well to understand how difficult life can be with insomnia. Physicians in the past had been traditionally trained to work with little sleep. Thus, they may be particularly unsympathetic to a person's complaints of insomnia. It can be difficult for a healthy sleeper to understand how nighttime insomnia affects the way a person thinks during the day. To understand better what it is like to live with insomnia, imagine that instead of sleep we were discussing food. There are many similarities between sleeping and eating. Typically, we expect to eat every day and anticipate our meals. For example, if you know in advance that you are going to your favorite restaurant or going to have an especially large dinner you might have only a light lunch in anticipation of the evening meal. Imagine what would happen if your access to food were instead unpredictable? Some days you may have a feast and sometimes there is no food available at all for many days in a row. How would you live your life if you could not predict when there would be food? It would constantly affect the way you eat and think about food. When you woke, food would be on your mind. Even if you were not hungry, if you walked past a plate of stale donuts you would grab some and put them in your pockets in case there was no food later. Insomniacs feel the same way about sleep. Since sleep is unpredictable to insomniacs, they grab any sleep or rest period they can. By sleeping at irregular hours their sleep patterns are further aggravated.

Just as it is possible to be obese and malnourished at the same time, it is possible to sleep a lot and feel tired. Insomniacs often talk about wanting to get 8 hours of sleep per night. However, the amount of sleep should not be the goal. It is not just the total hours of sleep that matters. What also matters more is the quality of sleep. A child that came to the sleep clinic illustrates this last point. Even though most insomniacs are adults, children can often provide the most

CHAPTER 15 Insomnia

instructive examples because of their inherent simplicity. Unlike adults, they are usually not as guarded in their responses or have accumulated as many other medical problems. This child, Eric was almost 4 years old and he was brought in to the clinic by his mother. Medical histories usually begin with a chief complaint. When Eric was asked why he was brought in he looked at the physician straight in the eyes, and earnestly answered, "Doctor, sleeping makes me tired." His mother went on to explain that she was perplexed. She said he slept more than any of his friends.

In preschool, he napped longer than the other children. Yet despite all this sleep, she thought he always looked tired. Put yourself in Eric's place. The reasons parents want their small children to sleep may be one thing and the reasons they give their children for needing to sleep maybe something else. Obviously, the most pressing reason that parents want their children to sleep is so that they themselves can sleep. However, children are told sleep is good for them. They might be told they are going to "recharge their batteries" so they may have more energy in the morning. If they are cranky or in any way irritable they are told, again rightfully, that if they sleep, they will feel better in the morning. Eric had heard all this before. He was baffled. He really was confused because the more he slept the worse he felt. So, sleeping did make him feel tired. It turns out Eric had obstructive sleep apnea syndrome (OSA), which will be discussed in detail in chapter 18. The more he slept the greater difficulty he had breathing and the more tired he became.

Does sleeping make you tired? If you go to an "all you can eat" buffet or your favorite restaurant, do you leave feeling hungry? Sleeping a full night should never make you feel tired. The concept that to have a good night's sleep we must sleep 8 hours through the night without awakening is a common misconception. No one sleeps through the entire night without awakening. In the previous chapters, we have described that we periodically briefly wake up throughout the night but do not remember the awakenings the next day. The sleep cycle for most of us is typically 90 minutes. At the end of every cycle, we have a major body movement and shift positions. We are briefly awake, but return to sleep without remembering the awakening. For insomniacs, these otherwise normal awakenings can lead them to remain awake.

We should not be expecting to sleep without periodic awakenings. It is essential for our survival to do without sleep at certain times. For example, how can we be 8 - hour sleeping mammals, but have babies that need to be fed every 2 - 4 hours throughout the night? We are hard-wired in our brains to be able to interrupt our sleep and later return to sleep. How else could a woman wake up, feed a baby and then go back to sleep. There are countless situations when people need to interrupt their sleep to take care of something and then go back to sleep. That is not considered insomnia. Most things that are essential for our survival we can do without temporality. We can skip an occasional meal without going into hypoglycemic shock. We can hold our breath while diving underwater. We are not such fragile creatures. Yet, insomniacs pressure themselves to sleep because they over emphasize how much their sleep will influence their next day performance.

Nobody sleeps well under pressure. A dramatic example is if we imagine somebody points a gun at our head and says stay awake or I will shoot you. You would of course stay awake. If somebody tells you fall asleep or I will shoot you, you will not be able to fall asleep. You might pretend to sleep but you will stay awake as long as that gun is pointed to your head! The point is insomniacs sometimes go to bed thinking their entire lives depend on how well they sleep that night. The pressure to sleep only adds to the hyper-vigilance to sleep.

Another example of how insomniacs are trapped by maladaptive thoughts and misinformation about sleep is revealed when they describe their sleep patterns as children. If you ask an insomniac how they slept when they were children, they might say they do not remember or it was not a concern. If you consider our sleep patterns throughout our entire lifespan, at what ages do we sleep the best? The most common answer is when we are infants. People announce a particularly satisfying night of sleep by saying, "I slept like a baby." However, it is a misconception that infancy is the best sleep of our lives. Babies in reality do not sleep so well. Would you really want to have your sleep interrupted by a need to feed every 2 - 4 hours during the night and wear diapers to sleep due to incontinence? Our best sleepers are not babies. Our best sleeping years are usually around 8 or 9 years of age. Think about the lifestyle of a typical second or third grade child. As a child, you have a set bedtime. Parents still tuck the child in to bed to various degrees. When you are 8 or 9 years old, you do not need to worry about the rent or mortgage. Safety of the home is taken care of for you. Somebody wakes you to go to school. Clothes are laid out for you. Someone gives you breakfast, checks your homework and takes you to school. How well do 8 and 9 year olds typically sleep? They fall asleep easily, apparently sleep through the night, and wake up refreshed. During the day, they are full of

CHAPTER 15 Insomnia

energy and do not take any naps. On weekends, 8 and 9 year olds do not sleep in. They have no need to catch up on sleep. On weekends, they may actually wake up earlier to play or watch cartoons. These children go to sleep in a state of serenity. This seemingly idyllic situation is the norm for many children. However, insomniacs do not go to bed in a state of serenity. Going to bed for them is a state of uncertainty. They are in bed wondering how bad the insomnia will be tonight. To help them sleep better we need to help them re-establish serenity at bedtime.

Insomniacs can sometimes pinpoint in their minds when they developed insomnia. They attribute the insomnia to a specific event and they think if they could only go back in time to change that event the insomnia would go away. In order to correct the insomnia, we cannot change the initial trigger for the insomnia. The event(s) is/are usually long past. The initial trigger for their insomnia is usually no longer relevant when the condition becomes chronic. What is relevant is identifying what is perpetuating the insomnia. If you want to put out a fire, focusing on the spark that started the fire is not the solution. Instead, you need to address what is fueling the fire. Imagine a mother of a nursing newborn. She routinely wakes up during the night to feed the child. She may complain to her family how tired she is of the infant's disruptions of her sleep. What happens the first night that baby does not cry out to be fed? The mother will still wake up and check on the child. What was initially waking up the mother was the child's need to feed; now the mother is waking up on her own. The cause for initial sleep disruption may end but the insomnia may persist and take on a life of its own. This can change over time.

Insomniacs can sleep so lightly that they think they are awake, when they are actually sleeping. They may complain of not having slept at all. On mornings that they wake up and recall a dream, they realize that they must have slept. They therefore anticipate they will have a good day. When insomniacs do have an occasional good night of sleep, they will spend the rest of the day thinking what was different about that night that caused them to sleep better. If, for example, an individual concludes they slept better because it was a quieter night than usual, the person will think that getting rid of noise will be answer to the insomnia problem. They convince themselves they need silence to sleep. That individual may then decide to sleep with earplugs. They will buy the most expensive earplugs they can find. They will buy custom-made earplugs if they are available. When bedtime approaches, they warn family members not to make any noise, and put in the earplugs and go to sleep. Since the insomniac feels they finally found an answer, they feel less apprehensive about sleeping. They may sleep better that first night which reinforces that idea that they need silence to sleep better. However, at some point the insomnia will often resurface. While using the earplugs, they become vigilant to any noise that might disrupt their sleep. But it is harder to hear with earplugs in, so to monitor for noise while they are sleeping they have to now sleep even lighter. The earplugs will work for a while and then they will stop working. Then they will need to add something else to their nighttime repertoire.

The above description of an individual with insomnia does not consider the complications that can develop when that individual's insomnia impacts family members living in the same household. Insomniacs rarely sleep in the same bed with another insomniac. There is usually only one insomniac per couple. The more severe the insomniac symptoms, the more likely their bed partner sleeps very well. Insomniacs may tell you they do not understand how they can be suffering staying awake at night and the person next to them is sleeping soundly, oblivious of their torture.

They may feel resentment towards their bed partner. Family members sensitize to their loved one's poor sleep will do anything they can to avoid disturbing the insomniac once they have finally fallen asleep. They will walk on eggshells. For example, a mother might tell her young children that tomorrow they can go to the zoo if tonight she gets a good night's sleep. The children will do anything possible to avoid disturbing the mother's sleep. If the mother has a poor night of sleep but also realizes that her children are making an effort to behave and not make noise she might feel guilty about her situation. In reality, the mother's sleep may not depend on the children's behavior, more accurately her sleep depends on her own insomniac's situation. The above scenario is simply an example of how a family member's sleep affects other family members. Another typical scenario is of the parent who feels that if they do not sleep they will underperform at work. The other family members will do anything possible to not disrupt the insomniac's sleep, which creates greater pressure on the insomniac to sleep. Family members often try to help the insomniac sleep better but actually reinforce the insomnia and the viscous cycle continues. There are many different combinations of a family's composition in our modern society. The entire family's situation must be taken into account when considering how an individual sleeps.

CHAPTER 15 Insomnia

Over time, insomniacs may describe a series of specific items or conditions they need to sleep. For example, a special pillow, blackout curtains, a new mattress, herbal supplements, the television or radio left on. The longer they have symptoms the more conditions they need to establish or control to sleep. Despite their efforts, their sleep remains unpredictable with a gradually worsening waxing and waning pattern. This is another example of how the paradox of sleep leads to insomnia. The more things you have to control to sleep, the more hyper-vigilant the person becomes.

The best way to address insomnia is to address the underlying hypervigilance to sleep. Insomniacs need to understand that the need for sleep is biological but the way we sleep is learned. We are taught how to sleep. Insomniacs have learned to sleep a certain way that undermines their own sleep. At the risk of the authors seeming to be food obsessed, as noted earlier, many correlations can be made when thinking about sleep and food. All newborn babies drink milk; yet, 5 year olds all around the world have different diets. The need for food is biological, but what you eat is cultural. You are taught what to eat. Sleeping is similar. There is a biological need for sleep, but the way you sleep is learned. Insomniacs have learned to be hyper-vigilant about sleep. Hypervigilance to sleep is a key concept to understanding the pathophysiology of insomnia.

It is easy to demonstrate that the way we sleep is a learned behavior. For example, if you have a regular bed partner, you and your partner will each have a preferred side of the bed. You do not randomly change the side of the bed you will sleep on every night. The longer you are a couple the more likely you will always go to your own side of the bed. While traveling, when you arrive at the place you will spend the night you might say, "this is my side and that is your side." If the couple has been together for some time, there is no need to exchange any words. You each already know which side of the bed you will each sleep on. Couples rarely if ever switch sides. One exception is in older couples. If one person gets up more often at night to use the bathroom, they may agree to switch sides while traveling so that one of them is closer to the bathroom. If you do not have a bed partner and want to prove to yourself that sleep has a learned component, imagine what would happen if you rotated your body in your bed so that your feet are where your head usually is. You will feel odd in bed despite being your usual bedroom environment. You might have trouble sleeping and you might even fall out of bed that night. Sleeping is clearly a learned behavior. Parents must teach their children when and where they will sleep.

Understanding that chronic insomnia is driven by a learned or acquired hypervigilance to sleeping, then a treatment approach that reduces the hypervigilance should be effective. This can be accomplished using technique called Cognitive Behavioral Therapy (CBT). CBT has been shown to be effective and will be further discussed in the treatment section of this chapter.

Evaluation of insomnia

In a typical doctor's office, insomnia is recognized and evaluated only from the subjective complaint of the patient and only if the patient complains. A physician usually accepts what a patient says, but may occasionally assume that the patient's sleep is not actually disturbed. In addition, the definition of "clinically significant insomnia" must include at least one negative daytime consequence of the nocturnal sleep problem. Someone who complains of insomnia but feels "great, wide awake, and energetic" all day long simply does not have a clinically significant problem. There are also very rare individuals who have been labeled "normal short sleepers." Such rare individuals, whose daily sleep need appears to be only a few hours, occasionally consult a physician because they have been told many times that they should sleep eight hours. Thus, even though they are wide awake and energetic throughout the day, they may worry obsessively that something must be wrong with their sleep.

Specific Types of Insomnia Complaints

People may complain only about being unable to fall asleep at the beginning of the night. They may state that they require one to several hours to fall asleep. Because most people expect sleep to come quickly once they decide to turn off the lights and go to sleep, even a sleep latency of only 15 to 20 minutes may be considered troublesome. Other individuals may fall asleep quickly, but complain that they wake up several times for varying durations throughout the night. A common complaint is what has been referred to as "middle-of-the-night" insomnia, which usually means one long awakening, and if it continues, increasing anxiety about getting get back to sleep in time to be rested the next day. Other individuals complain of awakening "too early" and being unable to get back to sleep. Finally, complaints may include any combination of the above.

Since the major purpose of obtaining adequate sleep is to make us alert in the daytime, clinically significant insomnia is generally associated with impaired alertness, fatigue, tiredness, or severe drowsiness in the daytime, but may also include other symptoms.

Severity of Insomnia

Severity is an important dimension because very severe insomnia means that the victim is experiencing considerable sleep deprivation which, in turn, creates a life-threatening hazard of impairment in daytime alertness that is associated with the progressively increasing sleep debt.

In general, we assume that an individual with severe stress-related insomnia can make a reasonably accurate judgment about the amount of sleep he or she obtains. Prior to the episode, such a person is presumably a normal sleeper with normal sleep processes. If the insomnia is severe, the individual is rapidly building up a large sleep debt and a very strong tendency to fall asleep at an inopportune time (recall the vignette at beginning of chapter).

When someone complains of chronic insomnia or of having been an insomniac for many years, the description of the severity of the degree of sleep disturbance is often exaggerated. It is simply not possible for a human being to go without sleep for months and even years as some claim. Whether the victim's description of the nocturnal sleep disturbance is truthfully accurate, it must always be assumed that the victim is suffering, and thus, it is the doctor's responsibility to deal with the problem and alleviate it, if possible.

The following are often associated with insomnia complaints: daytime fatigue, tiredness and/or drowsiness; trouble concentrating during the day and/or trouble remembering things; inability to handle minor irritations; depressed mood; anxiety; jitteriness; diminished ability to accomplish tasks; impaired family and social relationships; inappropriate sleep episodes; general sense of diminished well-being; and, most importantly, more frequent automobile accidents. In large population surveys, chronic insomniacs report auto accidents twice as often as individuals who did not have insomnia. Since a very high percentage of sleep related motor vehicle accidents result in death, it is likely that insomnia contributes heavily to crash-related tragedies. Studies which followed large numbers of people for several years have reported an increased death rate among those who complain of insomnia or who say that their habitual nightly sleep time is very low. Another major study showed that the treatment of insomnia improved symptoms. Despite the stereotypes and bad publicity about sleeping pills, the clear majority of insomnia sufferers who have used sleeping pills have found them beneficial and would use them again.

Clinical Vignette #3

When I was just getting started at Stanford University, a married couple I knew quite well left for a driving tour of France. Two weeks later my wife and I were shocked and dismayed to get a postcard from the husband tersely announcing that his wife had been killed in an automobile crash. When he returned, he told us between barely stifled sobs that he had fallen asleep at the wheel while driving in the French Alps. The car went off the road and crashed into a ravine. His young wife sustained fatal head injuries while he walked away with only minor cuts and bruises. He attributed his horrible lapse to severe jet lag. After a 12hour red-eye flight from San Francisco to Paris, they had rented a car at the airport and immediately begun their tour, not giving his body time to adjust to the nine-hour time shift. For five nights leading up to the accident he had experienced great difficulty in falling asleep. His sleep debt had become enormous by what tragically proved to be the last day of his wife's life.- WCD *The Promise of Sleep*

Prevalence of Insomnia Complaint

All of us have had an occasional bad night of sleep. How would one know a good night unless you had a bad one to compare it. There have been large-scale surveys studying the topic of sleep and insomnia. In one of these, the question was asked, "Have you ever had insomnia?" Ninety-five percent of the respondents answered "yes" to this question. (As the jokester said, the other five percent were lying). This particular survey included people of all ages who were able to speak on the telephone, and yielded the additional finding that children and adolescents also complain about disturbed sleep. Estimates of the prevalence of insomnia depend on the criteria used to define insomnia and more importantly the population studied. A general consensus has developed from population-based studies that approximately 30% of a variety of adult samples drawn from different countries report one or more of the symptoms of insomnia: difficulty initiating sleep, difficulty maintaining sleep, waking up too early, and in some cases, nonrestorative or poor quality of sleep. One of the surveys found that on any given night about 48 percent of the population report they had trouble sleeping. Of these, 75 percent reported intermittent episodes and 25 percent had chronic insomnia (every night or almost every night).

Another national survey found that 35 percent reported insomnia. It is clear that the prevalence of chronic insomnia increases with age and is more prevalent among women.

To paraphrase Mark Twain, everybody talks about insomnia, but nobody does anything about it. A gigantic number of Americans experience disturbed sleep, but do not recognize, except in a few cases, when it warrants the attention of an informed physician. When people do complain to doctors about insomnia, it is usually in association with, and apparently secondary to, complex medical, neurological, or psychiatric problems. In such cases, the sleep disturbance is expected to diminish when the primary medical, neurological, or psychiatric illness is effectively treated.

Insomnia Can Be Temporary or Persistent

Sleep specialists have found it very useful to organize the various manifestations of "difficulty sleeping" in terms of the duration of the problem. Thus, difficulty sleeping is usually either **temporary** or **persistent**. In the former case, the most common situation is that an otherwise normal sleeper is experiencing circumstances which produce an inability to obtain adequate amounts of sleep. The precipitating circumstances are usually short lasting, several days at most, and when they come to an end, the insomnia disappears. The major concern is that the reduced amount of sleep will cause a homeostatic increase in the tendency to fall asleep which eventually will become dangerously strong. Generally, temporary insomnia extends from one or two nights to occasionally one or two weeks. Bouts of temporary insomnia can be repeated frequently or infrequently, sometimes occurring intermittently at fairly regular intervals; or at other times occurring more or less at random. A longer period of insomnia may accompany a very severe psychological or medical problem.

A sleep disturbance as the result of anxiety related to events of the previous day invariably lasts only a single night with sleep apparently returning to normal the following night. No one would consult a physician the first time such an experience occurred. If, however, such episodes occur repeatedly and predictably, they should be mentioned to a healthcare provider, particularly if daytime sleepiness following the disturbed night is a problem.

An insomnia complaint may persist for months and even years. By far the most common manifestation of persisting insomnia that is considered troublesome is chronically intermittent episodes, where there are several bouts of severe insomnia every month or even every week. Although there is not a precise, widely accepted definition of when insomnia should be called "persistent" (implying that it will not go away by itself), difficulty sleeping that has endured for months or years clearly fits this definition. The International Classification of Sleep Disorders uses the clinical term chronic when insomnia has lasted at least more than 3 months. Yet many people would not think of insomnia as persistent if it only lasted 3 months. Persistent insomnia requires a completely different approach than temporary insomnia. It is usually caused by a specific, diagnosable sleep disorder, or has a chronic psychiatric or medical cause. Simply using sleep medications every night for long periods of time is rarely a suitable approach. The first step must be the identification of the specific sleep disorder that is causing the problem. This, in turn, will dictate a specific therapy. Specific therapies range from manipulations to reset the biological clock to a variety of behavioral approaches such as relaxation techniques, biofeedback, counseling, and psychotherapy. These are discussed further below.

Temporary Insomnia

Causes of Temporary Insomnia

There are three main categories of circumstances that temporarily disturb sleep. They are: (1) hyperarousal, (2) abrupt time zone and sleep schedule change, and (3) noise or other environmental disturbance. Hyperarousal can be produced by excitement, anxiety, worry, stress, anger, pain, other types of discomfort such as itching, and by ingesting stimulants such as caffeine or certain nasal decongestants at bedtime. The specific causes of disturbing or exciting emotional responses are, of course, almost infinite. Temporary circadian sleep disturbance is caused by abrupt time zone changes or schedule changes. In these instances, the circadian tendency to be awake occurs at times when individuals are trying to sleep. Persistent circadian rhythm disturbances will be considered later in the chapter. Finally, direct environmental disturbances must be considered. These include temperatures that are too high or too low, noise, light, a strange bedroom, or an uncomfortable bed.

A special example of environmental disturbance is the insomnia commonly experienced when a person is hospitalized. One of the authors was invited to a hospital formulary meeting to discuss what sleeping medications to have available for in-patients. The question raised was what medication could provide the patients with a normal night of sleep? The author's answer was "None. Why should you be sleeping well if you are hospitalized?" They never invited the author back to formulary meetings.

By far the most commonly reported cause of disturbed sleep is stress and anxiety. Though extremely common, the ways in which stress and anxiety occur are also very diverse. In addition, the tendency for any individual to respond to stress and anxiety with severely disturbed sleep can be highly variable. Some people may feel great emotional upheaval during the day, but have little or no problem falling asleep at night, while others experiencing only a slight aggravation will lie awake for hours.

Severity of Temporary Insomnia

The assessment of the severity of temporary insomnia should take into account the number of nights, amount of sleep lost per night, and the impact on daytime function. Individuals vary a fair amount in terms of the impairment of their daytime alertness in response to insomnia. For many people, one bad night is sufficient to cause daytime fatigue. To a large extent in cases of hyperarousal, the degree of the sleep disturbance will reflect the degree of the emotional distress or anxiety that produces the disturbance.

Consequences of Temporary Insomnia

Although many individuals find it extremely unpleasant to lie awake at night worrying about all their problems, the more serious consequence of stress-related insomnia is daytime fatigue and sleepiness. Sudden severe sleep loss can cause marked impairment of daytime function. The sleep loss is particularly dangerous because the anxiety, stress and aggravation frequently mask the mounting drive to sleep. The tendency to fall asleep can build up to a level which can overtake the insomnia victim with surprising and unexpected suddenness. Consequently, the danger of falling asleep while driving or in other dangerous situations greatly increases. Once again, the vignettes earlier in this chapter are tragic examples.

The effect of stress and anxiety on sleep is difficult to study in the sleep laboratory for at least a couple of reasons. First, to deliberately cause people to experience stress and anxiety strains the limits of ethical human research. Furthermore, it is very difficult to produce a predictable amount of stress and anxiety in a laboratory situation. Second, to bring people who are experiencing acute stress and anxiety into a laboratory situation is also very difficult because the events producing the stress are often unpredictable and would usually preclude departure from the home or work environment for purposes of having a sleep study.

Nonetheless, we must emphasize that severe sleep deprivation, daytime sleepiness can occur in association with transient insomnia, and this, in turn, produces impairment of a variety of cognitive and behavioral functions as well as increased risk for accident and injury. Severe sleep loss in normal sleepers always leads to tiredness, fatigue, drowsiness, and impaired performance in the daytime. This then interferes with work and other activities, and creates a danger due to inattention or falling asleep while driving or drowsiness in some other hazardous situation. Most people refer to the increased sleep tendency as exhaustion, fatigue, tiredness, misery, but rarely as excessive sleepiness. The increasing sleepiness is camouflaged by the ongoing stress and anxiety. For people behind the wheel driving is extremely dangerous.

It is difficult to know how sleepy an individual can become in the daytime solely as a result of an abrupt bout of insomnia because the amount of chronic sleep loss or sleep debt that the individual is already carrying when the insomnia occurs is never certain. A single night of severe insomnia in a person who already has a large sleep debt can be devastating. Doctors who are unaware of sleep debt will be skeptical that a single night of insomnia can produce such consequences. Nonetheless, physicians should accept as true the patient's description of exhaustion and misery in association with possibly only a few nights of insomnia.

Should Temporary Insomnia Be Treated with Medication? Many people, including physicians, feel that insomnia is trivial and that treating insomnia with sleeping pills is analogous to swatting a mosquito with a sledgehammer. People who say this are, of course, people who have almost certainly not experienced severe insomnia. The fact is that severe insomnia can be a very serious problem. As we have emphasized, with temporary insomnia it is not the sleep loss *per se* that is dangerous, it is the accumulating sleep debt. The Food and Drug Administration has approved the use of sleep medication for patients with complaints of insomnia even if it is only temporary insomnia.

If doctors have ever been guilty in the past of over prescribing sleeping pills, it is clear that there is also a tendency to err on the side of being overly cautious. Any person who consults a physician specifically because of stress-related temporary insomnia is probably desperate for help. Even so, physicians are very reluctant to prescribe sleep medication. In a survey of approximately 500 primary care physicians scattered across the country carried out in 2000, more than 90 percent of the doctors stated that they believe sleep medications are addicting and have serious side effects. Even if this were true, which it is not, the risk/ benefit ratio in many instances would favor treating a patient with sleep medication.

Clinical vignettes #1 and #4 (below) in this chapter exemplify such situations. Helen S. and Roger M., the two insomnia victims described in these vignettes, actually consulted physicians about their insomnia and asked for "something to help me sleep." In both of the above cases, the doctors could have prescribed sleeping pills, and moreover, they should be considered responsible for the tragic consequences of their failure to do so. At the same time, an inaccurate or misinformed understanding of the use of sleeping pills often prevents people from requesting medication.

Clinical Vignette #4

Roger M. was a hardworking young businessman. He was, as far as he knew, happily married. Upon returning from a business trip, he expected to be met at the airport by his wife. Instead, a complete stranger served him with divorce papers and gave him the keys to his car. When he arrived at his home, his wife was nowhere to be found and a number of his prized possessions were also gone. He called various acquaintances, but was unable to locate his wife. That night, he could not sleep at all, and the following day he was extremely upset and could not work. His agitation increased as repeated attempts to locate his wife met with failure. Again, he was completely unable to sleep. After the third night of no sleep and feelings of severe anxiety and depression, he began to question his own worth and began to feel that life was not worth living. Although he was exhausted, he did not feel sleepy. At the urging of friends, he consulted a physician for the purpose of obtaining sleep medication. The physician was unwilling to prescribe medication and suggested he see a psychiatrist. The earliest appointment the patient could obtain was for the following week. After five nights and days of no sleep at all, Roger M. experienced a somnambulistic amnesic episode which began in the afternoon and ended in the late evening when he "came to" and found himself standing in the bedroom of his burning house holding a loaded rifle. His shoes were on fire. He ran from the house and put the gun to his head. Fortunately, firefighters were already on the scene and were able to grab the rifle. He was then hospitalized for treatment of his burns. In the hospital, sleep was induced with a benzodiazepine hypnotic, and after several nights of lengthy sleep, his agitation and sleeplessness disappeared. He was then able to understand and confront his problems which now included a criminal indictment for arson.

Witness Testimony: National Commission on Sleep Disorders Research, 1990

Close Up

I have personally experienced another doctor's disdain when I asked her to prescribe sleeping pills. Until recently, I presented an annual lecture to undergraduates at Cornell University, and on my most recent trip I noticed a student who looked terribly sleepy. It turned out that she had recently endured a serious personal crisis that left her severely sleep deprived. Unable to sleep at night or to skip classes for a daytime nap, she appeared desperate. Knowing what horrific things can happen when someone is that sleep deprived, and fearing she might suffer some terrible accident, I placed a call to the student health service. I told the doctor who I was, and that this student needed sleeping pills for a few nights. The doctor seemed resistant, so the Cornell professor who was my host got on the phone and said I was one of the foremost experts in the world on the subject. Our pleading only strengthened the student health doctor's resolve: "I'm not going to be a pill pusher for you or anyone else," she said to me. It did no good to explain that no one was going to get addicted from a three- or four-day supply of sleep medication. By the scorn and indignation in her voice, you would have thought I was Joe Sleaze, the local drug dealer. All I could do was tell the student to try to find a doctor who would help her get the sleep she desperately needed. - William C. Dement

Insomnia Treatment

Insomnia is a very common problem in the general population; however the vast majority of these people will never seek medical attention specifically for their complaints of poor sleep. In part this may due to these people attributing their insomnia to other medical problems they may have. Many medical and psychological conditions are co-morbid with insomnia. Some of these include sleep apnea, restless leg syndrome, neurological problems, dementia, Parkinson, and chronic heart failure, which are all associated with insomnia. Other natural situations that are not disorders such as pregnancy and menopause are also associated with insomnia. Many internal medial conditions have co-morbid insomnia such as asthma, irritable bowel syndrome, ulcers, arthritis, diabetes mellitus and incontinences. Psychiatric conditions associated with insomnia are bipolar disorder, depression, generalized anxiety disorder and post traumatic stress disorder are all co-morbid with insomnia. Bulimia, anorexia, personality disorders also may be associated with insomnia. Given all of these possible co-morbid conditions, it is important to optimize these other medical problems to help improve the person's sleep. However even after the treatment of these other conditions is optimized the insomnia may persist. Fortunately, despite the high prevalence of insomnia, it is a condition that is readily treated. The majority of patients that seek proper medical attention for insomnia will improve. Keeping in mind that patients with insomnia might be suffering for years. It is amazing that any treatment is effective and even more surprising that more people do not seek out proper help for their insomnia. If we accept the pathophysiology of insomnia is based on a model of learned hypervigilance towards sleep, then a pathway to reverse this process can be established. Long-term improvements can be accomplished through behavioral changes. However, for this to take effect it may take some time. Typically, behavioral therapy is done over a 2-month period of time by psychologists, properly trained in behavioral sleep medicine. The American Board of Sleep Medicine created a specification certification for behavioral sleep medicine in 2003. These experts are not yet widely available. What is widely available is medication for insomnia. Sleep medications, collectively referred to as hypnotics, can provide immediate relief of insomnia. However, there are pitfalls with these hypnotics. The advantage of hypnotics is that the same medications are available throughout the country, unlike the behavioral treatment therapies. Hypnotics should be viewed as tools to improve sleep. They can be combined with behavioral techniques to help chronic insomniacs. The overall goal of any treatment is to have the patient sleep easily all throughout the night and wake up feeling refreshed in the morning without relying on medication. However, there are clinical situations where these medications are needed on a long-term basis in combination with behavioral approaches. For example, patients with Post Traumatic Stress Syndrome (PTSD) or chronic pain may benefit from long-term use of hypnotics. Further, in this chapter we will discuss both behavioral approaches and pharmacological treatment of insomnia.

Behavioral Treatment:

When a patient with insomnia is first seen at the clinic many of them will want to avoid hypnotics. Patients are quick to mention fear of addiction or dependence on the medications. These medications have earned a bad reputation. The reputation has been well earned when we consider the range of problematic medications used in the past such as barbiturates. The bad reputation is also built on years of stories of celebrities' deaths with sleeping pills such as Marilyn Monroe and Heath Ledger. (In fairness, these deaths were more properly attributed to polysubstance abuse in the context of significant psychological stress.) Nonetheless, in the general public's mind sleeping pills are dangerous. This explains an appeal towards a nonmedication approach to treat insomnia. However, a nonmedication approach requires a longer time and patient cooperation to be affective. The proper behavioral treatment for insomnia has been shown to be very effective. At least 2/3 of patients with chronic insomniacs will be able to correct their insomnia. The ideal behavioral treatments of insomnia are now typically described under the umbrella term of Cognitive Behavioral Therapy (CBT). CBT is based on the assumption that patients with insomnia have learned to sleep a certain way due to the experiences they have had in the past. These experiences have now led them to behave a certain way and in turn these negative experiences lead to negative thinking and the vicious cycle goes on. In medicine, most disease states can be modeled on a vicious cycle and insomnia is no exception. The goal of CBT is to give the patient new information about sleep to help them think differently about sleep. If they think differently about their sleep then this can lead to changes in their behavior. These changes in behavior can lead to initially small improvements in their sleep. These improvements will lead to further changes in their thinking and behavior towards sleep. This process can gradually reverse the vicious cycle of insomnia. Under the general term of CBT, there are a group of behavioral techniques and concepts that can be customized for the patient's individual situation. These techniques and concepts include sleep restriction, stimulus control, and sleep hygiene.

Adults with insomnia often complain of greater difficulty staying asleep as opposed to falling asleep. Sleep restrictions is a very effective technique to help patients stay asleep. The term sleep restriction must be distinguished from sleep deprivation. Sleep deprivation is (unfortunately) a sanctioned form of torture. In contrast, sleep restriction is a therapeutic tool. The concept is that insomniacs are inefficient in the way they sleep. They may spend 9 hours in bed but only sleep 5 hours. The wasted time failing to sleep increases their frustration. With sleep restriction, the patient is given a narrow window of time to spend in bed. The time allowed in bed is typically restricted to 5 ½ - 7 hours depending of the clinical situation. The patient is advised not to sleep at any other time except for this window of time. Using sleep restriction, the homeostatic drive will improve the ability to sleep. As the sleep improves, the

CHAPTER 15 Insomnia

time in bed is gradually increased over time until the patient is more satisfied with the amount of sleep they are attaining. These increments of time in bed are done in intervals of 15 - 30 minutes once a week. For example, if a patient reports that they spend 8 hours a day on average in bed but they only think they sleep 5 ½ hour, then they waste 2 ½ hours per night. Using sleep restriction, we first fix the patient's wake up time to the patient's ideal preference. For example, the patient wants to ideally sleep from 11 PM to 7 AM, but currently only sleeps 5 ½ hours, we have the patient lock in the 7 AM wake up time but not go to bed until 1:30 AM. The idea is that the patient will spend less total time in bed. Often when you tell an insomniac who has been going to bed at 11 PM to stay up until 1:30 AM they say they cannot stay up that late. You may need to remind them they have insomnia and of course, they can stay up later. The key feature of sleep restriction is locking in a wake-up time, which allows the patients to gradually widen the window they can sleep by going to bed earlier. This allows insomniacs to become more efficient sleepers.

Another behavioral therapy technique for insomnia is stimulus control. This refers to a set of instruction designed to reinforce the instruction between the bed and bedroom. Many insomniacs spend more time in their bed than sleeping in their bed. Their bodies no longer associate the bedroom with sleeping. Typical stimulus control instructions include going to bed only when they are sleepy and to get out of bed when they cannot sleep. If a patient is tossing and turning in bed and getting frustrated they are instructed to get out of bed. They should not sleep anywhere else except their beds. They should get up and do something else and then to go to bed, typically something not productive. They should not read, check email, or get a sandwich. Doing any productive or enjoyable activity is a positive reinforcement of their insomnia. Patients that get up in the middle of the night should not do anything interesting, for example they should not find a TV show to watch. Instead just wait until they feel drowsy and then go back to bed. Patients are instructed to not allow themselves to get frustrated; anticipate that tomorrow will be another chance to sleep. Part of the stimulus control recommendations is that the bed is for sleeping only. It is not for reading or watching TV in bed. Also, we recommend patients to get out of bed every single morning at the same time and discourage napping.

Another behavioral technique is relaxation training. This helps lower patients' anxiety to having to fall asleep at night. This includes things like muscle relaxation techniques. The aim is to lower the anxiety and muscle tension. Sometimes yoga type techniques are recommended. Patients often complain they cannot sleep because they say their mind is racing. They say they cannot "turn off" their brains. The reader should know that when we are sleeping the brain is never turned off. Another useful technique includes, scheduled thinking time, or scheduled worry time. This technique is effective in helping patients decrease the racing thoughts they are fighting before trying to fall asleep. First, think back to a typical day, if you had to do 10 things and accomplished 8 of them you had a really good day. However, the 2 things that you forgot to do, when are you likely to remember? You will likely remember them when lying in bed. When are you doing nothing else but thinking? For most of us, it is only when we are in bed. During the rest of the day we are constantly doing other things and our attention is diverted. Being alone with our thoughts in bed is altered with this 24/7 society there is always something else to do. To apply a scheduled thinking technique the patient is instructed to reserve some time away from the bedroom for approximately 15 - 30 minutes every evening. This time should be set aside after any other evening work or activity is finished. Since we typically combine this technique with sleep restriction, the patient will have more time in the evening. They are instructed to sit in a quiet place of their home, away from their bedroom with a paper notebook (not a laptop) and start a journal. Just an old-fashioned paper notebook is all that is needed. We do not want something with a light source staring in their eyes. They should write down the little things they need to do tomorrow, for example do laundry or buy groceries. Then write down things they have been meaning to do but have not gotten around to, for example sending a thank-you for a birthday gift from an uncle from 3 years ago. All of us have loose ends that we let slip. Patients are told to write these things down in addition to anything that is worrying them. Finally, the patient should write down those things they have always want to do for example, go skydiving, or visit the Himalayas. After all these things are written down and the patient has had some time to be alone with their thoughts, they then close the notebook and tell themselves "I am done with my day whatever is not down can wait till tomorrow." Say it and believe it. Put the notebook aside. It is no one else's business what is in that journal. After they have finished journaling, they should not immediately get into bed because their minds are still running. Instead, they should spend some time doing something relaxing they enjoy such as reading for pleasure or taking a bath. They should do something as a reward for having finished their day. When they feel drowsy, then they should get into bed.

If when they get into bed, and they start thinking about something remind yourself that it is already written down and that the day is over. In the morning when the patient wakes up they should take a moment and take a look at what they wrote so that their day is organized. Continuing with this technique of scheduled thinking/worry time for a few weeks and patients will report a less tendency to wake up with racing thoughts during the night. While patients are applying these techniques to their sleep, they are also at the same time tracking the amount of time they are in bed with a sleep diary. Along with the process of completing the sleep diaries they also record what kind of a day they are having. By doing this when they review their notes, they can see that their sleep is gradually improving over time. They may also notice that they have good days and bad days independent of how well they feel they slept on a given night. This will help them decrease the tendency to worry that tomorrow depends on how well they sleep tonight.

These behavioral techniques alone may help many people sleep better. They however may take several weeks to correct the insomnia. Therefore, when more immediate improvement is desired, medications may be incorporated into the therapeutic plan in a combination approach.

Pharmacological treatment of Insomnia

Even though behavioral treatments may be more desirable for chronic insomniacs, they are not always readily available. Sleeping medication, in particular over the counter types are readily available. Patients with chronic insomnia have ambivalent feelings about the use of medications. On the one hand, they fear if they take medications for their sleep they will become addictive and they think about all the problems that can arise. They also worry that if they take medication and it does not work then they are really stuck and they have no other options. When somebody has uncertainty about his or her situation it is very hard to sleep. Sleeping pills, as mentioned above have earned a bad reputation over the years. In the 1950's, we used barbiturates to help people to sleep. These substances had fatal complications especially when mixed with alcohol. In the current times, hypnotics have become safer. Before patients receive prescription medications for sleep they will usually try over the counter remedies. There are many herbal remedies available for sleep. Perhaps the best-known herbal substance is valerian derived from the valerian plant. Its uses are thought to go back over 1000 years. With any herbal agent or under regulated substance it is hard to know exactly what the patient has taken. Valerian is a commonly used substance, however in patients with chronic insomnia it has not been shown to be very effective. Valerian may have a role in mild cases of temporary or situational insomnia. Valerian has not been found to have any significant side effects. Given the uncertainty of valerian preparations it is not a substance that is routinely recommended by physicians. Patients with chronic insomnia often try herbal remedies such as valerian before making an appointment to see their physician. By the time, they seek medical treatment they are no longer satisfied with herbal or other over the counter remedies.

Histamine is a neurotransmitter with wake-promoting substance produced in the brain. Medications that block the effects of histamine cause sedation and are called antihistamines. The primary use of antihistamine medication is for the treatment of allergies. Their sedating properties also make them appealing as an over the counter hypnotic. The most commonly used antihistamine is diphenhydramine, which is the principal product in Benadryl. Antihistamines clearly cause sedation; however, they are fairly long acting medications. When a person uses these medications, they may wake up feeling groggy the next day. Medications with antihistamines are often sold in combination with other medication (Tylenol PM). It is ironic that patients often will think that these over the counter medications are safer than prescription medicine such as zolpidem (Ambien). However, that is not the case. It is easier to cause fatal liver damage specifically from an overdose of Tylenol PM than with zolpidem. Any emergency room physician is more worried about a Tylenol PM overdose than an Ambien overdose. Another over the counter agent to consider is melatonin. This is a naturally occurring hormone produced by the pineal gland in the brain that is released in anticipation of the evening. This is the hormone that tells the brain night is approaching. Melatonin may help people fall asleep faster but does not help in staying asleep or significantly increasing the total sleep time in most adults. Melatonin has been shown to help children fall asleep faster. However, the purity and quality of various melatonin preparations has been questioned.

Perhaps the most commonly used agent to help people sleep is alcohol. Alcohol is a sedating substance, however not a treatment for chronic insomnia. With alcohol you can drink enough to be completely passed out, however you will not wake up feeling refreshed the next morning. It does not provide the restorative sleep we need so it is not an effective hypnotic for chronic insomnia. Even small amounts of alcohol at bedtime may be disruptive to sleep. Since most insomniacs have problems staying asleep, alcohol at bedtime is strongly discouraged. It is a troublesome compound especially when mixed with other medications.

Marijuana is commonly thought of as a sleep promoting agent however the data supporting this is not clear. Studies have reported mild suppressive effects of REM and increased total sleep time however, after prolonged nightly use any therapeutic effects may be lost. Although marijuana may have therapeutic value for other medical conditions, its use for chronic insomnia cannot be supported without adequate research trials.

Prescription medications

The National Institute of Health (NIH) has concluded that there is valid scientific data supporting the use of prescription hypnotics in the treatment of both transient and chronic insomnia. The best-studied hypnotics are benzodiazepines receptor agonists. These hypnotics include many commonly advertised sleeping pills such as zolpidem (Ambien) and eszolpiclone (Lunesta). These medications have a common mechanism acting as allosteric modulators of the GABA receptors in the brain. GABA, gama-amino butyric acid, is the major inhibitory neurotransmitter of the brain. The benzodiazepines receptor agonists will open a chloride channel, which leads to the inhibitory effect of this substance. GABA agonists tend to not only have the ability to promote sleep but they also relax muscles, control seizures, and lower anxiety. Since these medications can lower anxiety they can become habit forming. Within this class of medication, the best-known medication is diazepam (Valium). All of these medications can be tools in the treatment of insomnia under a physician's direction. Overall, these medications are thought to be relatively safe to use, however due to the wide availability of these medications they can also be used incorrectly. Taking these medications too early in the evening or in the daytime will not effectively help a person fall asleep. These medications are meant to be taken when getting into bed. When taken earlier than appropriate the medication is fighting the clock-dependent alertness that was discussed in earlier chapters. Other prescription hypnotics are available but are less popular than the benzodiazepine receptor agonists. These alternative medications can be less likely to be habit forming. Sedating antidepressants such a doxepin are approved to help patients stay asleep. A melatonin agonist,

Ramelteon, has also been approved to help patients fall asleep but is ineffective in helping patients stay asleep.

Combination Therapy:

It is common to divide the treatment into both behavioral and pharmacological treatments. Patients may be prescribed medication to obtain immediate relief but behavioral treatments may be prescribed to help patients with long-term improvement of their insomnia. How these therapies are combined is based on the individual patient and what is available in the patient's local community.

Conclusion

There is extensive documentation that insomnia is associated with serious symptoms such as irritability, memory problems, apathy, lost days of work, and impaired interpersonal relationships. Both temporary and chronic insomnia can have negative consequences due to fatigue and drowsiness. These consequences should never be ignored. They can be devastating and tragic. Fortunately, effective treatment for insomnia is available. An effective treatment approach may involve a combination medication or behavioral techniques such as cognitive behavioral therapy.

"I'm beginning to lose sleep: one sheep, two sheep Going cuckoo ...I'm friends with the monster that's under my bed. Get along with the voices inside of my head."- Eminem & Rihanna

CHAPTER 16 Restless Legs & Sleep related movement disorders

"Yeah, because she is throwing off my whole sleep. She's got the jimmy legs." -Kramer

Clinical Presentation:

Restless leg syndrome (RLS) is a movement disorder that has been recognized for centuries. It was historically first described over 400 years ago, by Sir Thomas Willis, the same person who identified the Circle of Willis in the brain. Several centuries ago, people with RLS feared a curse had been put upon them. Those with RLS feel an urge to move when they are resting. The feeling goes away as long as they are working, but recurs when they try to rest, especially at night. When their legs are examined, nothing appears to be wrong with them. The disorder can seem designed to drive somebody insane. Those afflicted have to keep moving and cannot rest. Even today, patients with RLS still say it feels like a curse. The diagnosis of RLS is based solely on a patient's history. There are no physical signs or laboratory tests necessary to make the diagnosis

In 1995, a consensus emerged from the International RLS Study Group (IRLSSG) for uniform description and diagnosis of RLS. The group decided that a diagnosis of RLS required four essential features

1. An urge to move the legs usually accompanied by or caused by uncomfortable sensations in the leg. Sometimes the urge to move is present without the sensations.
2. A worsening of these uncomfortable sensations during restful activities such as lying or sitting down Partially or totally relieved by movement such as walking or stretching, as long as activities continues
3. The partial or total relief of these sensations from movement such as walking or stretching, and for as long as these activities continue.
4. A worsening of these symptoms in the evening/at night. They may also only occur in the evening or at night

Those who suffer from RLS often have a family history with the disorder. The prevalence of RLS among first degree relatives of RLS sufferers is 3 to 5 times greater than that of the general population. Additionally, treatment is usually quickly responsive to dopaminergic medications. Patients with RLS may also have periodic limb movements of sleep (PLMS) during a polysomnographic recording. The absence of PLMS does not exclude the diagnosis of RLS. RLS may begin at any age and can show up in children. The severity of the symptoms may fluctuate a great deal from patient to patient. Some patients may go for years without feeling any RLS symptoms. These symptoms tend to be more severe in middle age or in elderly persons. They often flare up during pregnancy, as many women first become aware of the symptoms during pregnancy. They may then forget they ever had the problem until they are reminded of it by the emergence of RLS symptoms in their children. The prevalence of RLS is reported to be as high as 15% among Canadians in the Montreal area. This makes RLS one of the most common sleep disorders. Since RLS is readily present in family members, there is strong scientific interest in the genetics of the disease. RLS is a genetic disorder with an apparent autosomal dominant inheritance pattern. This means it is uncommon to find a patient with RLS who does not have another family member afflicted by it. These family members might not realize they even have the condition. Sometimes, family members are aware of RLS symptoms but think it is normal because many of their relatives have it. They may then simply try to ignore it.

Pathophysiology

The exact cause of RLS is unknown, but it is strongly suspected that it is due to dysfunctions in the brain's ability to manage iron and produce the neurotransmitter dopamine. Dopamine is a neurotransmitter in the central nervous system involved in movement. Medications that block dopamine make RLS symptoms worse. Examples of medications that block dopamine activity in the brain are many of the same medications for the treatment of schizophrenia. Medications that increase dopamine levels in the body quickly improve symptoms of RLS. Patients may report dramatically improved symptoms in less than a week. Depletion of dopamine is classically seen in Parkinson's disease, a neurodegenerative condition of the central nervous system. RLS is a disorder independent from Parkinson's, meaning that having RLS does not predict the future development of Parkinson's. RLS symptoms may flare when a patient is deficient in iron. Iron is an important cofactor in the synthesis of dopamine. For the brain to make dopamine it must convert it from the amino acid tyrosine using the enzyme tyrosine hydroxylase which acts with iron as a co-factor. One hypothesis of the pathology of RLS is that it is due to a possible dysfunction in a way iron is used by the brain. A genetic modification of a protein that helps transport iron in the brain, transferrin, may be working inefficiently. Although RLS was presented as a "curse" at the beginning of this chapter, it does have a potential silver lining. Adults over the age of 50 are at risk for colon cancer. This cancer is especially deadly because it may grow silently within patients for an extended period of time Their only symptom may be blood loss in their stool. Patients with RLS are very sensitive to blood loss. The RLS will flare up if they develop colon cancer. RLS serves as an early warning system for any other medical condition characterized by slow blood loss.

Treatment

RLS is similar to headaches in the since that there is no objective test to determine whether the patient suffers from RLS or not. As awareness of sleep disorders and RLS in particular has spread in society, a market of treatment options have emerged. In 2005, the FDA approved for the first time a medication specifically to treat RLS. This led to a flurry of direct consumer advertising that included many television commercials. These commercials became the butt of many jokes, especially those from late night comics. Bill Maher provocatively quipped that 'they invented the pill before they invented the disease.' However, the false charge that RLS is a disorder fabricated by the pharmaceutical company was offensive to those who had suffered from the disorder for many years. The reader should recall that RLS was first identified in the 1600s.

In support for a role of iron in the mechanisms causing RLS is the fact that intravenous iron treatments decrease patient symptoms. Intravenous iron supplementation is not an easy treatment to administer. It can take several hours and can lead to other complications. Iron supplements can also be given by mouth. Since iron is slowly absorbed it may take a long time to see any beneficial results. Determinations regarding the use of supplemental iron in patients with RLS are guided by measurements of serum ferritin. Ferritin is a protein used by cells to store iron. When ferritin levels are in the low range of normal or lower, iron supplements are recommended. Excessive iron should be avoided, however, since it can cause liver damage among other problems.

In addition to iron supplementation, other nonpharmacological treatments are available. These may be preferred before starting medication, particularly in children. Patients are encouraged to develop good sleep habits and to especially avoid caffeine. Donating blood may exacerbate RLS due to the resulting iron loss. Sometimes, rubbing a cold cloth on the legs can help relieve symptoms. While moderate evening exercise may also be helpful but more strenuous exercise may worsen the symptoms.

The most common medications for RLS are dopaminergic agents such as levodopa/carbidopa (Sinemet). While these agents act as precursors for dopamine, this medication has side effects including nausea. Other dopamine agonists which are better tolerated have been developed. The most common medications used in this category are pramipexole (Mirapex) and ropinarol (Requip). These agents are very effective in the treatment of RLS. Low dosages of this medication often lead to dramatic improvements of RLS in just a few days. One of the drawbacks of dopamine agonists is that although they can suppress RLS symptoms in the evening, sometimes RLS symptoms come back even worse at unusual times such as the morning. This phenomenon is called augmentation. Augmentation forces changes in dosage or medication type.

Due to the possibility of developing augmentation of symptoms, other treatments have been sought. A highly effective alternative is the opiate medication class. Preparations with oxycodone hydrocodone (Vicodin) have been shown to be effective for this condition. Longer acting opiates such as methadone can also be an alternative. Of course, using opiates to treat a chronic condition can lead to other potential problems, and many patients are reluctant to use a potentially habit forming medication.

The other classes of medications that decrease symptoms are anticonvulsants. Specifically preparations with gabapentin are effective and have a lower side effect profile. An extended duration preparation of gabapentin, Horizant, is FDA approved for this condition.

Nocturnal Leg Cramps

Nocturnal leg cramps occur during sleep, typically in the calf but sometimes in the foot. They are typically quite painful and usually cause awakening. The pain is associated with objective cramping and muscle hardness or tightness in one or rarely both legs. Nocturnal leg cramps occur in an episodic, unpredictable fashion and typically last a few seconds to several minutes although they may persist for 30-45 minutes. The frequency of recurrence is variable but is generally of the order of once or twice a week to every night. Frequent occurrence of leg cramps may lead to a form of sleep maintenance insomnia and possible consequent daytime sleepiness and fatigue. The condition is most common in the elderly but can be experienced at all ages from young childhood on. Males and females appear equally afflicted. While the prevalence of the condition is not well-known, some studies suggest that nocturnal leg cramps affect at least 15-20 percent of individuals at some point during their lifetimes.

Several facilitating and etiologic factors exist. Genetic factors are not well documented although some familial cases have been reported. Pregnancy, diabetes mellitus, excessive exercise, metabolic diseases, various neuromuscular disorders, peripheral vascular disease, and disorders associated with reduced mobility (such as arthritis involving leg joints and the rigidity of Parkinson's disease) are all predisposing factors. Treatment consists of slow stretching of the leg and muscle, massage, or the application of heat, all of which may reverse the muscle spasm. Episodes may sometimes be avoided by taking either clonazepam (0.5-2.0 mg).

Bruxism

Bruxism is the repeated grinding or crunching of the teeth during sleep. The noise is usually very disturbing to a bed partner. Bruxism is frequently associated with a preexisting dental, mandibular, or maxillary condition. Repeated sleep-related bruxism may lead secondarily to excessive tooth wear and decay or periodontal tissue damage and is often associated with temporomandibular pain.

Tooth grinding is often associated with stress and may therefore be seen in otherwise normal individuals. However, it is particularly common in children who have some degree of mental deficiency. Tooth grinding in wakefulness as a habit or tic is often associated with bruxism. The condition has no gender preference. Polysomnographic testing shows repeated phasic increases of masseter muscle tone during instances of grinding. These repeats at a frequency of approximately 0.5 to 1.5 per second in bursts lasting a few to several dozens of seconds and occur mainly in NREM stages 1 and 2, as well as in REM sleep. Some cases in which phasic increases occur exclusively in REM sleep have been described.

Evaluation and treatment should include a comprehensive dental examination and correction of any causal anatomic anomalies. Rare patients have treatable CNS lesions. In most cases, no evident jaw or brain abnormality is found. Appropriate counseling or psychotherapy may reduce stress. In some cases, relaxation techniques or biofeedback have been useful. Many bruxers require a rubber mouth guard over the teeth to prevent further dental and jaw damage.

Conclusion

Restless leg syndrome is a common neurological disorder that can disrupt sleep by causing an unpleasant sensation to move the legs at night. The condition often affects multiple family members to various degrees of severity. Iron deficiency makes the condition worse. RLS symptoms improve quickly with a variety of treatment options. Other sleep related movement disorders occur that are also readily treatable.

"Restless legs is the Rodney Dangerfield of sleep disorders"-Mark Buchfuhrer

CHAPTER 17 Obstructive Sleep Apnea

"Laugh and the world laughs with you; snore, and you sleep alone!" - Anthony Burgess

Case History #1

A 59-year-old male was driving his 11-year-old grandson to school. According to the grandson, the car drifted off the road for no apparent reason, hit the concrete support of an overpass, bounced off, and skidded to a halt in the middle of the road. The grandfather was unconscious with minor head lacerations. His grandson was unhurt. The boy told police investigators that his grandfather's head was nodding just before the crash. The grandfather recovered consciousness in the ambulance. Upon arrival at the hospital, he was admitted to intensive care and observed for 24 hours while numerous neurological tests were conducted, including electroencephalogram (EEG), brain scan, and cardiovascular tests – all the modern tests for cardiovascular disease and epilepsy. Clinically ambiguous minor EEG changes were seen - occasional EEG spikes in the temporal lobe with some slow-wave activity. No neurological abnormality was found. The grandfather was discharged on a trial of seizure medication at a low dose with instructions not to drive for a period of two months and to report immediately if any further episodes of loss of consciousness occurred.

In the first few weeks, the grandfather became progressively apathetic and worried about his condition. He was seen by a psychiatrist who made a diagnosis of depression and started treatment with Prozac (fluoxetine), *a well-known antidepressant. His condition worsened as he lost all interest in things around him. He merely sat in a chair all day long, nodding and dozing. Repeat brain scans revealed nothing. The dose of Elavil* (amitriptyline) *was gradually increased to a high level with no improvement. He was switched to another antidepressant at a very high dose with still no improvement. He was then switched to yet another antidepressant, but no change was observed, except for a marked weight gain. At this point, the grandfather's employers asked him to accept early retirement.*

Two years after the accident, the grandfather's wife read an article on Obstructive Sleep Apnea (OSA) in the Reader's Digest which stated that sleep apnea is characterized by

loud, intermittent snoring. She realized, "That's my husband." She took the article to their doctor who dismissed it. She then called the local sleep disorders clinic and arranged an appointment for her husband. He fell asleep in the waiting room, and his loud snoring could be heard throughout the clinic offices.

The diagnosis of severe obstructive sleep apnea with severe nocturnal heart arrhythmias was conclusively established by a polysomnogram. Because the oxygen saturation of the blood showed severe drops of 50 percent during many apneic episodes, the grandfather was immediately placed on an effective treatment (Nasal Continuous Positive Air Pressure CPAP). In his first follow-up visit, he stated that he had never felt so good, and he had the energy of his (now 17-year-old) grandson. All medications were discontinued. A repeat EEG showed no slowing and no evidence of spiking

. Introduction

A report in 1997 estimated that 93% of women and 83% of men with moderate to severe sleep-breathing disorders (SBD) were undiagnosed. **Obstructive sleep apnea (OSA).** The most common terminal events induced by obstructive sleep apnea are heart attacks, strokes, and accidents. Despite these serious consequences, the general public and the medical profession have yet to deal adequately with this major health problem. The continuation of this state of affairs should be regarded as both unacceptable and reprehensible since effective treatment is readily available and, if instituted soon enough, can often reverse the cardiovascular complications, and provide dramatic relief of the behavioral and psychological symptoms. Since the obstructive sleep apnea problem is so important and public awareness is relatively low, we will describe the illness in detail. Given the high prevalence of the disorder, everyone probably knows someone who has sleep apnea. Obstructive sleep apnea can be easy to diagnose, and severe cases can be identified by asking a few simple questions. After completing this learning experience, we expect that readers will recognize obstructive sleep apnea in family members and friends, and will act to save their loved ones' lives.

What is Sleep Apnea?

The word "apnea" means absence of breathing. Sleep apnea refers to a situation when breathing stops during sleep. The official definition of a **sleep apnea episode** accepted by sleep specialists is a cessation of airflow lasting ten seconds or longer during sleep, and terminated by an arousal or an oxygen desaturation. The ability to wake up when breathing stops is why millions of people with this problem do not die in their sleep. Apnea episodes can be easily seen and heard by anyone who is observing the sleeper. In a sleep disorders clinic, apnea is detected and measured by a polysomnographic sleep test that includes continuous monitoring of airflow and other respiratory variables. In most patients, the number of apnea episodes may vary somewhat from night to night; upper respiratory infections (head colds) and the use of sedating drugs or alcohol can increase the number of apneas. The total number of apneas for the entire night depends also on the amount of time spent sleeping.

Two Types of Sleep Apnea

Polysomnographic testing has identified two main types of apnea. **Obstructive sleep apnea** is defined by the absence of airflow through the nose and mouth despite persistent and increasing efforts to breathe.

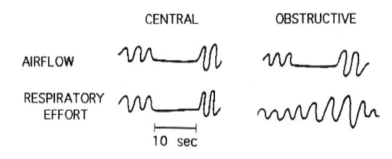

Figure 17-1: Comparison of Central Sleep Apnea and Obstructive Sleep Apnea. The upper tracing represents airflow with the upward direction indicating inspiration and the downward direction expiration.

Central sleep apnea (see Figure 17-1) is characterized by the absence of any effort to breathe and consequently an absence of airflow through the nose and mouth. Another pattern of apnea called mixed (central and obstructive) sleep apnea is simply a combination of the two types in a single episode in which a brief central apnea (no airflow and no effort) gives way to an obstructive phase. Predominantly central sleep apnea is less common; accordingly, the bulk of this chapter deals with the obstructive type.

Cardinal Signs of Obstructive Sleep Apnea

The two most common reasons that a victim of obstructive sleep apnea may consult a physician are (a) fatigue or sleepiness and (b) concerns about snoring, or snoring interrupted by pauses in breathing. Pauses in breathing during which the sleep apnea victim struggles for air are often very frightening to an observer. Sometimes the partner (and sometimes the victim) has learned about obstructive sleep apnea and its association with loud snoring from a newspaper article, a magazine article, or a television program. Almost everyday someone comes into the sleep clinic and claims that their spouse or bed partner snores. Why would they say that if it weren't true?

Symptoms

- Snoring
- Witnessed apneas
- Gasping for breath during sleep
- Sleepiness
- Bed wetting in children, nocturia in adults
- Mood, memory, or learning problem
- Impotence
- Recent weight gain
- Morning headache
- Dry mouth or dry throat in the morning

Signs

- Obesity
- Hypertension
- Crowded oropharynx
- Retrognathia

Prevalence of Obstructive Sleep Apnea

The population prevalence of obstructive sleep apnea has been established. A positive sleep test in obstructive sleep apnea was found in 24 percent of adult males and 9 percent of adult females. That same study found that the vast majority of people denied having any symptoms despite a positive sleep study showing they did have sleep apnea. Only 2% of women and 4% of the men had both admitted having symptoms and were shown to have sleep apnea from the positive sleep study. Since people do not understand or recognize their symptoms they do not realize they have sleep apnea. In 2005, the Institute of Medicine (IOM) stated that sleep disorders are among the biggest unmet problems in the United States.

The precursor of sleep apnea, snoring, is even more commonplace, occurring in over 50 percent of all adults. Habitual snoring is one of the most common symptoms of OSA. Yet the disorder of sleep apnea remains largely unknown to the general public.

While the prevalence of obstructive sleep apnea increases with age, it can occur as early as infancy. A non-random study of several hundred Stanford University undergraduate volunteers from the Sleep and Dreams course found that *nine percent* had obstructive sleep apnea.

Recognizing Obstructive Sleep Apnea and Saving Lives

Recognizing severe obstructive sleep apnea is a life or death matter. Accordingly, we will review how to recognize when someone, including yourself, has this serious and commonplace sleep disorder

Daytime sleepiness:

People with obstructive sleep apnea rarely complain of excessive sleepiness although they can be extremely sleepy. They are more likely to complain about feeling tired instead of sleepy. They may think that fatigue is a part of normal aging and ignore it. A simple question to ask someone you suspect of having sleep apnea is, "do you wake up feeling refreshed?" If somebody says she always wakes up feeling tired no matter how much sleep she gets, there may be something wrong with the quality of her sleep. Sleep apnea is one of the most common reasons for poor quality of sleep among all of us

Snoring

There is a misconception that snoring is normal. If, at the onset of inspiration, the brain fails to instruct the upper airway muscles to contract, or allows them to be too relaxed, airway resistance will be elevated, making the passage of air more difficult. When the throat narrows while we are sleeping but the lungs remain the same size, the work of breathing increases. When air rushes through the narrow throat it creates turbulence which in turn causes the membranes and tissues of the throat, particularly the soft palate and uvula, to vibrate. This vibration produces a sound that we all recognize and associate with sleep - snoring. However, snoring should never be considered normal. When we are awake our breathing is silent. We should not be making noise while we are breathing, including when we are asleep. Snoring simply indicates trouble in airway flow and airway obstruction. Not all snoring is due to sleep apnea, for example; somebody may snore when they have a cold or other causes of nasal obstruction. Some people may have obstructive sleep apnea and not snore at all because very little redundant tissue is in the back of the throat, but it is important for people to understand that snoring is not normal. In fact, the American Academy of Pediatrics recommended that for the care of all children in the United States, parents should be asked if their children snore. If they do snore, the nature of the snoring may require further investigation."

When someone snores so loudly that the noise disturbs others, some degree of obstructive sleep apnea is almost certain to be present. Oftentimes the loudness of the snoring is truly remarkable. For example, a patient once was brought to the Stanford clinic by his wife because a neighbor had come to their door to ask, "What kind of animal are you keeping in your house?" His wife had to admit, "That's no animal. That's my husband!" Unfortunately, while it may be very annoying to a spouse or bed partner, loud snoring, like sleepiness, is rarely regarded as a medical problem. In addition, victims typically see no link between their remarkably loud snoring and their complaints of severe fatigue and sleepiness.

Loud snoring is not only detrimental to the person snoring but also is a problem for others. In many instances, snoring is loud enough to seriously disturb the sleep of a bed-partner, roommate, or even someone sleeping in a different room. Either the snorer or the snorer's companion must repeatedly leave the bedroom to obtain adequate rest.

A recent study at the Mayo Clinic in Rochester, Minnesota has provided dramatic documentation that the sleep of the bed-partner is indeed very disturbed by snoring. In this study, the sleep of the patient and his or her spouse were recorded simultaneously while sleeping in the same room. During the first half of the night, the snoring patient was untreated. During the second half of the night the patient was treated with nasal CPAP (see later section on treatment) and his or her breathing was no longer noisy. During the first half of the night, the spouse's sleep was as disturbed, fragmented, and inefficient as the sleep of an "insomniac," whereas when quiet was restored in the second half, sleep was essentially normal. The investigators whimsically named this the "spousal arousal syndrome."

At best, these results suggest that millions of Americans are in a state of severe sleep deprivation because they share their bed with a victim of sleep apnea or loud snoring. Although daytime evaluations were not carried out, the degree of the nocturnal sleep disturbance certainly suggests that the bed partners of sleep apnea victims could have very impaired daytime alertness. A final thought is that successful treatment of a patient with a bed-partner means that you get two for the price of one.

Mood

Poor sleep can affect mood. People with OSA can be irritable when they wake up in the morning. Lack of sleep can aggravate any underlying mood disorder that a person may have. In addition to aggravating depression, symptoms of fatigue may be misdiagnosed as depression. Further complicating matters is the fact that if people are treated with antidepressant medications, these medications improve sleep apnea to a degree. As a result, someone who is incorrectly diagnosed will feel better after taking these medications because their sleep apnea has improved, further reinforcing the misdiagnosis

High Blood Pressure

An association between always or almost always snoring and arterial hypertension exists. Case controlled studies have shown that the prevalence of sleep apnea among patients with hypertension is 30%. In adults with drug resistant hypertension, the prevalence of OSA may be over 80%. There are now several studies showing that OSA is a reversible risk factor of hypertension. Up to 70 percent of patients with high blood pressure suffer from obstructive sleep apnea. Any clinical investigation aimed at identifying the cause of high blood pressure must include questions and tests to identify the presence or absence of obstructive sleep apnea.

General Appearance

Although obstructive sleep apnea can occur in individuals who are not at all obese or who may even be thin, sufferers commonly have a general appearance that includes a thick neck, large tongue, nasal adenoidal breathing and speech, and marked obesity with a large abdomen. Such an appearance should immediately make one think of obstructive sleep apnea. Jaw size is also a clue. Obstructive sleep apnea is more likely to occur in people who have a small jaw

Unexplained Changes in Behavior

Some individuals with obstructive sleep apnea begin to make more mistakes and to have more difficulty with tasks such as balancing their checkbooks. If these individuals are older, family members or doctors may immediately suspect Alzheimer's disease or another late onset neurodegenerative disease. In addition, individuals may show emotional changes that typically take the form of an increase in irritability and anger. At night, the struggle to breathe is associated with fairly violent thrashing, occasionally with the precipitation of bizarre somnambulist episodes. Occasionally, severe oxygen desaturation will precipitate a grand mal seizure.

Motor Vehicle Accidents

It is necessary to emphasize a simple stark fact as strongly as possible. In addition to cardiovascular complications, sleep apnea can kill in another way. Anyone who has survived an automobile accident due to a lapse of consciousness (most are fatal) is highly suspect for sleep apnea. If the lapse of consciousness episode has occurred in a middle-aged male, then sleep apnea is far and away the most likely cause, much more likely than fainting or epileptic seizures.

How We Breathe When We Are Awake

It is helpful to understand a few things about how we breathe in the waking state before we consider the problems of breathing when we fall asleep (see Figure 17-2). The human breathing apparatus can be divided into three functional components. The first component is the chest (thorax) which includes the lungs, pulmonary blood vessels, diaphragm, rib cage, intercostals muscles, and the accessory respiratory muscles. The second component is the upper airway. This component includes the nose and mouth, nostrils, buccal cavity, palate, pharynx, larynx, and most importantly from the point of view of obstructive apnea, the muscles of the tongue and throat. The third component is the control system - the machinery in the brain that regulates breathing and ensures the amount of air going in and out of the lungs is adjusted to the needs of the body for oxygen and for the elimination of carbon dioxide. Along with the brain, the control system involves nerves and sensory endings in various locations that transmit information about the respiratory cycle and blood gases to the brain, and motor nerves coming from the brain which cause the contraction of the muscles of the upper airway, the rib cage, and most importantly, the diaphragm.

The lungs operate like a bellow; when they expand, air must enter. This is called inspiration. The lungs are expanded by increasing the volume of the thoracic cavity, (which is accomplished by contraction of the diaphragm), causing it to descend while the intercostal muscles raise and steady the rib cage, very much like a bird holding its wings out. Structurally, the rib cage would collapse when the diaphragm descended if the ribs were flexible and did not have the additional support of the intercostal muscles.

When the volume of the chest begins to increase at the onset of inspiration, the air pressure inside the lungs becomes negative with respect to the outside atmosphere and air rushes in through the upper airway. If one attempts to inspire while holding the mouth and nostrils closed, a very large negative pressure is generated with respect to atmospheric pressure.

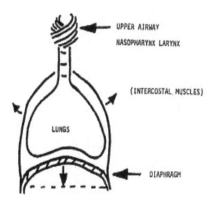

Figure 17-2: Contraction of the diaphragm substantially enlarges intrathoracic volume. The simultaneous increased tone of intercostal muscles tends to hold the rib cage in place. The tissues of the upper airway are not rigid and are capable of collapsing if intrathoracic pressure is sufficiently negative

When the inspiratory phase of the respiratory cycle is complete, the diaphragm relaxes and rises into the thorax. Because the lungs have elasticity, they contract and these two effects expel from the lungs essentially the same volume of air that was taken in during inspiration. This phase is called expiration. Inspiration and expiration are the two alternating phases of the respiratory cycle.

video on this

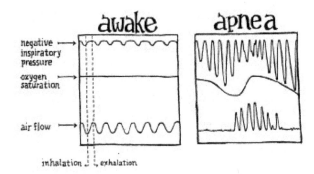

Figure 17-3: Intrathoracic pressure in changes in wake and sleep. In the awake state, the negative pressure required to achieve adequate inspiration is relatively small because airway resistance is low. The pressure tracing in the apnea diagram shows progressively increasing and much larger swings in negative pressure as the apnea victim struggles unsuccessfully to breathe. Airflow increases dramatically and the pressure swings become smaller when the apnea is broken.

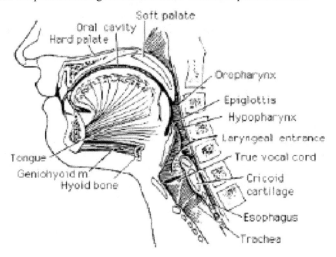

Figure 17-4: Schematic drawing illustrating key features of human upper airway in the sagittal plane. In addition to contraction of the diaphragm, there is a very complex and coordinated control of upper airway muscles during the respiratory cycle. Our tongue is tethered to our lower jaw. At the onset of inspiration, the tone in the tongue muscle (genioglossus) increases, pulling the tongue toward its attachment at the anterior mandible. Several muscles in the upper airway also increase their tone.

Blood flows through the lungs and red blood cells in millions of tiny capillaries adjacent to the air sacs (alveoli) pick up oxygen molecules by attachment to hemoglobin. At the same time, the red blood cells release carbon dioxide which then leaves the body in the expired air. If this gas exchange were not maintained, we would quickly asphyxiate because oxygen is required for proper cellular functioning and carbon dioxide in excess is poisonous to the body.

Air enters the lungs through the trachea and bronchi, which are essentially rigid tubes. Above the trachea and larynx, however, the airway is not rigid at all; it is a complex mixture of muscle, bone, and cartilage. The bones and cartilage of the skull, jaw, and neck form the framework upon which the tissues and musculature of the upper airway are woven. The muscles of the airway are constructed such that the contraction of certain muscles tends to enlarge the airway and decrease the resistance to airflow. These muscles contract at the beginning and throughout the inspiratory phase of the respiratory cycle to maximize airflow (Figure 17-4).

Obstructive Sleep Apnea: Why Breathing Is Blocked

Individuals with obstructive sleep apnea generally have smaller upper airway dimensions than do individuals whose breathing during sleep is entirely normal. Those with smaller upper airway dimensions generally have a smaller jaw along with a larger tongue. Airway size can also be reduced by the presence of tonsils, fatty tissue, edema (swelling), and a host of other anatomical factors. As the size of the pharyngeal airway is reduced, airway resistance increases. In order to maintain adequate air exchange, a greater effort must be made to inhale the same amount of air into the lungs. Smaller airway size constitutes the first and most important predisposing factor for obstructive sleep apnea.

As noted earlier, when human beings breathe air into their lungs (inspire), the tongue muscle contracts, which pulls the tongue forward. Other muscles, primarily fibers going from the skull to the sides of the throat, also contract during inspiration and hold the throat open, analogous to tent poles holding a tent. The vigor or tone of the throat muscle contractions is reduced during sleep, but more in some people than others. This reduced vigor or tone is the second important predisposing factor. Thus, an overly strong inspiratory effort through a smaller airway can suck the relaxed muscles and throat tissues closed, instead of sucking air into the lungs. This airway collapse initiates an obstructive sleep apnea episode. Figures 17-5 and 17-6 a, b, and c will help clarify the pathophysiological mechanism. On many occasions, the collapsing throat and tongue muscles do not entirely close the airway. Inspiration continues, but with a markedly reduced volume of air. The event is designated a **hypopnea** and is tabulated along with apneas if blood oxygen falls, inspiratory effort increases, and the episode is terminated by an arousal or oxygen desaturation. Thus, with regard to associated bodily change and interruption of sleep, a hypopnea has essentially the same clinical significance as an apnea.

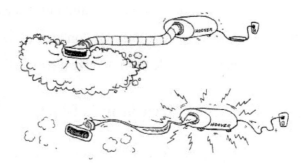

Figure 17-5: The smaller the tube, the higher the resistance; the greater the likelihood of collapse. The top vacuum cleaner has a large hose and is able to work in a relatively serene manner. The lower vacuum cleaner struggles and strains and the motor grinds and squeals because the hose keeps collapsing. The smaller the airway, the higher the resistance; the higher the resistance to airflow, the greater the likelihood of airway collapse.

An obstructive sleep apnea episode is identified on the polysomnogram when the recording shows that airflow through the nose and mouth stops while respiratory effort continues and progressively increases. Likewise, a hypopnea is identified when the polysomnogram reveals a limited airflow volume with increasing respiratory effort.

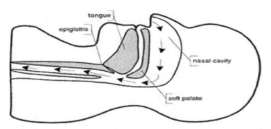

Figure 17-6a:

Schematic drawing illustrating normal breathing during sleep. Airway resistance is low.

Figure 17-6b:

Schematic drawing illustrating increased airway resistance.

The airway size is reduced, but not completely blocked. This partial blockage increases inspiratory effort and the increased velocity of airflow causes turbulence, which causes the soft tissue to vibrate creating the noise known as snoring.

Figure 17-6c

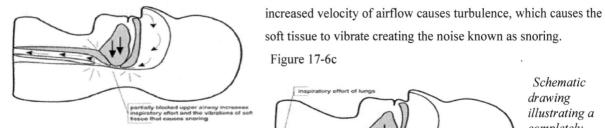

Schematic drawing illustrating a completely obstructed airway.

Degrees of Severity of Obstructive Sleep Apnea

In 1976 at Stanford University, Professors Christian Guilleminault and William Dement introduced the Apnea/ Hypopnea Index as a way of characterizing the severity of obstructive sleep apnea. The **Apnea/Hypopnea Index** (AHI) is the sum of the entire night's total number of apneas plus hypopneas terminating in either arousals or oxygen desaturation, divided by the hours of sleep. The AHI that was developed here at Stanford is now the universally accepted quantitative measure of sleep apnea severity. Taking into account that an occasional apnea can occur during a night of sleep in anyone and breathing is notably irregular in REM sleep, Guilleminault and Dement decided to establish an AHI of 5 as the minimal number for identifying sleep apnea as a diagnostic entity. For clarity, an AHI of 5 represents a nightly average of 5 apneas or hypopneas terminating in arousals per hour of sleep. It should be noted that the prevalence figures of obstructive sleep apnea in 24 percent of adult males and 9 percent of adult females utilize an AHI of 5 as the cutoff for identifying apnea victims.

Mild sleep apnea is defined by an AHI of 5 to 15, **moderate sleep apnea** is defined by an AHI of 15 to 30, severe sleep apnea by an AHI of 30 to 45, and finally, very **severe sleep apnea** is defined by an AHI of 45 and above. It is generally true, however, that as the apnea index increases, the number and severity of symptoms and complications also increases. In addition to characterizing severity, the AHI is also the simplest number to analyze in comparing the severity of apnea from night to night or from person to person.

Effects of Apneas on Sleep

Obstructive sleep apnea episodes are associated with EEG arousals immediately before breathing resumes. Typically, the arousal is very brief and is rarely recalled. It is common for a victim to complain that he or she sleeps nine or ten hours a night and is still very tired in the morning while being completely unaware that each night's sleep is interrupted by apnea episodes four to five hundred times. There is a tendency for obstructive apneas to worsen in REM sleep. This is because the general tendency to arouse is reduced in REM periods and the arousal response to decreasing blood oxygen is blunted. REM sleep is characterized by decreased muscle tone which causes airway muscle tone to be more collapsible and more predisposed to obstruction. In addition, the overall amount of REM sleep is often considerably reduced in patients with obstructive sleep apnea because returning directly into REM sleep after an arousal is much less likely than returning to non-REM sleep. Together, the increased duration of obstructive apneas during REM and the greater likelihood of entering into NREM after an awakening often greatly reduces the amount of REM sleep obtained.

Heart and Blood Vessels

Sleep apnea episodes are often associated with a drop in blood oxygen saturation followed by a sharp rise to near normal levels as breathing is resumed. The severity of these changes in blood oxygen depends upon the duration and the type of apnea - the longer the duration of the apnea, the greater the drop in oxygen saturation. Obstructive sleep apnea can be associated with life-threatening cardiovascular complications. Many patients will have some degree of high blood pressure. These blood pressure elevations are always present during sleep and frequently reach very high levels. Heartbeat irregularities are almost always present during sleep and range from mild to severe and life threatening. Blood pressure and cardiac abnormalities induced by the repetitive sleep apnea episodes are initially seen only in sleep but gradually begin to be present when the victim is awake in the daytime. Obstructive sleep apnea is now recognized as a reversible risk factor for hypertension.

Mortality Risk

Snoring alone is associated with sudden death. Habitual snoring quadruples the risk of sudden death in the early morning hours. Due to continuing failure on the part of the healthcare system to identify and diagnose obstructive sleep apnea and due to lack of sufficient federal public health presence, long-term outcome data on sleep apnea are very sparse. One study dividing untreated patients into those with an apnea/hypopnea index below 20 and those with an index above 20 has shown a highly significant difference in mortality over the course of an eight-year follow-up. In principle, all patients with sleep apnea should be given treatment as soon as possible.

In the early 1970s, chronic tracheostomy (an opening into the trachea just above the breastbone) was the only effective treatment for severe sleep apnea. Many patients who were diagnosed refused to accept this treatment. It is clear that these patients had a markedly increased mortality risk because nearly all were dead within ten years regardless of their age when they were first seen. Conversely, most patients treated with chronic tracheostomy were alive and healthy as long as the study could follow them.

Today, appropriate treatment options can effectively control the apnea episodes and may even reverse cardiovascular complications. Now that sleep centers are all around the country, there is no reason that someone with sleep apnea cannot be treated.

Case History #2

While I was investigating the facilities where a hearing of the National Commission on Sleep Disorders Research was to be held, the woman who was showing me the facilities asked what the hearing was about. When she heard that it was to be about sleep disorders, she said "Oh, I should attend, I fall asleep all of the time." She was obese and had high blood pressure. The sleepiness had begun insidiously ten years prior to the present encounter. Upon questioning, she admitted that she snored very loudly every night. Amazingly, eighteen months prior to our encounter, this woman had fallen asleep while driving and had been involved in a head-on collision on an urban highway. She was hospitalized in intensive care with severe lacerations to her left shoulder and a damaged brachial artery. Her arm was saved by vascular surgery. Because she had experienced a lapse of consciousness, she underwent numerous neurological and cardiovascular tests, all of which were negative. During her hospital stay, she was seen by perhaps 20 different physicians. Not one doctor asked about her sleepiness or snoring. She was discharged without medication. Although she continued to drive, none of her friends would ride with her. She remained, until encountering me, at high risk for another falling-asleep-at-the-wheel episode. She fell asleep frequently at work. I helped arrange an appointment for her to be seen at a local sleep disorders clinic. She was found to have severe sleep apnea and was immediately treated. Her sleepiness disappeared over the course of the first week. She was amazed at how alert and energetic she felt. - William C. Dement

Natural History of Obstructive Sleep Apnea

The time course of obstructive sleep apnea depends on a number of factors including the severity and type of apnea episodes, the patient's age and gender, and the presence and degree of obesity. In addition, the sequence in which symptoms appear may vary. The appearance of secondary cardiac failure often heralds the onset of a rapid deterioration. Although sleep apnea occurs at all ages, sparing neither infants nor the elderly, more than half of all patients are 35 years of age or older when first diagnosed. In most adult patients, however, abnormal respiration during sleep indicated by a history of loud snoring can precede the development of unambiguous daytime sleepiness and clinical diagnosis by many years. It is clear that severe obstructive sleep apnea does not develop overnight. In many cases, it is possible to pinpoint the beginning of severe sleepiness within a period of a few months or a year. At the present time, it is difficult to predict what will eventually happen to individuals who have either mild obstructive sleep apnea or only a tendency to snore very loudly without evidence of apnea. However, there is no question that such individuals have a much greater likelihood of developing severe obstructive sleep apnea than do individuals who breathe completely easily and noiselessly (normally) when they are sleeping. One possibility that has been suggested, but not conclusively proven, is a vicious circle mechanism in which repeated awakenings lead to severe sleep deprivation which worsens the apnea. This increased apnea then causes even more awakenings and further sleep deprivation.

Treatment of Obstructive Sleep Apnea

Treatment in the 1970s

There are several treatments for OSA. The first successful treatment for obstructive sleep apnea, introduced in 1969, was chronic tracheostomy. In recent years, this treatment has been largely replaced by other approaches. In the early days, however, when many patients had developed advanced, very severe sleep apnea and were literally at death's door, the reversal of the clinical picture after tracheostomy was nothing short of miraculous. Moreover, the availability of any effective treatment was a godsend. Today, there is only a very small number of patients for whom chronic tracheostomy may be necessary.

Tracheostomy is the surgical creation of a tracheal opening at the base of the neck that bypasses the upper airway where the obstruction occurs and permits air to flow freely in and out of the lungs. During the daytime, the opening (stoma) is closed with a cap, and air flows through the nose and mouth in a normal fashion as the patient inhales and exhales. Capping the tracheostomy also allows the patient to talk normally (the larynx is located above the tracheostomy site). When the patient goes to bed, the cap is removed to open the tracheostomy which allows breathing during sleep to be completely normal.

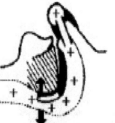

CHAPTER 17 Obstructive Sleep Apnea

Figure 17-7: This figure shows the mechanism of upper airway occlusion and its prevention by nasal CPAP. When the patient is awake (top), muscle tone prevents the collapse of the upper airway. During sleep (middle), the tongue and soft plate are sucked against the posterior oropharyngeal wall. The bottom figure illustrates how CPAP keeps the upper airway open.

At the time the only alternative treatment to tracheostomy was a conservative treatment approach where patients were encouraged to lose weight, sleep off their back and avoid alcohol close to bedtime. While these are standard recommendations for patients with very mild OSA, for most patients with significant OSA, these treatments are not adequate.

Positive Airway Pressure Therapy

In 1981, the method of **continuous positive airway pressure** (CPAP) was invented. CPAP applied through the nostrils was invented by Australian pulmonary specialist Colin Sullivan and his colleagues (see Figure 17-7.) They reported in 1981 that CPAP completely restored normal breathing during sleep. This has proven to be the most successful treatment for OSA to date. Nasal CPAP acts predominantly by providing a physical "pressure splint" for the upper airway. This prevents the inspiratory effort and associated negative pressure in the airway from causing the throat to collapse. The air pressure generated by the CPAP machine must be adjusted so that it prevents airway collapse in all sleep stages and throughout the entire night. It is particularly important to obtain a good sample of REM sleep because a higher pressure is often required to abolish apneic events during REM periods. Much developmental work has been done by vendor companies to improve the CPAP equipment, particularly the efficacy and comfort of the nasal mask.

Since their introduction, there have been many computer-based technological advances with CPAP machines. Computer chips and software have been introduced that allow some machines to react to changes in the airway over the course of the night. These so called "smart CPAP" machines can now continuously and immediately adjust the pressure to whatever level is required to maintain an open airway at any given moment. This immediate feedback ability results in lower, more comfortable pressures over the course of the night (rather than one fixed at the highest pressure needed). These lower pressures are associated with improved CPAP compliance. The machines now can connect via the internet to allow a two-way communication with the healthcare provider. The sleep data can be viewed and the machine settings can be viewed remotely.

The long-term use of nasal CPAP requires considerable understanding and com. mitment by the patient. Recent studies with computer-based treatment monitors concealed in the CPAP apparatus have shown that patients tend to use CPAP much less than they say they do. This has raised serious concerns about long-term compliance with nasal CPAP treatment. Non-compliance associated with continuing sleepiness is of great concern because of the many situations, e.g., driving, where one must be fully alert to be safe. At the present time, it is assumed that treatment with CPAP must continue on a nightly basis for the remainder of the patient's life.

The delivery and maintenance of CPAP equipment is customarily carried out by a home health care company or at a sleep disorders clinic. Companies will often demonstrate several CPAP machines including the nasal mask and headgear.

From its initial introduction in the 1980's, CPAP has dramatically changed in new modalities of pressure delivery using quieter and smaller machines. Modern CPAP machines are built for travel and include international voltage converters and automatic altitude adjustments. One of the first new modalities was a bi-level device. One of the early criticisms of CPAP was that it helped patients inhale, however, it was difficult to breath out against the pressure. The bi-level devices delivered a lower pressure on exhalation making devices more comfortable to use and more effective in patients with wide range of conditions including neuromuscular disease. The "smart CPAP" mentioned above is also referred to as autotitrating CPAP. This has become the new standard treatment for patients with OSA, delivering and adjusting pressure on a breath-by-breath basis. Autotitrating technology is capable of adapting to any changes in a patient's condition over time, such as weight gain/loss and pregnancy. Autotitrating technology has also been integrated with bi-level devices and now allows for autotitrating bi-level pressures. These devices can also include mandatory timed respiratory rates, which can act as non-invasive ventilators for home use. These devices would be used by patients whose medications cause both central and obstructive sleep apnea. Finally, the most recent advances in positive airway devices include servoventilation technology which can be used in complex cases of sleep apnea. These devices are particularly useful in patients with a combination of sleep apnea syndrome and congestive heart failure.

The Miracle of Treatment

The fountain of youth is in your bedroom. People, when they are treated, feel rejuvenated. Their memory may improve, libido may increase (more than one wife has asked us to turn off the machine), and patients overall feel better when their sleep improves. In many cases, obstructive sleep apnea victims are not aware of their illness, even when it is very severe. Because obstructive sleep apnea develops slowly, victims tend to have difficulty comparing their current condition of severe fatigue, lack of energy, and tendency to nod off to the experience of high energy and youthful exuberance of years earlier. They do not realize how much they have deteriorated. Typically, most patients are astounded at how much better they feel after treatment has been initiated. Once successfully treated, they can fully understand that they were indeed severely impaired.

The rapidity of the improvement and its dramatic nature are almost always tremendously rewarding to the patient. Quality of life is restored, the risk of accident is substantially reduced, and the risk of developing cardiovascular disease is reduced, or if disease is present, it is often partially ameliorated or entirely reversed. Who could possibly want to be implicated in any way in withholding this miracle deliberately or inadvertently from victims of obstructive sleep apnea?

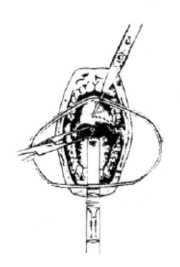

Firgure17.8 :Uvulopalatopharyngoplasty (UPPP), the oropharynx is exposed and incisions are made starting at the anterior tonsillar fossa.

Surgical Treatment: Uvulopalatopharyngoplasty (UPPP)

The treatment of obstructive sleep apnea by the surgical procedure, **uvulopalatopharyngoplasty** (UPPP), was reported by an ENT surgeon, the late Shiro Fujita and his colleagues, also in 1981. In subsequent years, UPPP, which had great initial popularity because it was publicized as a successful alternative to chronic tracheostomy, has not proven to be consistently effective. In general this surgery should be avoided as a standalone treatment.

However, a few surgeons have developed more sophisticated and effective surgical procedures (see Surgical Treatment at Stanford University in later section).

Surgical treatment is indicated for moderate and severely ill patients who, for whatever reason, cannot tolerate nasal CPAP. In the surgical procedure of uvulopalatopharyngoplasty (UPPP), the uvula and portions of the posterior soft palate, anterior and posterior tonsillar pillars are resected (cut out).

If the patient has not previously had a tonsillectomy, the tonsils and adenoids are also removed. Historically, UPPP was widely used until follow-up sleep laboratory studies began to accumulate results indicating that its efficacy was marginal.

The results of surgical treatment of obstructive sleep apnea must always be evaluated by an adequate follow-up sleep laboratory study or studies. Good medical practice requires that there be no exception to this principle. UPPP is considered successful when the apnea/hypopnea index is reduced to 10 or less, and ideally below five. This goal is typically achieved in only about 10 percent of UPPP procedures. In some instances, although snoring disappears, the patient's apnea condition may become worse! The worsening is thought to be a consequence of post-operative scarring. Amazingly, many ENT (ear, nose, and throat) surgeons still perform UPPPs as the sole treatment for obstructive sleep apnea patients. This surgery has gained a reputation of being painful and often ineffective.

Surgical Treatment at Stanford: Phased Airway Reconstruction

In the 1990's, Drs. Nelson Powell and Robert Riley, surgeons working with the Stanford University Sleep Clinic, successfully treated large numbers of patients utilizing a combination of ENT and maxillofacial procedures. These procedures require, in addition to ENT expertise, the surgical skills and experience of an endodontist. Collectively, the procedures are called airway reconstruction and range from simple resection of nasal polyps to major surgery on the jaws.

The Stanford approach called achieves very favorable results while at the same time requiring only two hospital days and limited surgical costs. This surgical treatment combines UPPP and advancement of the tongue (genioglossus advancement) into a single procedure. Advancing the tongue forward typically enlarges the anterior-posterior diameter of the upper airway. The advancement is accomplished by excising the small rectangle of anterior mandible to which the tongue muscle is attached (near the chin), pulling the rectangle of bone out of the mandible, rotating it so it cannot fall back, and finally, fixing and trimming it (see Figures 17-9).

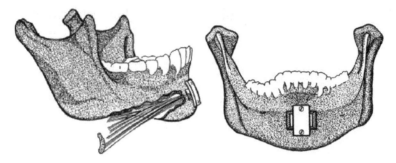

Figure 17-9: This shows the technique for moving the tongue forward by pulling the insertion point in the mandible forward. The bone fragment is rotated it cannot fall back.

Of the apnea patients who have had this surgery procedure performed by the Stanford surgical group, 60 to 70 percent improved. However, this procedure does not help all patients.

A more sophisticated surgery can be performed to increase the size of the upper airway called maxillomandibular advancement. By moving the upper jaw (maxilla) and lower jaw (mandible) forward, the entire airway can be enlarged. This procedure serves as an effective surgical treatment for obstructive sleep apnea. The lower half of the face is moved forward to create more space in the back of the throat. Maxillomandibular advancement surgery is at least 3-4 hours in duration. Hospitalization is usually two to three days and in general the patient can return to work in four weeks. This type of surgery is only available in very specialized centers.

More recent advances in sleep apnea surgery have been developed. The surgical techniques for both the genioglossus and maxillomandibular advancement have been refined. For many patients, the roof of the mouth (hard palate) is too narrow and the tongue does not fit into this space. The tongue subsequently falls back and obstructs the airway. So simply moving the tongue forward with the above techniques may not be sufficient. In these situations, a widening of the space may be done with a surgical technique called maxillomandibular expansion. These different surgeries may be combined in an individual. In these situations, a dentist and ENT may work together or the primary surgeon has gone through both dental and surgical training.

A new surgical approach has been developed and approved for patients with OSA that do not tolerate CPAP. A hypoglossal nerve stimulator can be implanted in a patient to help move the tongue forward. The device is like a cardiac pacemaker, but instead of connecting to the heart the device is connected to a nerve in the tongue, the hypoglossal nerve. The device is turned on via a remote control when the patient goes to bed. The device will synchronize with the breathing pattern to stimulate the tongue to move forward when the patient inhales. The device needs to be turned off when the patient wakes up.

Oral Appliances

Oral appliances are an established treatment option for snoring and mild OSA. It is a relatively simple and reversible approach to treatment. Oral appliances provide, a noninvasive approach to treatment and work by increasing upper airway space. A variety of dental appliances have been developed for the treatment of obstructive sleep apnea. Such an appliance is usually a mold of the teeth made in such a way that when it is in place, the lower jaw is moved forward and held in this position while the patient sleeps. Sometimes a tongue-retaining variant is used that simply pulls the tongue forward. In every case, the goal is to enlarge the airway by moving the mandible or tongue forward.

With collaboration and good communication between the dentist and the sleep clinician, many patients with snoring or obstructive sleep apnea-hypopnea can be treated effectively with oral appliances.

Other Treatments

Some patients with sleep apnea remain sleepy despite the appropriate use of CPAP. In this situation, the FDA has approved as an adjunct medication the use of modafinal to help the patient stay awake. Modafinil is not a substitute to CPAP; it is to be used in addition to it.

Due to the inherently cumbersome nature of CPAP devices, oral appliances and the risks of surgery there is a constant search for treatment for sleep apnea. Given the potential of billions of dollars involved in developing a successful treatment, it is not surprising that Silicon Valley entrepreneurs are aware of sleep apnea. Newer technologies that have been developed but are not mainstream treatments at this time. The most notable of these low technology treatments is playing the aboriginal instrument the didgeridoo. Also, various devices are available to force a person to sleep off their backs. These positional devices may be worthwhile in people who only

CHAPTER 17 Obstructive Sleep Apnea

have mild sleep apnea or are only focused on snoring reduction. We can look forward to future innovation in the treatment of sleep apnea as society's awareness of the condition grows and as its treatment is more greatly appreciated. Ultimately, the ideal treatment would prevent sleep apnea from developing.

After Diagnosis and Treatment: The Internet and/or Patient Support Groups

When there is a strong presumption that a person has obstructive sleep apnea, he or she should be referred to a sleep disorders clinic. A thorough explanation of the problem is important. It is also key to emphasize the progressively serious consequences if ignored. Once diagnosed and treated, the most common outcome is that an individual will be living with nocturnal CPAP for the foreseeable future. However, it is not always possible for a person who has the symptoms of sleep apnea to go directly to a sleep disorders clinic. In rural communities, the nearest sleep clinic may be hundreds of miles away. An alternative is to explain the situation to one's primary care physician and to insist on a referral.

The presumptive sleep apnea victim should also inquire at the sleep clinic if there is a sleep apnea patient support group in their vicinity. The improvement in the lives of these patients can be so dramatic that once they are treated they want to help others with sleep apnea. Such groups have names such as AWAKE, http://sleepapnea.org/support. html. This acronym stands for Alert, Well, And Keeping Energetic! The volunteer members exchange information, share experience and tips on living with nasal CPAP, and host speakers from the sleep research and sleep disorders community. When someone who has been diagnosed feels uncertain about treatment, talking with other apnea patients about their experience can be very helpful. In recent years, communication via the Internet has to some extent reduced the need for patient support groups.

In addition to responding to the needs of patients, groups like AWAKE can help increase community awareness, foster favorable insurance reimbursement policies, and advocate for sleep-related policy measures. For example, in years past, the AWAKE groups exerted political advocacy for Congress to implement the recommendations of the National Commission on Sleep Disorders Research. A new agency, the National Center on Sleep Disorders Research, now exists in the National Institutes of Health in part as a result of this type of grass roots advocacy.

Case History #3

Thomas W. was a corporate executive whose salary had been in the high six figures. He had begun to suffer from tiredness and fatigue about ten years earlier. His initial effort to obtain medical help resulted in a diagnosis of depression. Various antidepressant medications were prescribed, but his condition only worsened. Shortly thereafter, he began to have an increasingly difficult problem with high blood pressure that did not respond to very aggressive treatment. His fatigue finally reached a point where he was unable to meet the demands of his high-level job. He could not get things organized, his thinking was impaired, and he frequently canceled meetings because he was too tired. He was finally asked to leave his company (fired). After he had been unemployed for three years, he was at the end of his rope. His marriage was on the rocks because of his irritability and apathy and he was beginning to wonder if life was worth living. Although he had endured ten years of suffering and countless visits to various medical specialists, his illness remained unrecognized.

In 1996, he had the extraordinary good fortune to be one of five patients who were selected at random from a large group of patients being seen by Stanford Internal Medicine specialists to try out an ambulatory sleep-testing device. We were quite certain that we had identified someone with severe sleep apnea after looking at him and asking three or four questions. The ambulatory device confirmed our suspicions. Treatment was quick and easy, and his physical and mental status improved dramatically. He was interviewed after he had been on CPAP treatment for three weeks.

"How are things going on the CPAP?" he was asked. "Absolutely terrific," he replied. "I mean, the difference in the quality of my life is amazing. I don't know how I can possibly measure it. I really just thought that I was getting older, and that's the way life was going to be. But now it's really not."

"Looking back at the quality of your life before treatment," he was asked, "how impaired do you think you were?" "Really a lot. I would say 50 percent, sometimes nearly 100 percent. The quality of my life before this treatment was really awful. How can I describe it? It was just so difficult to make decisions and to get things done. I just didn't have the power or the energy to deal with my life. Now my blood pressure is controlled with about a quarter of the medication I was on previously, with no diuretic at all. It's just terrific I used to spend a lot of time watching

TV and dozing off. Now I have the energy to pursue other interests. I just bought my first computer, and I'm reading again now that I can stay awake to enjoy it. It's really terrific."

We must never forget that for every sleep apnea victim who has been restored to normal functioning and health, there may be a hundred who are disabled and dying without knowing why. This is unacceptable. We have always felt that patients who have lost years of their lives unnecessarily because their obstructive sleep apnea was not recognized should be extremely angry.

The medical profession finally got a good taste of this justifiable anger from Thomas W. "Your diagnosis of sleep apnea was made completely by chance," he was reminded. "You probably had apnea for eight to ten years. And every one of your doctors failed to recognize the symptoms. How do you feel about this?" "Really angry," he immediately replied. "Bitter. I mean, it's inexcusable and gross incompetence. I'm mad as hell. I really don't understand how any doctor couldn't diagnose this. It's easier to diagnose than a broken arm for heaven's sake. You ask a few simple questions and you don't need an X-ray. The thought that I spent so many years with such a low and deteriorating quality of life, and feeling really miserable, really, really aggravates me." "And to think about how many other people are walking around right now seeing their physician for these symptoms and not being recognized is inexcusable. If I had to mortgage my house to pay for this treatment, I would have done it in a second. It would have been an easy decision. The quality of my life in the past was really terrible, and now it's really terrific." In addition to everything else, Thomas W. had lost about $2 million dollars of salary. He decided to initiate a lawsuit against Stanford University putting Stanford sleep specialists in the very uncomfortable position of maybe having to choose sides. As it turned out, Stanford dodged the bullet. The angry but miraculously improved patient moved to the East Coast to accept an outstanding job offer. He was soon too busy to pursue his desire for justice.

Summary

Obstructive sleep apnea is so prevalent in adults that it must always be one of the first considerations when confronted with the conditions described in the foregoing material. The only reason that physicians can say they do not see this very often, or that it is not a problem, is because they do not think of it and do not look for it. Remember, more than 30 million Americans have sleep apnea and many millions more internationally. The overriding goal of this chapter is to put obstructive sleep apnea on the front burner in the minds of everyone, and we do mean everyone! We hope that the general public will eventually become more aware of this highly prevalent condition, and will know that snoring and fatigue are its cardinal symptoms. Compared with any single disease, obstructive sleep apnea could well be the number one treatable healthcare problem in America. Sleep apnea should never be overlooked by anyone! Any person who says that unintended sleep episodes are interfering with his or her life and who is a very loud snorer is almost certain to have this problem. Their lives can be saved!

"The most important item in your makeup bag is a good night's sleep." Tyra Banks

CHAPTER 18 Narcolepsy

"It's a normal thing happening at an abnormal time." -Mali Einen

Introduction

Led by Dr. Emmanuel Mignot, Stanford University, is the leading center for the study of narcolepsy in the world. Narcolepsy is a neurological sleep disorder characterized by excessive daytime sleepiness, cataplexy, sleep paralysis, and hypnagogic hallucinations. Patients have disturbed nocturnal sleep. The most unique of these symptoms is cataplexy, which is the sudden loss of muscle tone associated with emotions such as laughter. Attacks of cataplexy can result in the person falling down. These symptoms seemed to be related to REM sleep and are further described below.

The cause of narcolepsy has been a mystery for many years but work led by Stanford has found some possible explanations. Although the exact cause is not entirely clear, research at Stanford and beyond supports the theory that narcolepsy is an autoimmune condition. An autoimmune condition is a condition in which the body's immune system, which normally fights infections, has incorrectly attacked a normal part of the body. There are several other well-known autoimmune conditions. For example, juvenile diabetes, rheumatoid arthritis, and lupus are all considered autoimmune conditions. In the case of narcolepsy, the body seems to attack a tiny portion of the brain located in the hypothalamus. This portion of the brain produces a neurotransmitter that is alternatively called hypocretin or orexin. The impairment of the brain's natural hypocretin/orexin activity from an autoimmune process seems responsible for narcolepsy's symptoms. Our body's cells have a complex system of superficial cellular markers that the immune system uses to distinguish normal cells from foreign cells. These cellular markers include a group of markers referred to as **human leukocyte antigens (HLA)**. A specific variant of HLA called DQB1*0602 is often present in patients with narcolepsy. A discovery at Stanford found that the majority of patients with narcolepsy have abnormally low or absent hypocretin levels in the cerebral spinal fluid. These low levels may be due to the autoimmune attack previously discussed. Further evidence of an autoimmune mechanism's role in narcolepsy is supported by a cluster of narcolepsy outbreaks in Europe associated with a specific flu vaccine in the fall of 2010.

Historically, the word narcolepsy was first coined by Gélineau in 1880 to designate a pathological condition characterized by irresistible episodes of sleep of short duration recurring at close intervals. In the same article, he wrote that attacks were sometimes accompanied by falls or "astasias". Many years later the combination of cataplexy, hypnagogic hallucinations, sleep paralysis, and excessive daytime sleepiness was referred to as the "tetrad" of narcolepsy. Narcolepsy is not a rare condition. The prevalence of narcolepsy has been calculated to be approximately 0.02–0.05% of the general population. Narcolepsy affects more than 250,000 Americans. Age at onset varies from childhood to adults in there 50's, with a peak in adolescents and young adults. This is a condition that may be first diagnosed when people are in college. It generally appears in early adolescence and endures throughout the life of the patient. Males and females are equally affected. First degree relatives of narcoleptics have a moderately increased risk of developing the disease. However, the genetics are complicated. For example, if a narcoleptic has an identical twin, the twin sibling will probably not develop narcolepsy. Genetics predisposes a person to narcolepsy but additional factors are involved for the disorder to become manifest.

A capsule description of narcolepsy is that the victims are (a) sleepy all the time, day after day, month after month, year after year and (b) have episodic attacks of muscle paralysis (cataplexy) typically triggered by strong emotion.

The discovery of REM sleep and a description of its normal physiology, particularly muscle paralysis, were crucial developments in understanding the manifestations of narcolepsy and their pathophysiology. The involvement of REM sleep processes in narcolepsy was first demonstrated in 1960 when a REM sleep abnormality, the **sleep onset REM period**, was described in narcoleptic patients. This discovery led directly to the important realization that cataplexy, hypnagogic hallucinations, and sleep paralysis are manifestations of REM sleep processes occurring in an abnormal manner.

Case History

Joyce T. was a 16-year-old junior in high school. During an argument with a fellow student she felt a wave of anger. At the same time, she became aware of a strange weakness in her knees as if they were going to buckle and cause her to fall. A few weeks later while she was talking with several friends, someone said something hilariously funny. She started to laugh and she lost control of her muscles. She fell to the ground and was unable to move. She and her friends were very frightened. The paralysis disappeared after a few seconds, however, and she was able to get up, apparently quite normal.

She mentioned this incident at her next doctor's appointment. His response was that young women have a tendency to faint and that she should not worry unless it continued to happen. Around this time, she also had the very embarrassing experience of falling asleep in her morning psychology class. Although she seemed to be more tired than usual, she had never completely fallen asleep in class. The instructor made a big issue of it, which was very upsetting to Joyce. She began to notice that when she was angry, or when something was funny, she seemed to lose control of her facial muscles and sometimes her jaw would drop. At other times, her arms and legs would feel very weak, but she would not actually fall to the ground. Once, when she started to laugh at a joke, she dropped the soft drink she was holding. In addition, although she was worried about impending exams, she seemed unable to get things done and to stay motivated as she had in the past. One night, when she was falling asleep, she realized she could not move and saw a stranger coming into the bedroom. She tried to scream. She was utterly terrified. Suddenly, she was able to move and she did scream. Her parents came running into the room and were very concerned. When Joyce told several of her teachers about the frightening things that were happening to her, one of them immediately recognized that her problem was a sleep disorder. (This mysterious constellation of problems could have gone on for years while Joyce T.'s performance in school and quality of life steadily deteriorated. Fortunately, her high school was close to Stanford University). An appointment at the Stanford University Sleep Disorders Clinic was arranged. Following sleep tests, Joyce T.'s problems were diagnosed as narcolepsy and she received effective treatment. Her schoolwork returned to its previous high level and she eventually graduated with honors. She was accepted by the college of her choice and as of this writing has graduated from law school and is practicing law.
-William C. Dement

Sleepiness

Sleepiness is the most troublesome symptom of human narcolepsy. The patients' constant sleepiness is extremely disabling. The sleepiness can usually be momentarily relieved by activity or other stimulation. The sleepiness is increased by alcohol, inactivity, and lack of stimulation.

Sleepiness is usually the first symptom of the narcoleptic tetrad to appear, commonly in adolescence, and it is typically gradual and insidious in its onset. Narcolepsy can begin at a later age, but less commonly does it begin earlier. Sooner or later, the patient with narcolepsy will fall asleep in an extremely unusual situation such as in the middle of a meal, or while actively conversing with someone which cannot be dismissed as normal sleepiness. Driving is also a major problem for someone who has untreated narcolepsy. If the ability to stay alert while driving is threatened, the symptom must not be ignored.

Cataplexy

Cataplexy is characterized by the sudden loss of muscle tone while awake, typically triggered by a strong positive emotion such as laughter or surprise. It can also be triggered less commonly by anger or fear. Cataplexy is virtually an exclusive symptom of narcolepsy. When an experienced clinician witnesses a cataplectic attack, the diagnosis of narcolepsy can be established without waiting for a confirmatory sleep study. Narcoleptic patients remain awake during the attack and are able to remember the details of the event afterwards. The episodes are typically brief and may last only a few seconds. Some patients can have other narcoleptic symptoms manifest during an episode of cataplexy, such as hypnagogic hallucinations and sleep paralysis, or they may simply fall asleep.

Cataplexy may involve only certain muscles or the entire voluntary musculature. Typically, the jaw sags, the head falls forward, the arms drop to the side, and the knees buckle. The severity and extent of cataplectic attacks can range from a state of absolute powerlessness, which seems to involve the entire body, to no more than a fleeting sensation of weakness. The patient may complain of blurred vision. Respiration may become irregular during an attack, which may be related to weakness of the abdominal or intercostal muscles. Complete loss of muscle tone, which results in a fall with risk of serious injuries, including skull and other bone fractures, may happen during a cataplectic attack. The attacks may also be subtle and not noticed by nearby individuals. An attack may consist only of a slight buckling of the knees. Patients may perceive this abrupt and short-lasting weakness and may simply sit or stand against a wall.

Speech may be slurred owing to intermittent weakness of the throat muscles. If the weakness involves the arms, the patient may complain of "clumsiness," reporting activity such as dropping cups or plates or spilling liquids when surprised or laughing. A patient, particularly a child, may have repetitive falls that cannot be easily explained. A clinical suspicion of seizures may lead to a misdiagnosis. The duration of each cataplectic attack, partial or total, is highly variable and usually ranges from a fleeting few seconds to full minutes. Rare episodes lasting longer than 30 minutes may occur. The term "status cataplecticus" can be applied to these prolonged attacks. Emotion, stress, fatigue, or heavy meals can elicit attacks. Laughter and anger seem to be the most common triggers, but emotions triggered while listening to music, reading a book, or watching a movie can also induce the attacks. Merely remembering a funny situation may induce cataplexy. In children, it can occur when playing with others.

A mechanism similar to the one leading to rapid eye movement (REM) atonia is suspected in cataplexy. REM sleep is associated with increasing activity in the brain of the neurotransmitter, acetylcholine. Animal research has shown that these cholinergic mechanisms are also important in cataplexy. Physostigmine is a medication that has cholinergic properties and increases cataplexy. Cataplexy is blocked by the anticholinergic medication, atropine, in narcoleptic animals. Neurons in the part of the brainstem called the locus coeruleus have been shown to decrease activity in tight correlation with loss of muscle tone in narcoleptic research animals. Cataplexy is associated with an inhibition of the tendon reflexes that are typically elicited by tapping a person on the knees. These same reflexes are normally fully suppressed during REM sleep. This emphasizes the relationship between the motor inhibition of REM sleep and the sudden loss of muscle tone and areflexia seen during a cataplectic attack.

Another important neurotransmitter involved in REM sleep and cataplexy is norepinephrine. Noradrenergic neurons in another area of the brainstem, the locus coeruleus, stop discharging immediately prior to and during a cataplectic attack in dogs. Reports of decreased activation of the locus coeruleus may explain cataplexy.

The most amazing and mysterious aspect of cataplexy is the way in which attacks are triggered. Eruptive expressions of laughter, anger, and exhilaration are typical precipitants. Patients whose cataplexy is severe literally cannot experience strong positive emotions without becoming paralyzed. Imagine being an adolescent or young adult whose friends realize they can make you fall down if they can get you to laugh! When cataplexy is very mild, often only one curiously unique situation will trigger an attack. For example, one patient had cataplectic attacks only when he tried to pull the trigger of his hunting rifle. Another had attacks only when trying to slice rare roast beef. You can try to trigger a cataplexy attack but asking a person to tell a joke.

The anticipation of telling the punch line seems to be a somewhat reliable trigger. Many patients learn to stifle their emotions which may result in a decreased frequency of cataplectic episodes. Overall, the frequency of occurrence varies widely from less than once a week in some patients to countless attacks in a single day in others.

Isolated cataplexy without narcolepsy is extremely rare. It may occur in patients with discrete structural lesions involving the pontomedullary region of the brainstem. This may occur in multiple sclerosis or in some brain tumors. The treatment of cataplexy in patients with narcolepsy is discussed below.

Sleep Paralysis and Hypnagogic Hallucinations

Sleep paralysis and **hypnagogic hallucinations** are seen separately or in combination in about half of all victims are often referred to as auxiliary symptoms. Sleep paralysis refers to a situation in which a person is unable to move for a minute or so at the entry into sleep or the emergence from sleep. Hypnagogic hallucinations are vivid, perceptual, dreamlike experiences occurring at sleep onset. Occasionally, similar visualizations occur upon awakening at which time they are called hypnopompic hallucinations. The accompanying emotion is often fear or dread. Since all narcoleptic patients have sleep onset REM periods, the presence of the symptoms of sleep paralysis and hypnagogic hallucinations may be a function of whether or not these experiences are recalled. Furthermore, as noted earlier, these "auxiliary symptoms" occur occasionally in persons who do not have narcolepsy.

Abnormal REM Processes

It is strongly suspected that cataplexy, sleep paralysis, and hypnagogic hallucinations are dissociated manifestations of REM sleep occurring during wakefulness as a result of impaired control of REM sleep processes. In normal individuals, REM sleep is strongly suppressed during wakefulness. It is allowed to occur, in effect, only after substantial intervals of non-REM sleep have occurred. In patients with narcolepsy, REM sleep is poorly controlled and occurs too readily - as waking is subsiding at sleep onset, or for reasons that are not well understood, such as when wakefulness is invaded by strong emotion. Normal REM sleep usually develops sequentially. First, motor inhibition (paralysis) occurs, followed by increased brain activity and dreaming. When a cataplectic attack occurs, the brain is initially awake. The experience of cataplexy depends mainly on the length of the attack. First there is the occurrence of paralysis, then intrusive hallucinations, and finally, a complete entry into the dream world. On very rare occasions, patients may feel they are awake and dreaming simultaneously. The majority of cataplectic episodes, however, are very brief and experienced as being awake and unable to move.

Pathophysiology

The investigation of the pathophysiology of narcolepsy was greatly advanced when it was reported that 100% of Japanese narcoleptic patients studied were positive for human leukocyte antigen (HLA) known at the time as DR2. Researchers in other countries confirmed this. Further research narrowed this down to the currently named HLA marker DQB1*0602. The link between this HLA marker and narcolepsy is stronger than that of any other known link to an HLA marker to a disease. This has led to a vigorous search for an autoimmune mechanism in narcolepsy. Narcolepsy patients do not have the typical serum markers of an autoimmune disease. However, the discovery of specific antibody levels now supports an autoimmune pathology.

A better understanding of the genetics and pathophysiology of narcolepsy has been possible due to the discovery of the disease in animals such as dogs and horses. This has allowed for the development of an animal model for research. A dog model of the disease has led to the discovery of a narcolepsy gene. Canine narcolepsy is caused by a defective receptor to hypocretin. Hypocretin, also known as orexin, is a neurotransmitter located in the lateral hypothalamus with wide projections throughout the brain. Different subtypes of hypocretin were identified. Subsequently, it has been found that many narcoleptics are deficient in hypocretin 2. Many narcoleptics have decreased levels of hypocretin in the cerebral spinal fluid.

Diagnosis

The most widely accepted diagnostic criteria are published in the second edition of *International Classification of Sleep Disorders Diagnostic and Coding Manual.* Unless cataplexy is unequivocally present, an overnight polysomnogram and multiple sleep latency test (MWT) are necessary to establish an accurate diagnosis of narcolepsy. Even if cataplexy is present, since narcolepsy is a life-long diagnosis any uncertainty should be eliminated as much as possible with the MSLT. The MSLT was described earlier in this text. In summary, an MSLT is a daytime study, usually done following a polysomnogram, where the patient is recorded during 5 separate nap opportunities at 2-hour intervals. The number of sleep periods, the mean latency to sleep, and the number of REM periods that occur are documented. The overnight polysomnogram should last at least 6 hours and confirm that no other sleep disorders are present and untreated. An MSLT with a mean sleep latency (MSL) of ≤ 8 minutes plus ≥ 2 sleep-onset REM periods (or SOREMs, defined as periods of REM sleep within 20 minutes of sleep onset) is considered to be consistent with the diagnosis of narcolepsy.

However, there is ongoing debate about the value of the MSLT results. Some patients with other sleep disorders have been shown to have MSLT results similar to those of narcoleptics, and occasionally even normal subjects can present with 2 or more SOREMs, even without clinical symptoms of sleepiness.

Additionally, not all narcolepsy patients meet MSLT criteria. The MSLT may sometimes need to be repeated before the diagnosis of narcolepsy is established. The MSLT requires the patient's motivation and cooperation in order to obtain valid results. The MSLT requires the patient to not take any antidepressants or stimulants for 3 weeks before the test is conducted. If the MSLT cannot be properly performed or has ambiguous results, the diagnostic guidelines recommend a lumbar puncture be performed for measurement of CSF hypocretin levels.

The International Classification of Sleep Disorders includes diagnostic categories for narcolepsy with cataplexy, i.e. the classic form of the disorder, narcolepsy without cataplexy, which is based on the presence of the clinical symptom of sleepiness, and MSLT mean sleep latency criteria but without (or without yet) existence of the cardinal symptom of cataplexy.

Secondary forms of narcolepsy have also been reported. These cases can be associated with any neuropathology affecting the lateral and posterior hypothalamus. Conditions attributed to secondary narcolepsy include head injuries, multiple sclerosis, arterial venous malformations, Neimann Pick Type C, and CNS tumors such as craniopharyngioma.

Treatment

Although great progress is being made, there is no cure for narcolepsy at the present time. Treatment consists of relieving the narcoleptic symptoms with lifestyle changes and medications. The goal of all therapeutic approaches in narcolepsy is to control the narcoleptic symptoms and to allow the patient to continue full participation in familial and professional daily activities. However, drug prescriptions must take into account possible side effects, especially considering that narcolepsy is a lifelong illness and that patients must take medication for years. Some compounds may lead to tolerance or addiction. Treatment of narcolepsy must thus balance avoidance of secondary side effects, avoidance of tolerance, and maintenance of all activities.

The management of cataplexy is ideally part of a comprehensive narcolepsy treatment plan. Successful treatment typically must combine both behavioral and pharmacological treatments. The situation is analogous to other chronic conditions, such as juvenile diabetes mellitus, in which a combination of diet with medication can control the disease. Patients with narcolepsy-cataplexy will benefit from the healthy sleep habits referred to as sleep hygiene. Some patients with narcolepsy with relatively mild cataplexy may prefer not to take medication for their cataplexy, in part to avoid medication side effects. Also, some patients learn to anticipate the attacks.

Historically, the drugs most widely used to treat EDS are the CNS stimulants (Table 1). Amphetamines were first proposed in 1935. The alerting effect of a single oral dose of amphetamine is maximal 2 to 4 hours after administration, and many patients require a single daily or twice-daily dose. However, a number of side effects, including irritability, tachycardia, nocturnal sleep disturbances, and sometimes tolerance and drug dependence, may occur. The use of methylphenidate was later encouraged because of faster action and lower frequency of side effects. Pemoline, an oxazolidine derivative with a longer half-life and a slower onset of action, was used in the past but it is no longer available. Pemoline is associated with severe hepatotoxicity. Use of modafinil, a newer medication, may also result in substantial improvement. Modafinil's mechanism of action is not entirely clear, but modafinil does have important advantages over traditional stimulants. It has a safer cardiovascular profile and a lower risk of substance abuse. Usage of a modafinil isomer with a longer halflife allows for once-a-day dosing, and it does not typically interfere with nocturnal sleep. Modafinil is currently the preferred medication for newly diagnosed narcoleptics. Modafinil's most common side effect is transient headaches, particularly with high starting dosages. This may be avoided by starting with a lower dosage. Modafinil has important drug interactions to keep in minds. For example, it lowers the efficacy of oral contraceptives in woman. This is an important consideration since narcolepsy can develop in young adults.

Cataplexy does not usually respond to the stimulant medications used to treat the sleepiness of narcolepsy. Narcoleptics in the past would typically take a medication to improve alertness and a different medication to avoid cataplexy attacks. This is remains a viable treatment option. However, the FDA has approved sodium oxybate for treatment of both excessive sleepiness and cataplexy. This further discussed below. Cataplexy seems to respond best to medications with noradrenergic reuptake blocking properties. Medications used effectively include tricyclic antidepressants, such as clomipramine, protriptyline, and imipramine. Selective serotonin reuptake inhibitors, such as fluoxetine and venlafaxine, are effective and have fewer undesirable side effects than the tricyclic antidepressants.

The novel agent sodium oxybate (γ-hydroxybutyrate/ GHB) is very effective and well tolerated in the treatment of cataplexy among narcoleptics. This is a precursor to γ-aminobutyric acid. This medication also improves daytime sleepiness. The medication increases slow-wave sleep without changing the amount of REM sleep. The dosage is usually approximately 2 or 3 grams given in divided doses at bedtime and 4 hours later. Keep in mind that most medications are prescribed in smaller units such as milligrams. In the United States, GHB is a very controversial compound. It has become a popular drug of abuse among some segments of society and has been given the notorious nickname of the date rape drug. The medication has strong sedating properties when mixed with alcohol. It has been shown to have medical benefit in narcolepsy, but it should be used with caution in patients with a known history of substance abuse. Behavioral therapeutic approaches must be considered in the treatment of narcolepsy. These approaches include taking short daytime naps and support groups. A 15- to 20-min nap taken once to twice a day may improve alertness. Undoubtedly, narcolepsy can be a disabling disorder leading in many instances to employment difficulties. It is also a disorder that can be poorly understood by patients, family members, and peers that can result in rejection from families and other social entities, divorce, loss of self-esteem, and depressive reactions. For these reasons, and in consideration of age of onset, it is important to put narcoleptic patients in contact with support groups and to help with the creation of regional narcolepsy associations and patient groups.

The aim of narcolepsy treatment is maintenance of the patient's wakefulness and alertness throughout the day. However, the medications currently available often have significant side effects. Treatment plans must therefore be tailored to individual preference and tolerance, with good communication between the physician and the patient.

Close Up: An Animal Model of Narcolepsy and the Pathway to a Cure. My constant efforts to make movies of cataplectic attacks throughout the late 1960s yielded a number of good clips. In 1972, I showed a couple of the films at the annual convention of the American Medical Association. Afterward, a neurologist approached me and stated that he knew about a dog that had similar attacks. The dog's veterinarian offered both bad news and good news. The bad news was that he had euthanized the dog because he mistakenly assumed the animal had intractable epilepsy. The good news was that he had made a movie of the dog having several attacks. What I saw on the film certainly looked a lot like cataplexy - but only by observing the dog directly and doing some tests could I be certain.

The next year at the annual meeting of the American Academy of Neurology, I showed both the human film and the dog film. Another neurologist informed us about a female French poodle in Saskatoon, Canada that had identical behavior. The owner was persuaded to donate the poodle (whose name was Monique) to Stanford for tests. Monique did indeed have narcolepsy. In addition to excessive sleepiness, she had very frequent attacks of cataplexy which were triggered by strong emotion, such as the excitement of eating. At this point, I was determined to find more narcoleptic dogs and breed them with one another. I brought Monique with me on numerous tours of veterinary schools and animal medical centers all over America. Monique demonstrated unique attacks of cataplexy and I lectured about narcolepsy. I asked those in the audience to keep an eye out for any dog that exhibited similar abnormalities. During the next few years, veterinarians from around the country sent me about a dozen dogs that had unambiguous narcolepsy. In 1977, the successful mating of two narcoleptic Doberman pinschers produced a litter of five pups, all of whom developed narcolepsy when they were about two months old.

As with humans, canine narcolepsy does not involve pain: the dogs merely collapse when they become excited, for example when they are given a new toy or a favorite snack. A few seconds or a minute or two later they are up again as if nothing had happened. As far as I can tell, the dogs are all healthy and happy and have no idea they are different from normal dogs.

By breeding homozygotes (2 affected genes), heterozygotes (1 affected and 1 unaffected gene), and normals (2 unaffected genes), our group at Stanford was able to show that the cause of canine narcolepsy was a single autosomal recessive gene. Using selective breeding and genetic analyses of blood samples from the dogs, my close colleague at the Stanford Sleep Center, Dr. Emmanuel Mignot began his work to isolate the gene we now call canarc-1 (from canus - Latin for dog - and narcolepsy). The first step was to search for linkage markers and then to sequence DNA in these regions. By sheer brute force, Mignot and his coworkers were able to clone much of the physical region in the vicinity of the gene in bacterial artificial chromosome clones constructed from a heterozygous canarc-1 Doberman. The breakthrough came when they were able to identify a very small region in the target area of the dog genome that corresponded with a segment in human chromosome 6 and used homology mapping with expressed sequences. In May 1999, Mignot and his colleagues reported the isolation of the narcolepsy gene and the identification of the gene product. The canarc-1 locus encodes for a receptor in the brain which was shown as recently as 1998 to be one of two receptors that bind the recently discovered neuropeptide, hypocretin. The molecule was first isolated in 1996 from a hypothalamic preparation and the structure was formally described early in 1998. There are two peptide ligands for the hypocretin receptors that arise from a single polypeptide precursor. The cell bodies of the hypocretin neurons are located in the lateral and posterior hypothalamus and their axons project widely throughout the brain.

These results from narcoleptic dogs were immediately applied to the human disorder. Late in 1999, Stanford sleep scientists reported that hypocretin is absent in the spinal fluid of narcoleptic patients and present in controls. Subsequent studies of the hypothalamus in human brain bank material showed greatly reduced numbers of hypocretin neurons in patients with human narcolepsy.

A daunting puzzle remains, however, regarding the primary cause of human narcolepsy. Whereas the cause of canine narcolepsy is a single recessive gene, in human narcolepsy identical twins are more often than not discordant for the illness. This means that the cause of the illness in humans cannot be completely genetic; environment and/or developmental factors must also play a role. Nonetheless, the discovery of the crucial role of the hypocretin system augurs well for the likelihood that new treatments will be developed and that the pace of research will quicken.

In both human and canine narcolepsy, an impairment of the hypocretin system is certainly involved. In dogs, the hypocretin-2 receptor is abnormal and presumably cannot respond to the hypocretin-2 ligand. In humans, there appears to be an absence of the peptide in cerebral spinal fluid and this is very likely to be related to the greatly decreased numbers of hypocretin neurons in the hypothalamus. It is possible that there is an autoimmune attack on these cells earlier in the life of the patient. Replacement therapy may be the answer. In any case, these discoveries collectively are a true breakthrough in our understanding of narcolepsy.

Close Up #2: Narcolepsy In Other Species

I am frequently asked if narcolepsy exists in species other than dogs and humans. I identified several ponies that had unambiguous cataplexy. I assume that wild animals with narcolepsy would not survive very long, since any tendency to collapse when frightened would quickly lead to the animal becoming dinner for a predator.

In 1977, a short piece in Sports Illustrated magazine mentioned that the bullfighting industry in Spain was in trouble. The bulls would enter the arena and suddenly fall down. They would immediately recover and fall again. The Spaniards considered such a bullfight a travesty or a joke, so this was a serious problem for the industry. Sports Illustrated reported that a group in Madrid was offering a reward of several million pesetas to anyone who could solve the problem. I was dazzled by the possibility that I could be the one with the answer and possibly fund my research for the rest of my career. I was therefore very disappointed when I found that this seemingly huge sum of money was only about $25,000 U.S. Nevertheless, my professional curiosity was greatly aroused and I was convinced that the falling bulls had cataplexy fostered by excessive inbreeding. I sent a close friend who had previously lived in Spain on a mission to obtain movies of the bulls. Video cameras did not exist at this time, and since the longest rolls of Super-8 movie films available were 8 minutes, the camera could not be run continuously. There was no way to predict exactly when a bull would fall, so my friend was always two or three seconds too late. He attended countless bullfights and had lots of movie footage of bulls on the ground, but he never actually captured the cataplectic attack *itself. Although the Spanish veterinarians never identified the cause, the problem was finally solved by eliminating the bovine strains in which the falling tendency occurred. My friend did manage to turn the escapade into a hilarious magazine article entitled "El Toro Tanglefoot, or Sitting Bull Revisited." -William C. Dement*

Summary

Narcolepsy is a neurological condition apparently caused by auto immune mechanism. The pathophysiological mechanism for this condition was unraveled in part of the Stanford K9 narcolepsy colony. The condition is readily treatable. For individual patients, the major disability associated with narcolepsy is usually daytime sleepiness. The patient cannot remain awake in soporific classroom or job situations, and interpersonal relations become difficult. Without treatment, a patient with narcolepsy may be doomed to fail in school and will be unable to hold any job which requires sustained intellectual effort or attention. Cataplexy can also be very disabling when it is severe. Fortunately, in the many patients, cataplexy is relatively mild.

"How many narcoleptics does it take to change a zzzzzzzzzzzzzzzzzzzzzzzzzzzzzzzzzzzzzzzz."

CHAPTER 19 Parasomnias and Sleep Related Movement Disorders

"The scariest thought in the world is that someday I'll wake up and realize I've been sleepwalking through my life." -George Saunders

Introduction

Parasomnias consist mainly of inappropriate behaviors that intrude into the sleep process. They do not directly involve an overall increase or decrease of the normal sleep tendency or relate to abnormalities of circadian sleep-wake regulation. Since parasomnias are primarily manifestations of sleeping brain activity aberrantly transmitted through skeletal muscle or autonomic nervous system pathways, their existence adds to the evidence that the sleeping brain is an active brain.

The third edition of the *International Classification of Sleep Disorders* (ICSD) lists and characterizes over a dozen distinct parasomnias. Parasomnias are divided into 3 main groups: NREM-Related Parasomnia, REM-Related Parasomnia, and Other Parasomnias. It also includes a category for isolated symptoms and normal variants, under which sleep talking is included. Some of the parasomnias rare such as exploding head syndrome (yes, you read that right!) and others are very common. In this chapter, we will also discuss some conditions that were previously described as parasomnias but are now reclassified as Sleep Related Movement Disorders.

Parasomnias are thought to occur through a dissociated state. Usually we think of the brain as only functioning in three modes: awake, dreaming (REM), and sleeping but not dreaming (nREM). In reality the brain is complex to the point that parts of the brain are awake while other parts are in another state. We refer to this phenomenon as a dissociation. For example, during sleepwalking, which will be discussed in more detail later, the frontal lobes to be "offline" while other parts are activated.

The material in this chapter is restricted to the more common behavioral parasomnias including violent behavior during sleep, which has important psychiatric and legal implications. A large percentage of the general public is intermittently affected by one or another of the parasomnias, and should therefore be thoroughly familiar with the most common of them, if not with the entire group.

NREM Related Parasomnias

Case History

John M's roommate was startled out of his sleep by a series of blood-curdling screams. "What's the matter? What's the matter?" he called. John was sitting up in bed babbling about someone coming in the window which was closed. His eyes were wide open and his pupils were dilated. The roommate was very disturbed and shouted again, "What's the matter?" But John did not respond. After a minute or so, John lay back down and apparently resumed sleeping. The next morning, John remembered nothing when his roommate described the incident. A week later, the nocturnal screams occurred again; the next week, they happened twice. The sleep of some of the other students in the residence was also disturbed and they were worried about John. John M. was subsequently seen by a sleep specialist who found that he had a history of frequent sleepwalking and sleep terrors as a child. The episodes ceased during his adolescence with one or two exceptions. John had a younger brother who also walked in his sleep. It was determined that the reappearance of the night terrors and screaming in this predisposed student was a side effect of a medication he was taking. When the medication was changed, the screams stopped. Had John not been predisposed, it is likely that the medication would have been innocuous.

Confusional Arousals (Nocturnal Sleep Drunkenness)

Episodes of marked confusion during and after arousal from sleep, but with no or minimal sleepwalking or sleep terrors, have been referred to as **confusional arousals**, nocturnal sleep drunkenness, and excessive sleep inertia. Such episodes arise most typically from a deep sleep state in the first part of the night.

During episodes, the victim awakens only partially and exhibits marked confusion with slow mentation, disorientation in time and place, and perceptual impairment. Behavior is often inappropriate. Cognition during episodes is typically different from cognition during full wakefulness in that it features confused thinking, misunderstandings, and errors in logic. The confusion may last several minutes to a quarter of an hour or more. Memory for any associated mental activity or for the episode itself is typically totally absent. A not uncommon medico-legal case involved an on-call physician who, when called in the early morning hours approximately one hour after sleep onset, gave incorrect recommendations with serious negative repercussions for the patient. It is clear that he would never have made such recommendations during normal wakefulness. Confusional arousals are very common in children and may be almost universal before approximately five years of age. In the case of children, they are usually relatively benign and disappear with time. Predisposing factors include anything that deepens sleep or impairs ease of awakening. The major factors are young age, sleep after severe sleep deprivation, fever, and CNS-depressant medications, including hypnotics and tranquilizers. There is also often a family history of confusional arousals.

Polysomnographic recordings (PSG) have shown that the episodes typically have their onset during deep NREM stages 3-4 (slow wave sleep). They most commonly occur during the first third of nocturnal sleep but have also been reported to occur later at night during light NREM sleep and NREM/REM transitions, as well as during afternoon naps. They rarely, if ever, accompany arousal from REM sleep. During confusional episodes, the EEG may show some residual slow-wave activity, stage 1 theta patterns, repeated microsleeps, or a diffuse and poorly reactive alpha rhythm, all indicating incomplete awakening.

The differential diagnosis of confusional arousals includes four other parasomnias that occur during sleep and have amnesiac characteristics: (1) sleep walking; (2) sleep terrors in which there are signs of acute fear, usually including screams; (3) REM sleep behavior disorder associated with explosive movements such as fighting or diving out of bed; and (4) nocturnal epileptic seizures of complex partial type associated with epileptic EEG discharges. Aggressive treatment is usually not necessary for confusional arousals in children. Triggers for these parasomnias include sleep deprivation, CNS depressants, stress, fever illness and alcohol in adults.

Confusional arousals, sleepwalking, and sleep terrors all primarily occur during incomplete arousals from slow wake sleep (SWS), and all can be experimentally induced in some individuals by simple attempts at forced awakenings during SWS in the first third of the night. Arousal from SWS is therefore by itself sufficient to produce these attacks in predisposed individuals.

Sleep Terrors (Incubus Attacks, *Pavor Nocturnus*)

In a characteristic **sleep terror** episode, the victim sits up suddenly during sleep and emits a scream. He or she appears to be in a state of great terror; showing wide open, staring eyes, dilated pupils, rapid breathing and heart rate, and increased muscle tone. During the episode, communication is difficult if not impossible; the victim cannot be consoled or comforted. The usual attack duration is several minutes and the termination is usually spontaneous. Sleep terrors characteristically begin during deep sleep in the first third of the night. However, when episodes are very frequent, they may be diffusely distributed across the sleep period and occur in any NREM stage. Attacks may also occur during daytime naps. Accordingly, the term "sleep terror" is preferred to "night terror." Once awake, the subject often remembers having difficulty breathing or marked palpitations, but recall of detailed mental activity or dreaming is very rare. If present, memory consists almost always of a static scene like a single photograph, rather than the progressive continuity of images that characterize a typical dream experience.

Attacks in children are sometimes referred to as *pavor nocturnus* (Latin for sleep terror) and in adults as incubus attacks. They have also been referred to as nightmares, but this creates confusion with the true terrifying dream or REM nightmare. The words incubus (Latin: in, upon; cubare, to press) and nightmare (Teutonic: mar, devil) reflect the medieval belief that sleep terrors are caused by a devil sitting or pressing on the chest of the sleeper, which leads to the breathing difficulties and acute terror.

The attacks can occur from early childhood on and many children experience an occasional episode. The prevalence of frequent attacks is low, affecting approximately three percent of children and one percent of adults. Sleep terrors are sometimes familial and are associated with increased personal and family incidence of sleepwalking and confusional arousals. Males are more frequently affected than females. In children, psychopathology is very rare, and the attacks tend to disappear before adolescence. On the other hand, psychopathology is commonly associated with sleep terrors in adults. If present, adult attacks are most frequent in the 20- to 30-year-old range and less frequent in old age.

Sleep recordings have shown that sleep terrors occur mainly during arousals from SWS. When many episodes occur during a single night, they may occasionally arise in stage 2 sleep as well. Marked tachycardia and hyperventilation are all present and testify to an acute autonomic discharge. In adults, a wide range of activities including sexual behaviors are included with these parasomnias. In predisposed subjects, attacks may be elicited experimentally by forced arousals during monitored SWS, as mentioned previously. In the interest of making a correct identification, sleep terrors must be distinguished from nightmares. The latter typically occur in the last (rather than the first) third of the night; they do not have the same intensity of terror, autonomic arousal, or motor activity; and they are not followed by such intense mental confusion. Moreover, a person who has had a nightmare typically recalls an organized dream in which evolving events became frightening and personally threatening.

Treatment is often unnecessary when episodes are rare. Parents of young patients must be reassured that these dramatic episodes have no psychopathological implications, that they seldom cause injury and that they are almost always outgrown.

Sleepwalking (Somnambulism)

Sleepwalking consists of recurrent episodes in which the subject arises from a deep sleep, typically in the first third of the night, and, without awakening, leaves the bed and walks for some distance. Communication with sleepwalkers is difficult or impossible, and it appears that the behavior must be allowed to run its course. A pattern of repetitive behavior that has obvious symbolic meaning may occur, such as a child repeatedly crawling into his or her parents' bed or an adult trying to prepare meals. Mumbling or even comprehensible speech may occur as well. Rarely, a sleepwalker may become aggressive. This is most likely to occur as a response to attempts to restrict the sleepwalker's mobility. It is very important to emphasize that injury may arise from sleepwalking into dangerous situations or from cuts/burns.

The usual duration of a sleepwalking episode is brief, but prolonged complex behaviors lasting more than an hour, including driving, have occasionally been documented. Rarely, a sleep terror immediately precedes and evolves into a sleepwalking episode as a combined or hybrid attack. The episodes may terminate either spontaneously or by forced awakening. The sleepwalker is usually extremely hard to awaken and, if actually aroused, often exhibits mental confusion with amnesia for the event. Such amnesia varies in degree, however, and a few individuals may have clear recall of the events. One student stated that she was aware of the environment and of her movements during a sleepwalking episode but simply could not change her behavior while it was occurring. As noted above, violent or aggressive behaviors may occasionally occur and a few well-documented cases of apparent homicidal somnambulism have been reported. Sleepwalking can occur at any age after a person learns to walk. It is most common in children aged 4-6 years and usually disappears during adolescence. Adult cases, however, are not infrequent. A strong family history is common, and there is often a family or personal history of other arousal disorders from SWS, specifically sleep terrors and confusional arousals.

Facilitating and precipitating factors that increase the probability of an episode on a particular night have been identified. These are mainly factors that deepen sleep, particularly sleep deprivation and depressant medication. The presence of factors that disrupt SWS, including stress, pain, or bladder distension, can also trigger an attack. In chronic sleepwalkers, episodes can at times be externally precipitated by a simple forced arousal during deep sleep in the early part of the night.

Polysomnographic recordings have shown that episodes arise in stages 3 or 4 of deep sleep, most often at the end of the first or second cycle of NREM sleep. EEG during sleepwalking has been recorded by using long cables or telemetry that allow mobility during the episodes. These recordings have typically shown lightening from SWS to NREM stage 1 patterns. The marked autonomic activation characteristic of sleep terrors is absent. Episodes occasionally occur in light stage 2 sleep or even during NREM-to-REM transitions.

Diagnosis is seldom a problem, although REM sleep behavior disorder or "escape" behavior that sometimes occurs during a sleep terror episode must be distinguished. If eating or drinking is a common behavior during sleepwalking, the disorder must be differentiated from the nocturnal eating (drinking) syndrome in which the consumption-related behaviors consist of recurrent awakenings with the inability to return to sleep without food or drink.

Beyond stressing the importance of safety, treatment of young children for sleepwalking is usually unnecessary. Parents may be reassured that children often discontinue the behavior when they are older. When the behavior seriously distresses the individual or the family or presents the risk of injury, countermeasures should be instituted. Precipitating factors should be carefully avoided. Efforts should be made to minimize possible injury by locking doors and windows, removing sharp objects, and so forth. Above all, there should be no possibility of falling down stairs or off a high balcony. Medication may be indicated if episodes are extremely frequent. The medication clonazepam is sometimes prescribed. Adult cases are typically more difficult to treat than childhood cases are.

REM Related Parasomnias

Nightmare disorders

The term **nightmare** is best reserved for dreams with content that is frightening to the sleeper and leads to awakening. The term should never be applied to the previously described sleep terror, a condition that is a distinctly different form of parasomnia. Nightmares occur mainly in the second half of the night within REM sleep. Movements, mumbling, or vocalizations may occur during sleep before awakening. The awakening is seldom accompanied by the cry typical of an episode of sleep terror. Palpitations and labored breathing, if present, are much less marked. The sensorium is relatively intact with little of the confusion and disorientation that are characteristic of sleep terrors. The most important distinction between nightmares and sleep terrors is that a nightmare can almost always be recalled in detail while a sleep terror cannot. Nightmares usually last 4-15 minutes, although shorter and longer episodes do occur. Subjective perception of imminent injury or death is not infrequent.

Nightmares are common in children and have been said to occur in 10-50 percent of 3- to 6-year-olds. Nightmares may persist into, or initially appear in, adolescence or adulthood. Approximately 70-80 percent of adults report at least occasional nightmares. Borderline personality disorder, schizoid personality disorder, and schizophrenia appear to be associated with lifelong nightmares. People with weak ego boundaries and strong artistic tendencies are often vulnerable. In military personnel with combat fatigue and emotional trauma, nightmares may be particularly intense and recurrent, and the relationship of those nightmares to life stresses is evident. Nightmares are frequently associated with so-called REM rebound during a period of recuperation after REM sleep deprivation from stress, drugs, or other causes. A number of medications also predispose a person to nightmares.

Sleep Paralysis

Sleep paralysis consists of episodes of inability to perform voluntary movements either at sleep onset (called hypnagogic or predormital) or at awakening (called hypnopompic or postdormital). Although sleep paralysis is one component of the narcolepsy tetrad, it occurs much more commonly in a completely isolated form. Characteristically, limb, trunk, and head movements cannot be executed, but eye movements can be made and individuals have some control over breathing. The paralyzed individual is usually conscious of the environment and feels vulnerable, and is often quite frightened. Fragments of dream imagery may be recalled; when present, they are occasionally superimposed on the environment as hypnogogic hallucinations. Episodes typically last 1-3 minutes. The paralysis typically disappears spontaneously or is aborted by being touched by someone else or having one's limbs moved.

Sleep paralysis is not infrequently experienced by rotating shift workers maintaining irregular sleep-wake patterns, individuals with recurrent exposure to jet lag, and medical students and interns. Surveys suggest that sleep paralysis is experienced at least once during the lifetime of more than half of the general population. Both sexes are equally affected.

Polysomnographic recordings have been carried out during sleep paralysis only in patients with narcolepsy. These records show that sleep paralysis occurs during direct transitions into or out of REM sleep. Muscle tone is absent, tendon reflexes cannot be elicited, and the EMG is silent. At the same time, brain waves show waking patterns along with waking ocular movements and blink patterns. Occasionally, drowsy NREM stage 1 EEG patterns with slow eye movements occur. Electrically induced H-reflex studies during sleep paralysis episodes have shown total suppression of this monosynaptic reflex which indicates loss of anterior motor neuron excitability. Such loss of excitability is always present in normal REM sleep episodes. Sleep paralysis is thus considered a dissociated REM state in which the motor atonia component of REM sleep is present in isolation. The correct identification of sleep paralysis seldom poses a problem, as the episodes are so characteristic. The presence of narcolepsy can be excluded by absence of its other components.

Multiple attacks of sleep paralysis are not uncommon. Since the public is typically uninformed about this problem, it usually causes great concern and anxiety. After all, to suddenly become completely paralyzed can be terrifying or at least greatly puzzling. When a person learns about the fundamental nature of sleep paralysis and that it is a manifestation of normal REM sleep as well as its benign nature, there is always great relief. In most cases, the sleep paralysis is ameliorated by avoiding the tendency to become sleep deprived and by lowering one's sleep debt.

REM Sleep Behavior Disorder

In REM sleep behavior disorder (RBD), the sleeper physically acts out a dream. In RBD the intermittent absence of the normal muscle atonia of REM sleep allows vigorous motor activity to occur in relation to what is happening in the dream. The movements are often sudden and can be violent. Patients may dive from the bed and run about, suffering injuries with furniture. On awakening, recall of a dream consistent with the behavior is typical.

Episodes may repeat only once every several weeks. Alternatively, several attacks may occur each night in cyclic fashion approximately every 90 minutes after sleep onset. There is, however, a very marked tendency for the frequency of behavior episodes to fluctuate. Duration of episodes is typically 2-10 minutes.

RBD may also occur within the context of neurodegenerative disorders such as Parkinson's Disease or Lewy Body dementia. It actually may predict the onset of these disorders and allow for earlier recognition of them. Sleep tests show a characteristic feature of an absence or interruption of the usual REM atonia. Decades ago it was found that experimental lesions in the brain stems of cats will replicate RBD.

Treatment of RBD may utilize medications such as clonazepam. Breakthrough attacks can occur on medication and therefore common sense safety measures should be maintained. These include removing weapons or any object that can cause injury from the sleeping environment.

Sleep Enuresis

Sleep enuresis, or bed-wetting, consists of involuntary micturition (urination) during sleep. It is often classified into primary and secondary types. In primary enuresis, the full nocturnal urinary continence, which should occur by five years of age, has never been achieved. In secondary enuresis, after learning bladder control for at least 6 months, children lose control and become enuretic.

Bed-wetting is also often divided into idiopathic and symptomatic forms, the latter being associated with genitourinary pathology such as small bladder size, increased renal output, or other medical conditions. Idiopathic primary enuresis is by far the most common type. Bedwetting tends to occur in the first third of the night during a partial or complete arousal. Dreaming may be recalled in association with the enuresis, but it has been shown that dreams about fluids represent post-event incorporation of the wet bed stimuli rather than being causal. Bed-wetting often leads to embarrassment and secondary psychological trauma, especially in older children, adolescents, and adults. Bladder control during daytime wakefulness is typically normal.

The reasons for the perpetuation or reappearance of bed-wetting after age five vary. In some children toilet training is not fully provided or acquired. A small functional bladder or an insensitive bladder may be present without demonstrable organic disease. The most common causes of secondary nocturnal enuresis in children are obstructive sleep apnea or upper airway resistance syndrome.

Idiopathic enuresis is more common in males, in people from lower socioeconomic classes, and in institutionalized children. In adults, idiopathic enuresis is much rarer, and an organic cause, an associated metabolic or endocrine disorder, or coexisting sleep apnea syndrome is more common. A hereditary factor appears to be present in primary enuresis and in symptomatic cases with urogenital malformations. Polysomnographic sleep testing has shown that enuresis episodes can occur in all stages of sleep, with or without concomitant arousal. In NREM sleep episodes there is a tendency for increasing episode duration (from first signs of state change to micturition) associated with lighter sleep-wake patterns at the moment of bed-wetting. Sleep cystometry has shown increased bladder reactivity to environmental stimuli, increased detrusor contractions in sleep before onset of micturition, and waves of elevated bladder pressure during sleep at the onset of the enuretic event.

Treatment of the secondary form involves treating the underlying disease. This may include surgical correction of genitourinary pathology, treatment of diabetes insipidus, and so on. If sleep disordered breathing is implicated, this condition should be specifically treated. Usually tonsillectomy is sufficient.

In the case of the much more common idiopathic primary enuresis, a number of approaches are helpful: avoidance of fluids in the late evening may be sufficient. Bladder training exercises during wakefulness, in which the bed wetter is taught to hold increasingly larger amounts of urine in the bladder may help, as may sphincter training exercises in which the patient repeatedly interrupts the stream of urine while voiding. Bed wetters may respond positively to a conditioning device in which a bell or alarm triggered by the release of urine stimulates the sleeper, who becomes more aware of the enuresis over time. This apparently leads eventually to learning to sense the detrusor contractions during sleep and, consequently, awakening in time to void.

Family support and avoidance of teasing may be crucial. Medication may be necessary, particularly for older children who are under much social embarrassment and stress, and for adults. The most effective substance is imipramine hydrochloride, 10-75 mg of which is taken at bedtime. Its effect does not appear to relate to any change in sleep structure, but to a direct anticholinergic effect on the bladder myoneural junction.

Sleep-Related Panic Attacks

Sleep-related panic attacks are a parasomnia due to a medical disorder. Most or all of the attacks of **sleep-related panic disorder** occur in association with sudden awakenings from sleep. During the attacks patients are very frightened ("scared to death") with feelings of impending doom or death as well as a variety of somatic symptoms. These symptoms may include dizziness, palpitations, trembling, chest discomfort, choking, sweating, and so on. After the attack, the patient remains in an excessively aroused condition for a significant period of time and has difficulty falling back to sleep.

In the pure condition, which is rare, attacks occur only during arousal from nocturnal sleep. The most common type of sleep related panic attack, however, occurs in association with daytime panic attacks in persons who have a psychiatric disorder involving panic and phobias. The daytime panic attacks are usually more common than the sleep panic attacks and the phobias most commonly involve agoraphobia (the fear of being inside enclosed spaces, such as a car, a plane, a meeting room with the doors closed and in which escape is difficult or impossible).

The etiology and precipitating factors of panic attacks are variable. The number of people suffering from panic attacks only during sleep is unknown, but the condition does not appear to be common. Most patients are young adults and women are three times more commonly afflicted than men. There are no known genetic factors. Separation anxiety from parents in childhood is common in the patient history. Depression is seen in approximately one-half of the patients. Alcohol abuse is often associated. Triggering usually occurs by exposure to the particular feared objects during the daytime preceding the attacks.

Sleep recordings show panic attacks during sleep to be most common during transitions from NREM stage 2 to early stage 3 of SWS and, to a lesser degree, in light NREM stages 1 and 2. Occasionally, attacks are seen during sleep onset at the start of the night. There are usually some tachycardia and tachypnea during a nocturnal panic attack but their intensity is much less than one might expect and certainly less than during sleep terrors. Outside of the attacks, the overall sleep architecture tends to show increases in wakefulness with a corresponding reduction in sleep amount and increased movement time. There is usually little complaint of daytime sleepiness as patients tend to be hyperaroused.

Sleep panic must be distinguished from sleep terrors. Treatment is directed toward the underlying clinical panic disorder and agoraphobia and can include psychotherapy, desensitization programs for specific phobias, relaxation therapies, and use of anti-anxiety medication.

Rhythmic Movement Disorder (*Jactatio Capitis Nocturna*)

Though rhythmic movement disorder has previously been classified as a parasomnia, it is now considered a sleep-related movement disorder. It consists of repetitive stereotyped movements involving large body areas and typically occurs just before sleep onset and persists into light sleep. The most common movement patterns are of the head and neck (head banging) sometimes called *jactatio capitis nocturna*, and less often of the entire body. When the disorder takes place in head banging form, the patient often lies prone and repeatedly lifts the head or entire body, then bangs the head down onto the pillow or mattress. The head may also be struck rhythmically against the headboard. At times, a sitting posture is associated with backward banging of the occiput. In typical body rocking, the entire body is most usually rolled forward and backward from a sitting position or side-to-side from a supine position. Such movements are repeated rhythmically with a frequency of 0.5-2.0 seconds in long clusters. Rhythmic chanting or other vocalizations may occur during head banging or body rocking. As discussed, rhythmic movement disorder was traditionally categorized as a parasomnia. In the ICSD it was reclassified as a movement disorder along with RLS. For the purposes of this text we will retain its original classification as a parasomnia. The behavior is relatively common in infants and young children, the majority of affected children being otherwise normal. The rare persistence into, or appearance during, later childhood or adulthood is more often associated with significant psychopathology including autism and mental retardation. When particularly persistent or intense, significant complications can occur. These include scalp and other body wounds, subdural hematoma, retinal petechiae, skull callus formation, and significant family and psychosocial problems.

Polysomnographic testing has shown the presence of rhythmic movement artifacts, primarily in the immediate pre-sleep period during light NREM stages 1 and 2. In very serious cases, persistence into deep SWS has been noted. A typical feature is the absence of any EEG signs of arousal during, or immediately after, intense rocking movements. Diagnosis seldom poses a problem because of the stereotyped nature of the movements. In infants and in very young children, it may have to be distinguished from bruxism (tooth grinding), thumb sucking, and other such disorders. Periodic limb movements in sleep are easily differentiated, as they appear only after sleep begins, involve the legs primarily, and, although "pseudorhythmic," show several or dozens of seconds between individual contractions. Treatment in the majority of infants and children is not required and the parents should be reassured about this. Padding the bed area or a protective helmet is sometimes indicated. Restraining affected individuals is generally ineffective. Behavior modification procedures have been used with varying degrees of success. If the condition appears in later life, psychiatric and neurologic evaluation may be required. The possibility of seizures should be strongly considered.

Isolated Symptoms and Normal Variants

Sleep Starts (Hypnic Jerks)

Sleep starts, previously classified as a parasomnia, are now considered an isolated symptom of sleep-related movement disorder. They are variously referred to as hypnagogic jerks, or hypnic jerks. Almost everyone has experienced one or more sleep starts. Sleep starts are commonplace throughout the general population. They consist of bilateral and sometimes asymmetric brief body jerks that occur during the process of sleep onset, usually in isolation but occasionally several in succession. Sleep starts principally involve the legs but may also affect the arms and head. Asymmetry of the jerks may occur due to an asymmetric body position in bed.

Sleep starts may be spontaneous or evoked by stimuli. They are almost always associated with or in response to a feeling of falling. They may also be accompanied by sensory symptoms such as a flash of light, or more organized hypnagogic hallucinations. The intensity of sleep starts varies widely among individual victims. The intensity of the contraction may cause an abrupt expiratory cry. Particularly intense and frequent jerks can lead to a form of sleep-onset insomnia with consequent sleep deprivation. Rarely, mild injury may occur.

The cause of the intensification of this otherwise normal physiologic phenomenon to clinically significant levels is unknown. It is important to recall that during sleep onset, hypnogenic brain structures actively inhibit those responsible for wakefulness, thereby momentarily creating an unstable state before sleep maintenance mechanisms become fully established.

The best way to treat hypnic jerks is to simply go to sleep sooner.

Sleep Talking (Somniloquy)

Sleep talking is almost always a benign phenomenon and seldom represents a significant sleep disorder. It consists of speech or sound uttered during sleep without awareness of the behavior. The duration can vary from a few seconds to many minutes. Sleep talking typically becomes a problem only when it is sufficiently frequent or loud enough to disturb the sleep of others.

Occasional episodes are almost universal. Sleep talking is usually spontaneous but can sometimes be elicited, resulting in a dialogue with others. The occurrence of two sleepers having a dialogue, none of which is recalled by either on awakening, has also been reported. The course of sleep talking is typically benign and may occur only for short intermittent periods, then disappear completely. Chronic sleep talking may be related to significant psychopathology or to intercurrent sleep disturbances. The phenomenon shows no gender preference. Occasionally, it appears to be familial. If you sleep talk, tell your partner that whatever you say may or may not be true! The presence of sleep talking can pose problems in a marriage or in other intimate relationships. Offensive content can occur with sleep talking and may have adverse interpersonal consequences.

Polysomnographic recordings have shown that sleep talking most commonly occurs in NREM sleep stages 1 and 2, REM sleep, and only rarely in SWS. Speech utterances typically are associated with increased muscle tone and a transient and incomplete EEG arousal. Autonomic activation during the EEG arousals is minimal. Is not classified as a parasomnia that requires treatment but rather is considered as a normal variant.

Violent Behaviors in Sleep

It is clear that several parasomnias can give rise to aggressive and violent behaviors in sleep. This topic is reexamined here because of its important personal, social, and forensic implications.

These aggressive and violent behaviors are extremely variable and range from kicking/hitting to outright homicide. Most commonly, violent or sexual behaviors develop during the occurrence of REM behavior disorder or one of the arousal disorders. In the latter case, the term violent somnambulism is frequently used. Somnambulistic behavior can occasionally be quite vigorous, with affected individuals running and walking quickly despite not being violent or angry.

The prevalence for all parasomnias in which violent behaviors in sleep have been studied is higher in males than in females. As with most parasomnias, documentation of episodes is best with video telemetry, ambulatory monitoring, or other means that permit full, unimpeded movement of the patient.

Treatment involves several approaches. The first consists of buffering the violent behavior while it is happening. Physically impeding the behavior of a person with a sleeping, or a partially sleeping and confused brain usually enhances the level of aggression, and should therefore be avoided. Typically, the best approach is to let the episode run its course while attempting to awaken the person by talking to him/her or using other nonphysical methods. Efforts to protect the person from self-injury may also be necessary. Most such behaviors stop after 1-3 minutes. Another approach involves treatment of the underlying parasomnia triggers, which, if successful, will prevent further aggressive episodes. Counseling and reassurance of the patient and family to help them understand the unconscious nature of the behaviors can be helpful. Medicolegal implications of violent behaviors must be acknowledged and understood. In general, unintentional violent behavior in sleep is considered an automatism. This unintentional behavior must be distinguished from malingering. If violent behaviors are chronic and not under medical control, the law generally considers the patient to be a risk to society. Probably more common, however, is that victims of this sleep problem have already been incarcerated for assault and occasionally murder without any recognition of the true cause of the victim's behavior.

There are established forensic criteria to consider when evaluating whether a parasomnia may have a valid defense for a violent act:

1. There should be reason (by history or by formal sleep laboratory evaluation) to suspect a bona fide sleep disorder. Similar episodes, with benign or morbid outcomes, should have occurred previously. (It must be remembered that disorders of arousal may begin in adulthood.)

2. The duration of the action is usually brief (minutes).

3. The behavior is usually abrupt, immediate, impulsive, and senseless – without apparent motivation. Although ostensibly purposeful, it is completely inappropriate to the total situation, out of (waking) character for the individual, and without evidence of premeditation.

4. The victim is someone who merely happened to be present, and who may have been the stimulus for the arousal.

5. Immediately following return of consciousness, there is perplexity or horror, without attempt to escape, conceal or cover up the action. There is evidence of lack of awareness on the part of the individual during the event.

6. There is usually some degree of amnesia for the event; however, this amnesia need not be complete.

7. In the case of sleep terrors/sleepwalking or sleep drunkenness, the act may: a. occur upon awakening (rarely immediately upon falling asleep) – usually at least 1 hour after sleep onset. b. occur upon attempts to awaken the subject. c. have been potentiated by sedative/hypnotic administration or prior sleep deprivation.

There is some controversy over how intoxication with alcohol or other substances should be factored into an evaluation of a violent parasomnia. Some have argued that the intoxication with alcohol precludes the use of the sleepwalking defense as a defense since it may be hard to retrospectively differentiated the amnesia of sleepwalking from an alcohol induced black out state. However, alcohol is known to fragment sleep which can also be a trigger for a sleepwalking parasomnia. Therefore, the amount and timing of alcohol consumed may need to be factored into an individual situation if that information is reliably present.

Summary

As noted earlier, when readers have the opportunity to view videotapes of the extraordinary behaviors associated with some of the parasomnias, they are invariably amazed. As with many other sleep disorders, the victim is not fully conscious of what is going on and when awake has poor recall of experiencing sleep terrors and/or violent somnambulism as well as most of the other parasomnias.

To see the actual behaviors, however, and to realize that some of one's fellow human beings do these things is very eye-opening. Awareness of these behaviors also helps to foster a new viewpoint about the importance of adequate healthy sleep in the life of every person. Parasomnias have legal consequences. Finally, as pointed out in the introduction, the existence of this panoply of behavioral manifestations once again dramatically emphasizes that the sleeping brain is an active brain.

"The list of fun and easily fixed brain diseases is very short." Mike Birbiglia

CHAPTER 20 Pediatric Sleep Disorders

"People who say they slept like a baby usually don't have one." -Leo Burke

Introduction

Sleep disorders in children are relatively common and readily treatable. This chapter will review the various sleep disorders common in children and the management of these sleep disorders. For children and adults our nighttime sleep affects our daytime behavior. Sleep disorders mimic behavioral and psychiatric conditions. Pediatric sleep medicine has evolved into a major field of study. Both the American Academy of Pediatrics and the American Academy of Sleep Medicine have published relevant clinical practice parameters. Sleep disorders should be considered whenever a child is being evaluated for any behavioral problem. Insufficient quality or quantity of sleep may be associated with learning difficulties and attention problems. When a child does not sleep well, it can impact the entire household leading to significant stress for the family. Unrecognized and under treated sleep disturbances can carry over into adulthood.

A question commonly posed by parents is, "How much sleep does my child need?" This common question is interesting because one must question why that question is even raised. Parents do not ask how much oxygen their individual child needs. Oxygen and sleep needs should be self-regulated in a healthy child. The reason parents ask the question is because they have concerns about the child's behavior while awake, or perhaps in part parents want to know how much time they can have to themselves while the child sleeps. We should keep in mind that the best opportunity for parents to have some time to themselves is when their child is sleeping. This may add to pressure to get the child to sleep as quickly and for as long as possible. The parent's focus should not be solely on the timing or amount of sleep; ultimately, the overall quality must be considered. If a child is getting quality sleep at an appropriate circadian time, the child should awaken spontaneously feeling refreshed. "Good sleep" can be simply described as the ability to fall asleep easily, sleep through the night, and wake feeling refreshed, without relying on hypnotics.

The Stanford Sleep Disorders Clinic has a long history of working with children with sleep disorders. Even before the clinic opened, children played an important role in the development of modern sleep research. Dr. William Dement, as a medical student, was interested in finding a neurological basis to understand Freud's dream theories. Infants and small children were an early part of Dement's research into rapid eye movement (REM) sleep. Rapid eye movements in sleep were first described in infants. Since the size of an infant's eyes is so large relative to the rest of their head, the eye movements were more easily identified. It was speculated at the time that perhaps these infant eye movements might be related to dreaming. In 1952, Dr. Dement helped discover the relationship between electroencephalographically measured rapid eye movements in sleep and dream recall in adults based on earlier work in infants.

In the early 1960's, Dr. Dement joined the department of Psychiatry at Stanford University and studied the relationship of sleep and dreams to mental illness. In 1972, he along with Christian Guilleminault, opened the first clinic devoted exclusively to the treatment of sleep disorders. Not surprisingly, children were among the first patients seen at the new clinic. This landmark sleep clinic served as a worldwide model for the comprehensive multidisciplinary care. The first medical textbook on sleep disorders in children was published based in large part on the work done at the Stanford Sleep clinic. This book included a description of the Stanford Summer Sleep Camp. For ten consecutive summers the same group of children returned to the same dorm on the Stanford campus to participate in sleep research. The data collected from these children has been used throughout the world. In that dorm, now called the Jerry House, a plaque commemorating this historic research was installed.

The spread of clinical sleep medicine knowledge eventually led to the American Council of Graduate Medical Education recognizing it as a unique medical specialty. Stanford supports the largest ACGME training program in sleep medicine in the country. The study of normal sleep and sleep disorders in children is an integral part of the development of modern sleep medicine.

EPIDEMIOLOGY

It would seem self-evident from speaking to parents that sleep problems are common among children. The National Sleep Foundation's (NSF) yearly surveys of sleep in America confirm parents are dissatisfied with their child's sleep habits. In 2004 the NSF reported that America's children sleep less than experts recommend for their age group, but parents do not always know how much sleep their child needs. More than two-thirds of all children experience one or more sleep problems at least a few nights a week according to parents. The NSF found that nearly one-third of children 10 and younger wake up at least once a night needing attention. In addition, about one-quarter of infants, toddlers and preschoolers appear sleepy or overtired during the day, according to their parents/caregivers. About 26 percent of children between the ages of 3-10 drink at least one caffeinated beverage a day. In 2006 the NSF reported that over 25% high school students fall asleep in class at least once a week. Many of the high school students reported falling asleep doing homework and arriving late for school because they overslept. In 2011, the NSF reported about half of teens say they send, read or receive text messages every night or almost every night in the hour before bed. That same year it was reported that 9% of teens surveyed said they are awakened after they go to bed every night or almost every night by a phone call, text message or email. A larger group, 18%, said this happens at least a few nights a week. It seems many teens and young adults are reluctant to turn off their cell phones at night.

In this chapter, we will first review normal sleep in childhood. Then we will review the most common sleep disorders starting with sleep-disordered breathing. Sleep-disordered breathing is the most common reason polysomnograms are performed in children. Narcolepsy is reviewed, which is a condition that commonly begins in childhood but is not typically diagnosed until adulthood. Insomnia is perhaps the most common sleep complaint in primary care. Restless leg syndrome is often not considered in the evaluation of sleep disorders in children but seems to be more common than previously thought.

NORMAL CHILDHOOD SLEEP

Prior to discussing specific sleep disorders, it is important to review normal sleep in children. Sleep needs and patterns differ with age. The basic principles of normal sleep are similar for children and adults. Normal sleep can be simply described as the ability to fall asleep easily, sleep through the night, and wake up feeling refreshed. A certain degree of serenity is needed to sleep well.

Long-term longitudinal data has been published by Iglowstein and colleagues to illustrate the developmental course and age specific variability of sleep patterns. They followed 493 children for 16 years. Total sleep duration decreased from an average of 14.2 hrs (SD: 1.9 hrs) at 6 months of age to an average of 8.1 hrs (SD: 0.8 hrs) at 16 years of age. Between 1.5 years and 4 years of age, there was a prominent decline in napping habits. At age 1.5 years 96.4% of children had naps; by 4 years of age only 35.4% napped. However, napping behavior must be put into a cultural context. It would be less concerning if napping or "siestas" are part of the child's cultural environment.

In adults and most children, REM periods occur in cycles of approximately 90 minutes throughout the night. At the end of every REM cycle there is usually an arousal or brief awakening. In infants REM cycles are shorter (approximately 40-60 minutes). Parents may be concerned that their infant seems to "wake up every hour." These brief awakenings may be part of the child's normal rhythm but an overly attentive parent may inadvertently reinforce and prolong the awakenings. On the other hand, it is possible that in a child the awakenings are due to a condition that is exacerbated by REM sleep such as obstructive sleep apnea. The percent of the total sleep time spent in REM decreases from birth when it may take up 50% of the total sleep time to approximately 20-25% by the time the child is 3 years of age. It remains mostly at this percent throughout adulthood with some slight decreases in the elderly.

During drowsiness or light sleep (stage 1) children may not realize they are asleep and think they are awake. They may also have fragments of auditory or dream-like imagery called hypnagogic hallucinations. Similar episodes upon awakening are called hypnopompic hallucinations. These episodes occur more commonly with sleep deprivation and in conditions such as narcolepsy. They should not be confused with psychosis.

Delta or slow-wave sleep represents our deepest sleep as measured by the arousal threshold (amount of stimulation needed to wake up the sleeper). Delta sleep obeys homeostatic principles; it increases in duration and intensity in response to sleep deprivation. The amount of delta sleep is about 20% of the total sleep time in children and decreases after adolescence. Delta sleep clusters in the first third of the night. During delta sleep, it is extremely difficult to arouse children. For example, smoke alarms will not reliably wake up a child during slow wave sleep. If aroused, they often appear disoriented and cognitively slowed. Parasomnias such as sleepwalking and sleep terrors usually emerge from slow wave sleep and in children commonly occur in the first third of the night.

The circadian modulation of alertness and sleepiness is an important part of sleep physiology. Part of our circadian physiology is our tendency to have decreased alertness in the afternoon and to have a surge of alertness ("second wind") in the evening. If a child is having trouble falling asleep the imposition of an earlier bedtime or start of the nighttime routine may make the problem worse by creating greater frustration. The circadian system and normal adolescent sleep patterns are further discussed below in the section on the most common circadian disorder in children - delayed sleep phase syndrome.

GENERAL CLINICAL CONSIDERATIONS

One of the characteristic idiosyncrasies of pediatric training is starting a medical presentation with short blurb on the birth history. For example, "the patient is 15-year-old girl born normal through spontaneous delivery to a gravid 3 para 2 mother with a chief complaint of …". If the clinician taking the sleep history of a child uses the birth history as a starting point, they have started too late. The history should begin with the history of parents' sleep habits and patterns (or other caretaker) before they were parents. Being a parent is arguably one of the most rewarding experiences for many adults; however, there is little argument that being a parent is one of the most impractical things for adults to ever experience. The only opportunity for many adults to have a semblance of what their pre-parenthood life was like is while their child is sleeping. Knowing what the parents' expectations are for their own sleep has to be taken in consideration when evaluating a child with a putative sleep disorder. As a simple example, if an infant who sleeps seven hours in a row without requiring the parents' assistance is born to a woman that, before she was a mother, used to be happy with getting six and half hours at night, that mother will proudly tell all who will listen what an easy baby she has and by inference, what a wonderful mother she is. She will readily give advice to other mothers on how to get their children to sleep. Take that same seven-hour sleeping infant with a mother that was in the habit of sleeping nine hours at night before she was a mother and she will bitterly complain that her baby does not sleep enough. Another example is a parent, say the father who had a subclinical case of obstructive sleep apnea. The sleep apnea may become unmasked with additional nighttime requirements of taking care of an infant. Financial factors may influence the parents' sleep. If both parents were receiving outside income before they were parents and one of them gives up that income to stay home with the infant, there will be greater pressure on the parent still receiving income not to lose that income source. The parent that gives up the income source to stay with the infant may fall into a trap of trying to quiet the infant immediately at night so as not to disturb the sleep of the parent that still has to leave the house in the morning to go to work. It is easy to conceptualize how resentment and increased stress can develop if the infant is perceived as not sleeping well. Given the near endless variations on what constitutes a family, factored in within the different cultural backgrounds of the individual family members, the importance of understanding the family's sleep expectations should be part of the evaluation of a child's sleep.

When taking a history, it is important to keep in mind that the need for sleep is biological. However, the way someone sleeps is learned. A family can only place a child in a crib in a separate room if they can afford a crib in a separate room. Early humans did not place their infants in separate caves or huts. Having an infant sleep in a crib in a separate room is clearly a cultural phenomenon. It is also a cultural decision whether to pierce an infant's ears or have them wear shoes. The clinician should avoid the temptation to make value judgments over cultural differences.

All of us sleep best when we go to sleep feeling safe and comfortable. We learn to feel this way by forming associations with our sleeping environment. These associations are formed starting from infancy. Just like in adults, these associations may be maladaptive which can lead to sleep difficulties. When evaluating school age children for a sleep problem, it is important to directly ask the children if they are scared of the dark or of being alone. The parents may previously have minimized these concerns and the children may be too embarrassed to volunteer the information. It is important in any evaluation to keep in mind the interaction between the physiology (the need for sleep) and the psychology of sleep.

When evaluating a child's sleep problem, it is crucial to realize what wakes a patient up may not be what keeps the patient awake. For example, you could be awakened at night because of the sudden loud noise coming from a party next door. The fact that you were not invited to the party could bother you and keep you awake! It is not uncommon in a clinical situation to have two or more sleep disorders interacting in the same patient. A child may have an awakening due to difficulty breathing from sleep apnea, but then be unable to return to sleep. The child has not learned to settle back down without the parents' intervention.

Since the person who is being evaluated cannot provide us with information about their sleep once they are asleep, it is important to get as much information as possible from not only the child but also all the adults that are part of the household.

When parents bring in a child for evaluation, a common question asked is the sleep problem due to a physical or behavioral cause. Although many times it is a combination of both, a useful rule of thumb is that a behavioral sleep problem, such as fear of the dark, may improve when the parents share sleeping space with the child but a physical condition such as obstructive sleep apnea will not improve without diagnosis and treatment.

SLEEP DISORDERED BREATHING

Sleep disordered breathing (SDB) is a clinical syndrome ranging from simple snoring to potentially life threatening obstructive sleep apnea (OSA). This clinical spectrum can occur at any age. Many clinicians learn that OSA emerged from the study of Pickwickian syndrome. However, it is important to point out that in Charles Dicken's first novel, *The Posthumous Papers of the Pickwick Club*, the classic description of snoring with arousals and excessive daytime sleepiness was not in Mr. Pickwick (although he probably did have OSA too) but instead it was of a boy, Joe, who constantly falls asleep in any situation. The first medical description in English of children with abnormal breathing in sleep is attributed to William Osler in 1892. Osler wrote, "Chronic enlargement of the tonsillar tissue is an affection of great importance, and may influence in an extraordinary way the mental and bodily development of children…. At night, the child's sleep is greatly disturbed, the respirations are loud and snorting, and there is sometimes prolonged pauses, followed by deep, noisy inspiration. The child may wake up in a paroxysm of shortness of breath. In longstanding cases the child is very stupid-looking, responds slowly to questions, and may be sullen and cross."

In the modern era, there has been a realization that patients may be symptomatic in the absence of frank apneas. This has led to the use of the term "sleep-disordered breathing" to better describe the clinical spectrum which includes snoring, upper airway resistance syndrome (UARS,) and obstructive hypopnea syndrome.

The most obvious nocturnal symptom is snoring. Snoring indicates turbulent airflow and is not normal in children. The American Academy of Pediatrics has recommended all children should be screened for snoring as part of well-child care. Not all snoring is due to obstructive sleep apnea. It may be due to other forms of obstruction such as nasal allergies or a cold.

Many of these children mouth breath. Regular mouth breathing should always lead to suspicion of SDB. Children with disordered breathing may avoid going to bed at night due to hypnagogic hallucinations. Upon awakening these children may report morning headaches, dry mouth, confusion or irritability. As mentioned, daytime sleepiness may not be obvious depending on the age. It may translate only as a complaint of daytime tiredness.

Clinical signs include increased respiratory efforts with nasal flaring, supra-sternal or inter-costal retractions, and abnormal paradoxical inward motion of the chest may occur during inspiration and sweating during sleep. Information regarding the sleep position is helpful. Typically, the neck is hyper-extended and the mouth is open.

A physical finding that may be overlooked in a child with SDB is a narrow and high arched palate. Interestingly, the Diagnostic and Statistical Manual of the American Academy of Psychiatry, DSM IV's, description of attention deficit disorder mentions that minor physical anomalies such as high arch palates may be present. Since both conditions may have similar daytime behavior in the same age group, a child with SDB could be misidentified as having attention deficit disorder. The possibility of a sleep disorder being present should be considered in any child being evaluated for attention deficit disorder. This is particularly important since treatment of SDB may improve behavior and academic performance. Parasomnias maybe triggered or exacerbated by SDB. Individuals identified with SDB have a much higher incidence of nightmares, with reports of "drowning," "being buried alive," and "choking." SDB leads to sleep fragmentation or disruption. Any condition that disrupts slow wave sleep may lead to sleep terrors. and sleepwalking in children. SDB should be included in the evaluation of any child with parasomnias.

Epidemiology

The prevalence of SDB in children was studied by Rosen et al who performed a cross-sectional study of schoolaged children in a Cleveland cohort. The cohort consisted of 829 children, 8 to 11 years old all of which had unattended in-home overnight cardiorespiratory sleep recordings. SDB was found in 20% of the cohort. The remaining 80% had neither snoring nor OSA. Functional outcomes were assessed with 2 parent ratings scales of behavior problems. Children with SDB had significantly higher odds of elevated problem scores in the following domains: externalizing, hyperactivity, emotional liability, oppositional, aggressive, internalizing, somatic complaints, and social problems. The authors concluded that children with mild SDB, have a higher prevalence of problem behaviors, with the strongest, most consistent associations for externalizing, hyperactive type behaviors. An interesting finding in this study was that only 55% of the parents of children diagnosed with a polysomnogram with obstructive sleep apnea (OSA) reported loud snoring. If pediatricians and surgeons screen for OSA by asking the parents/caregiver if the child snores loudly, they may miss close to half of the cases. The term snoring may be misunderstood. It is more helpful to instead of asking parents does their child snore, to instead ask can they hear the child breathing at night.

Pathogenesis

The pathogenesis of sleep-disordered breathing is based on the inherent complexity of the throat. The throat is the only area of the body where food and air mix. We need a rigid enough structure in the negative pressure of the lungs to allow the passage of air. It also has to be able to have the muscular strength to close and allow us to swallow food. In addition, it has to allow for the intricate movements of speech, singing and laughter. There is clearly conflicting activity occurring in a relatively small space. When we fall asleep, our throat reflexes relax causing the throat to be more collapsible. Any fixed obstruction in the throat such as enlarged tonsils will narrow the throat. Since the lungs need to obtain the same amount of air through a smaller passage, the work of breathing will increase by generating more negative pressure. This increase in negative pressure may further collapse the airway with resultant turbulent airflow. The result will be the ubiquitous snoring sound being generated. If the patient is sleeping supine, the tongue can slip back further occluding the airway. This can result in partial or complete obstruction called a hypopnea or apnea respectively. These events can result in multiple either partial arousals or full awakenings from sleep throughout the night. While sleep-disordered breathing in children has many important similarities to the adult version of this disease, there are also marked differences in presentation, diagnosis and management. The abnormal daytime sleepiness may be recognized more often by schoolteachers than by parents of young children. An increase in total sleep time or an extra-long nap may be considered as normal by parents. Nonspecific behavioral difficulties are mentioned to the pediatrician such as abnormal shyness, hyperactivity, developmental delays, and rebellious or aggressive behavior. Chervin et al found that conduct problems and hyperactivity are frequent among children referred for sleep-disordered breathing SDB during sleep. Bullying and other specific aggressive behaviors were generally two to three times more frequent among children at high risk for SDB. Other daytime symptoms may include speech defects, poor appetite, or swallowing difficulties. Nocturnal enuresis or bedwetting accidents should raise suspicion of possible SDB.

An important similarity is that both morbidly obese adults and children are at risk for obstructive sleep apnea. Children do not need to be overweight to have SDB. Children with sleep apnea tend to be of normal weight. Very young children and infants may actually be underweight. OSA may be a risk factor for overall growth failure. The increased work of breathing through the narrow airway, along with the increased total sleep time in children and possible disruption of nocturnal growth hormone secretion from sleep fragmentation may all factor in to difficulty gaining weight. Ironically, children have become obese after being treated for obstructive sleep apnea. Behavioral problems in children with obstructive sleep apnea seem to be independent of obesity. The clinician should not assume that a child with SDB must be overweight.

A study from Israel found that children with SDB had lower scores on neurocognitive testing compared to controls but the scores improved after treatment. This prospective study of 39 children aged 5 to 9 years underwent a battery of neurocognitive tests. Children with SDB had lower scores compared with healthy children indicating impaired neurocognitive function. Six to 10 months after adenotonsillectomy, the children demonstrated significant improvement in sleep characteristics, as well as in daytime behavior. Their neurocognitive performance improved considerably, reaching the level of the control group. The authors concluded neurocognitive function is impaired in otherwise healthy children with SDB. Most functions improve to the level of the control group, indicating that the impaired neurocognitive functions are mostly reversible. An abrupt and persistent deterioration in grades must also raise the question of abnormal sleep and SDB

In schools the tiredness and sleepiness may be labeled as inattentive or daydreaming. The association between SDB, learning problems, and attention deficit disorder has been established.

Diagnosis

The diagnostic criteria used for adults with OSAS cannot be used reliably in children. The diagnosis of sleep-disordered breathing is based on the history, physical findings, and supportive data. Laboratory testing should be, ideally, tailored to the clinical question.

The polysomnogram in a child uses the same technology and the same type of information that is recorded as in adults. Airflow, respiratory effort, and pulse oximetry make up the breathing measurements usually monitored.

Along with the absence of controlled studies, another problem with understanding of pediatric sleep-disordered breathing is that definitions for key terms vary. Obstructive sleep apneas are defined as lasting at least 10 seconds in adults. However, since children have faster respiratory rates clinically significant apneas can occur in less time. Apneas as brief as 3 or 4 seconds may have associated oxygen desaturations. There is no universally accepted definition of hypopneas in children. The clinician needs to know how apneas and hypopneas are defined and scored when interpreting a polysomnogram report. The most recent edition of the International Classification of Sleep Disorders defines sleep apnea in children as having an AHI > 1. In adults, a higher AHI of 5 is required.

Controversy exists over whether a diagnosis of OSA or the larger spectrum of sleep-disordered breathing should be routinely made without a formal polysomnogram. Some have suggested that this diagnosis can be made in patients using either the history and physical examination alone, or in combination with an audio or videotape of the child sleeping. Others have found an inability of clinical history alone to distinguish primary snoring from OSAS in children. A sleep study is the most definitive test for SDB. Currently, some otolaryngologists who treat sleep-disordered breathing children may make the surgical recommendation based on clinical findings of airway obstruction, sometimes reviewing an audio or video tape. Further controversy exists about the value of doing the sleep study in an attended sleep lab setting or allowing the child to sleep in his or her own bed with a home setting. The clinicians must be aware of the potential pitfalls to these different options. If the sleep study is done in an attended in-lab setting the staff must be trained how to work with children. The bedroom must be able to accommodate a parent or other adult that will accompany the child. SDB events tend to cluster during REM sleep in the last third of the night or early morning. The child should be allowed to awaken spontaneously. If on the contrary the sleep study is terminated abruptly and the child is force to wake up earlier than usual, then the true severity of the SDB may be underestimated.

SDB may occur more often in special populations. Any condition or syndrome associated with cranial facial anomalies may be associated with SDB. Pierre Robin's, Apert's and Crouzon's are among these syndromes. At least half of all children with Down syndrome have SDB. However, symptoms of daytime sleepiness and sleep disruptions at night may be due to non-neurological factors such as maxillofacial abnormalities, large tonsils or adenoids, small jaw, or other abnormalities. Such factors often lead to sleep fragmentation and daytime sleepiness. Sleep disorders often occur in patients with neuromuscular disorder because of associated weakness in respiratory muscles, which is further exacerbated by hypotonia during sleep. Diagnosis and treatment of sleep-disordered breathing in these patients can be an important part of comprehensive management.

Treatment

Not only are the diagnostic criteria different in children from adults but also the treatment. SDB in adults have different treatments options, which may be combined. Sleep specialists and ENT surgeons typically provide treatment. As the technology improves, the treatment may be provided by a wider range of healthcare providers. The most common treatment for adults is continuous positive airway pressure (CPAP) to help split open the upper airway. When CPAP is used correctly snoring should be absent during sleep. As a conservative measure, adults with SDB are advised to sleep off their backs, lose weight and avoid alcohol before sleeping. Unlike adults, for children surgery is the most common treatment for SDB. Adenotonsillectomy is the most common initial treatment for SDB in children. This procedure can be extremely effective and result in dramatic improvements (and very grateful parents). When surgery is being entertained, as a general rule, the adenoids and tonsils should both be removed during the same surgery. It is tempting in very small children to only remove the adenoids if the tonsils do not appear overly enlarged since this allows for less post-operative pain and lower risk of adverse events such as bleeding. This practice should be discouraged since even though the tonsils do not seem enlarged, the surgeon must keep in mind that they are examining a child that is awake and sitting. If only the adenoids are removed there is the risk of residual breathing difficulties remaining needing further surgery.

The anesthesiologist should be familiar with sleep apnea since postoperative pulmonary complications can occur. Children with sleep apnea are often thinner than expected. This may be due to multiple factors including the greater caloric demand of breathing through a narrow airway and possible disruption of growth hormone secretion. Children after sleep apnea surgery may unexpectedly increase their weight.

Surgery does not always completely cure the child's SDB. If surgery is not a viable option for the child, then continuous positive airway pressure (CPAP) therapy should be considered. CPAP uses a small air compressor attached to a mask via a hose. CPAP is effective but can be cumbersome to use. Over time the CPAP devices have become smaller and quieter. The masks have also been improved with many more styles and sizes available. Newer CPAP masks specifically for small children have been created. The most recent advance in positive airway pressure has been the development of machines that can adjust the pressure required to keep the airway open on a breath-by-breath basis. These so-called "smart CPAP" units are quickly becoming the standard treatment option.

An alternative to CPAP for residual SDB is the use of orthodontics for rapid maxillary expansion. The general concept is that expansion of the maxilla will allow more space for the tongue in the hard palate. The tongue is then less likely to fall back into back of the throat and cause obstruction. This treatment option is promising but not yet readily available. It does bring the orthodontist into the increasing large sleep medicine tent further demonstrating the multi-disciplinary nature of sleep. Another optional treatment is the use of inhaled nasal steroids and aggressive treatment of any allergies.

In summary, it is important for clinicians to be aware that SDB and snoring are not normal. Difficulty breathing while asleep may impact the daytime behavior of children. SDB is readily treatable.

NARCOLEPSY

Narcolepsy is a chronic autoimmune neurological disorder in which the boundaries between the awake, sleeping, and dreaming brain are blurred. The awake narcoleptic will feel sleepy. The sleeping narcoleptic will have disturbed sleep due to arousals. The neuropeptide hypocretin, also called orexin, is often dysfunctional. The cardinal features of narcolepsy are daytime somnolence, cataplexy, sleep paralysis and hypnagogic hallucinations.

Clinical Presentation

Narcolepsy is described in detail in chapter 19. To review, it is characterized by abnormal sleep, including excessive daytime sleepiness with often-disturbed nocturnal sleep later in life and pathological manifestations related to rapid eye movement sleep (REM sleep). REM sleep-related abnormalities include early REM sleep onset and cataplexy. It is important to consider narcolepsy, especially in young patients, because it can take up to 20 years between the initial onset of the first symptom, commonly sleepiness, and development of the full clinical syndrome. During this lapse of time, the patient may be mislabeled with wide variety of diagnoses. The patient may be considered lazy or depressed. Prior to being correctly diagnosed, the teenagers and young adults may turn to illegal drugs such as "crank" amphetamine to combat the sleepiness.

Narcolepsy is under-recognized. Since the symptoms may begin gradually in childhood it may be several years before the diagnosis is made. The clinician's lack of suspicion may cause the symptoms to be misidentified as manifestations of psychosis or another mental disorder. These experiences are so frightful that the child may resist going to bed or talking about them. Primary care providers and psychiatrist will often see these patients before the sleep specialists and therefore should consider the possibility of narcolepsy being present in a child with new onset of either sleepiness or hallucinations.

Treatment Strategies

If a primary care provider or psychiatrist suspects a child has symptomatic narcolepsy, they may consider initiating empirical treatment while awaiting a consultation with a sleep specialist if the clinical situation warrants it. However, most medications would be discontinued prior to formal diagnostic testing. Successful treatment for narcolepsy needs to include both behavioral and pharmacological treatments. The situation is analogous to juvenile diabetes mellitus where a combination of diet with medication can control the condition. Treatment strategies must include conservative behavioral treatment such as usage of short naps, emotional support to parents and their child, and education of school authorities. Children with narcolepsy should be encouraged to take one to two brief naps, typically less than 30 minutes, every day. Brief naps are typically refreshing. Only after the importance of naps and proper sleep habits is emphasized should medication be discussed.

Narcolepsy is a lifelong illness and this must be taken into account when considering drug therapy along with possible side effects. Central nervous system stimulants have been the drugs most widely used to treat narcolepsy. The American Academy of Sleep Medicine practice parameter for the treatment of narcolepsy recommends modafinil as a first line treatment. However, there are no guidelines for the use of this medication in children with narcolepsy. A novel agent to treat both the excessive sleepiness and cataplexy symptoms of narcolepsy is sodium oxybate, also referred to as gammahydroxybutyrate (GHB). Illegal use of this substance for recreational purposes has been of great concern. Sodium oxybate has powerful central nervous system depressant effects and can increase slow wave sleep. The medication is a liquid preparation taken twice at night. The product's package insert does not recommend using the medication in patients younger than age 16. In younger patients the off label usage of this medication may necessary.

BEHAVIORAL INSOMNIA OF CHILDHOOD

The symptom of insomnia is defined as "repeated difficulty with sleep initiation, duration, consolidation, or quality that occurs despite adequate time and opportunity to sleep and results in some form of daytime impairment." The clinician must distinguish between insomnia symptoms and insomnia syndromes. The International Classification of Sleep Disorders lists several different insomnia syndromes but there are dozens of other conditions listed that may have insomnia symptoms.

Clinical Presentation

Behavioral insomnia of childhood is a relatively new term that combines two previously described pediatric sleep disorders: sleep onset association disorder and limit setting sleep disorder. Both of these disorders have in common inconsistent enforcement of bedtime routines. In normal infants, brief awakenings throughout the night are common. In reality nobody, neither adult nor child, sleeps through the night. Awakenings lasting several seconds occur at the end of every REM period accompanied by large muscle movements. Small arousals lasting for only approximately 3 seconds may occur up to 10 times per hour. These awakenings and brief arousals are all part of normal sleep. The child and the parent are unaware of most of these, and unbeknownst to the parents, the child may simply drift back to sleep. The concern should not be the awakenings or arousals per se; the real problem for parents is does the child need their assistance to return to sleep. It is only when the child's awakenings disturb the parents' sleep that a problem arises. If a brief awakening is followed by a full arousal in an active, alert child, and the child is unable to fall asleep, the child may simply become agitated. The child might only return to sleep after being comforted by the caregiver.

Although the need for sleep has a biological basis, the way we sleep is learned and influenced by the cultural environment. As the child is readied for bed, the child may associate sleep onset with particular events. The child may associate this process with increased attention and activity rather than getting ready for bed. If continued, the child may then develop an association with parental activity that requires persistent parental involvement in order to fall asleep. The awakenings in the middle of night often necessitate similar actions. What initially were brief awakenings may develop into lengthy struggles. What wakes up the child may not be what is keeping the child awake. Now the parental actions became the focus of the problem. If returning the child to sleep requires frequent parental assistance the parents' sleep may be very disturbed. The entire family may start to dread the nighttime instead of looking forward to going to sleep. The nighttime routines are no longer pleasurable but become a source of frustration. A vicious cycle may form which is exacerbated by middle-of-the night awakenings. Parents may report the infant wakes up crying every hour. The normal sleep cycle length in an infant is approximately one hour and is followed by a brief arousal prior to initiating the next sleep cycle. Therefore, when a parent complains of hourly awakenings with prompt return to sleep with the parents' intervention the possibility of a sleep-onset association disorder should be considered.

Limit-Setting Sleep Disorder is the term used to describe stalling behaviors or refusal to go to bed at the desired time associated with inadequate limit-setting for a child's behaviors. This sleep disorder is usually seen in a child who is sleeping apart from the parents and is old enough to get out of bed or climb out of the crib. The child then may come to the parents' bedroom and disrupt the parents' sleep. Usually there is no enforcement of consistent bedtime rules. There may be possible recurrence of behaviors after nighttime awakenings. When the child does sleep, it is usually of normal quality and duration. The parents may lack an understanding of the importance of setting limits and inadequate knowledge of limit-setting techniques. There may be underlying psychosocial factors limiting the parents' ability to set limits. The limit-setting difficulties may also be manifest in the child's daytime behavior such as tantrums.

Insomnia characteristics fluctuate with age. With infants, the problem usually involves trouble settling to sleep without the parents' assistance. In toddlers, the problem may also involve inconsistent limit-setting with regard to bedtime expectations. By the time children are 8 or 9 years old, they seem to have fewer problems with insomnia. When problems with sleep are present at this age there may be some unresolved fear of being alone at night. This might be expressed as a "fear of the dark". This fear may not be initially volunteered during the clinical interview but a description of the sleep routine and environment may reveal a need for nightlights or hallway[1] lights to be left on at night. Adolescents have a characteristic insomnia pattern associated with sleep onset insomnia without sleep maintenance insomnia called delayed sleep phase syndrome (DSPS). DSPS is such a common sleep disorder among adolescents that it will be discussed in Circadian Disorders chapter.

All of these insomnia types may emerge in different ways in a given individual. In addition to this clinical heterogeneity there are children who are described as being lifelong poor sleeper without any apparent cause. This condition is referred to, for lack of a better term, as idiopathic insomnia and fortunately is considered to be rare.

These insomnia patterns are of a behavioral nature and usually do not need to be treated with medication. Antihistamines and clonidine may be tried in some of these children but there is little evidence to suggest that these preparations provide anything but short-term sedation. In particularly difficult situations, medication can be used as part of a comprehensive treatment program with behavior modification at its core.

Behavioral techniques have been helpful and applied to a variety of clinical situations. The first issue to be addressed in a behavioral sleep problem in an infant or small child is where do the parents want the child to sleep. The child will typically prefer to sleep with the parents; the question is what the parents' preference is. There are many variations possible to where a child will sleep, such as in a crib in a separate room, a crib in the parents' room, in the same bed with the parents, and in the same bed or bedroom of a different relative. The sleep situation may change based on multiple factors such as new siblings or the size of the home. Cultural issues will also play a part in this situation. Once all parents and caretakers agree, then a plan to teach the child what to expect can be developed.

A well-known behavioral approach was popularized by Dr. Richard Ferber at Harvard. This approach has not been without some controversy and the clinician needs to be familiar with it since, undoubtedly, the parents are advised once the child is in bed, to leave the room. If the child cries, wait a certain amount of time before checking in on the child. The suggested

---

[11]

waiting time should be based on how comfortable the family is with the situation and can be charted to help track the progress. When the parent returns to the child's room, they are to soothe the child verbally but avoid picking up, rocking, or feeding the child will gradually increase the amount of time that passes between parental interventions. The time interval should be predetermined and not shortened during the night. Ideally, within a week, the child will learn to sleep without the parents' help.

This method is not desired nor effective for all families. A discussion of different approaches and counter arguments about helping a child sleep at night are beyond the scope of this chapter. Suffice to say, that parents will typically seek professional help only after they have tried one or more of the different popular approaches. The relationship between the parents may be suffering due to frustration and sleep deprivation. Given this situation, a pragmatic approach that takes into account the individual situation is necessary. A judgmental or patronizing attitude from the healthcare provider will not be appreciated. From the authors' personal experience at Stanford usually when an infant or child is brought in for behavioral insomnia, the parents are often isolated without the assistance to extended family members such as the child's grandparents. In some families, grandparents may take on the role of helping the infant or child sleep and decrease the pressure on the parents. In other families, there may be a conflict with extended family members that are playing a role in this behavioral pattern. When getting a history of a putative behavioral problem, it is important to obtain a history of how the child sleeps with other caretakers or environments, such as nannies or daycare centers.

The clinician should keep in mind physical causes for the poor sleep. Obstructive sleep apnea, food allergies, or gastrointestinal reflux may mimic behavioral insomnia of childhood. As a rule of thumb, if the child sleeps better when they share a bed (for example, while traveling) with the parents a behavioral component is likely. If the child sleeps poorly even when given the opportunity to sleep with the parents from the beginning of the night, then a physical factor needs to be explored. This may require a polysomnogram.

The parents may need additional direction to be able to implement firm limits for the child. For example, reassure the parents they have not abandoned the child if they decide to sleep separately or that they are spoiling the infant if they decide to share their bed with the infant. If the child has learned to sleep in the parents' bed and the parents want to change this arrangement they need to realize the infant has no incentive to change. One approach is for the parents to start the night sleeping in the child's room. The amount of time that the parent spends with the child in bed could be decreased gradually until the child learns to sleep alone. These gradual interventions are not meant to be punitive, and require patience, persistence, and reassurance to both the parents and the child.

CHILDREN AND HYPNOTICS

Children with pervasive developmental disorders and mental retardation may be particularly prone to an irregular sleep wake rhythm type of circadian sleep disorder. It is important to keep in mind that waking up is a biological process but getting out of bed is a volitional event. Ask any person why they got out bed on a particular morning and they will almost inevitably say, "because I had to." But what happens when there is no incentive to get out bed? For many of our more impaired children there may not be any motivation to get out of bed in the morning. The principal reason to attempt to regulate their sleep is help the other members of the family sleep better. The importance of helping other family members sleep better cannot be minimized. If the child's caretakers are exhausted due to frequent interruptions of their own sleep, they will not be able to provide optimal care. If the patient's sleep patterns disrupt their sibling's sleep it will further exacerbate a very difficult situation. Before embarking on any pharmacological approach, it is important to determine that sub-acute OSA is not present since many sedatives can worsen sleep apnea.

There is relatively little information available on the pharmacological management of sleep disorders in children. Most pharmacological guidelines were developed for sleep disorders in adults and must be empirically extrapolated to children. The medications are typically neither FDA approved for the specific sleep disorder nor for the pediatric age range. The physician is often forced to prescribe medications as an "off label" indication. This may result in frustrating insurance reimbursement delays or denials for the family. These reimbursement problems may affect the availability of a specific medication, the family's compliance with the medication or force the physician to prescribe a less desirable alternative medication. The medication may not be available commercially in an easily administered format. Certain age range children may not be able to swallow pills, or ingest chewable tablets, would require the local pharmacist to compound the medication into a suspension. In addition, due to the natural aversion among both parents and physicians to use medications for pediatric sleep disorders, medications are usually prescribed as a last resort or in the most refractory situations.

At times, a decision to use medication in a child may be made not necessarily to assist the child as much as to help the parents or other family members sleep better. It is not unusual that parents may finally seek help for a child's longstanding sleep problem when the parents feel they can no longer put up with interruptions to their own sleep.

Further complicating the pharmacological treatment of sleep disorders in children is the general lack of specialized training in sleep disorders available to the health care providers who are working with these children. Failure to consider or properly apply non-drug treatments as part of the comprehensive management of the child may also lead to unsatisfactory results for the patient and the family. These factors result in children with sleep disorders that are not properly managed due to either under-dosing or overdosing of medication or incorrect medication selection.

There may be an over-reliance on medications by both the parents and healthcare provider without adequate attempts at the application of behavioral techniques to help improve the child's sleep. A common scenario in clinical practice is a parents' complaint of a child's paradoxical reaction to a hypnotic medication. "It made it worse" or "he became hyper" may be the parent's complaint. This may occur because the timing or the dose provided was incorrect. The parents may expect that once a medication has been finally prescribed to help their child sleep that it will "knockout" the child. It is important that the healthcare provider advising this family take into account the circadian modulation of alertness. Humans will typically experience an enhanced alertness in the evening, which is often referred to as a "second wind". During this circadian phase, it is harder to fall asleep. If the hypnotic medication is given during this circadian time window, the medication may not work or the child may be frightened by hypnagogic hallucinations.

If a medication has been previously shown to shorten sleep latency by only 20 or 30 minutes, giving this medication 2 or 3 hours prior to the usual falling asleep time could elicit this common scenario. This same medication given at a more appropriate circadian time could be effective. A similar lack of efficacy or paradoxical reaction could occur if the medication dose is inadequate.

CHAPTER 20 Pediatric Sleep Disorders

RESTLESS LEG SYNDROME

Restless legs syndrome (RLS) is discussed in chapter 17. RLS can occur in children. It is a chronic autosomal dominant neurological disorder characterized by leg discomfort, most common in the evening, that makes the patients want to move their legs. Discomfort is relieved with movement. The leg discomfort may be hard to describe and in children may be characterized as "growing pains." In patients with restless leg syndrome, sleep may be disrupted by hundreds of involuntary kicking movements of the legs during sleep called periodic limb movements of sleep.

Restless leg syndrome can result in significant daytime drowsiness as a result of poor sleep quality, which may lead to daytime behavior that mimics Attention-Deficit/ Hyperactivity Disorder. Children with RLS may present with nonspecific symptom such as growing pains, restless sleep, insomnia, and daytime sleepiness, but most often ,these issues go unnoticed by their parents.

PARASOMNIAS

Parasomnias are discussed in chapter 20. As a group, many parasomnias are present in children. Per the International Classification of Sleep Disorders (ICSD), parasomnias are groups of clinical disorders in which undesirable events or experiences occur during entry into sleep, within sleep or during arousals from sleep. The ICSD describes over a dozen different parasomnias, most which can occur in children. In medieval times, the phenomena of sleep paralysis, night terrors, and sleepwalking may have been interpreted as supernatural events. The arousal parasomnias are more common in children and include the clinical spectrum of confusional arousals, sleep terrors, and sleepwalking. These conditions are thought to arise from impaired arousal from sleep, typically slow wave sleep. Since slow wave sleep (stages 3 and 4) dominates in the first third of the night, the arousal parasomnias also occur at that time. Anything that increases the amount of slow wave sleep, such as recovery from sleep deprivation, may increase the likelihood of these parasomnias occurring in susceptible individuals. If necessary, clonazepam may help decrease these arousal parasomnias but should be avoided in the presence of sleep disordered breathing.

Confusional arousals may be partial manifestations of sleep terrors and sleep walking events. The episodes are described, as the name implies, as confusion during and following arousals, usually in the first third of night. The individual is disoriented, speech and mentation are slowed, and response to questioning is confused. These behaviors usually last only a few minutes in children and can be precipitated by forced awakenings out of slow wave sleep. The condition is benign and tends to decrease over time. However, a child with a tendency to have these events may also be at risk for sleepwalking, and parents/caretakers should be warned of this possibility. The events can be minimized by avoiding situations that can increase slow wave sleep or sleep disruption.

Sleep terrors are characterized by a sudden arousal from slow wave sleep with a characteristic "blood-curdling" scream accompanied by intense fear. Pronounced autonomic nervous system discharge may occur with tachycardia and sweating. The child may be described as wide-eyed with an intense look of fear staring past the parents. The prevalence of sleep terrors in children is 3% and decreases with age to about 2% in adults. Arousal is difficult, and if successful, the patient may be confused and disoriented. The child may try to describe fragments of images with poor coherence. These images will not have the rich detail of a nightmare. Waking the child 15 minutes after they have fallen asleep has been a popular treatment. Awakening the child at the point that they would be expected to be entering slow wave sleep seems somehow to decrease the tendency for these parasomnias to occur. When the child returns to sleep the homeostatic mechanism driving slow wave sleep may raise the arousal threshold and decrease the likelihood of the recurrence of the sleep terror. The parents should be warned that the child with sleep terrors might also sleepwalk.

Sleepwalking is also called somnambulism. The term sleepwalking may be a misnomer; perhaps "sleep fleeing" would better characterize the dramatic behavior witnessed. The range of behaviors witnessed may range from simply sitting up in bed to running out of the home and driving an automobile. During the sleepwalking episode, pain thresholds are elevated. The child will be hard to awaken and may be confused and disoriented. The motor activity may terminate spontaneously or the child may simply return to bed without reaching alertness. Sleep talking may occur during this event. The treatments mentioned above for the other arousal parasomnias may be helpful. Safety precautions should always be taken. Specifically, the child should sleep on a ground floor, if possible, to avoid injury while navigation stairs. If the child has a bunk bed, they should use the lower bunk. Drapery should be kept over the bedroom windows to protect the hands of the child in case the child tries to punch a window. A door alarm is helpful to warn the parents if the child walks out of the home. Weapons should be kept away from adults who may sleepwalk. It is helpful to keep precipitants of sleepwalking at a minimum. Thus, sleep deprivation should be avoided. Obstructive sleep apnea may be an aggravator and should be treated if present.

Parasomnias such as sleepwalking and sleep terrors need to be distinguished from nocturnal seizures. The timing of seizures is different from sleepwalking and sleep terrors. Nocturnal seizures do not cluster during slow wave sleep in the first third of the night. The motor activity of generalized tonic-clonic seizures is very different from these parasomnias. Tongue biting and urinary incontinence are not characteristically seen in sleepwalking and night terrors.

Patients with generalized nocturnal seizures have a low risk of daytime seizures. The motor activity of a partial complex seizure may be more difficult to distinguish from parasomnias and may resemble a confusional arousal. Usually, the overall clinical picture can help with the diagnosis. If a patient has been injured during an apparent parasomnia, a comprehensive neurological evaluation should be considered including a sleep-deprived electroencephalogram. Some forms of epilepsy may be misidentified as a benign parasomnia and as narcolepsy in children. Frontal lobe epilepsy is poorly understood and often unrecognized by health care workers dealing with children.

Conclusion:

The field of pediatric sleep disorders needs much more research. We lack sufficient normative and epidemiological data in children. There is a paucity of therapeutic clinical trials and health policy outcomes research. With greater awareness of sleep disorders and their impact in the daytime behavior of children, it is crucial that the field progress in a direction that allows non-sleep specialists to recognize and treat these children. Sleep disorders are so common that if a primary care provider or psychiatrist is not routinely identifying sleep disorders in their clinical population they are probably being missed totally. Sleep disorders in children are fairly easy to treat and their treatment can have a positive impact on the entire family.

> *"Hush little baby, don't say a word*
> *And never mind that noise you heard*
> *It's just the beast under your bed,*
> *In your closet, in your head"-Metallica*

CHAPTER 21 Circadian Rhythm Sleep Disorders

"If I get home before daylight I just might get some sleep tonight"-Friend of the Devil

It is essential for most animals to distinguish the day from the night and behave accordingly. The world becomes a very different place at night. As the seasons change, diurnal creatures such as us have to predict when the sun will rise and when it will set. An active circadian system is necessary in a world where the timing of dawn and dusk fluctuates gradually every day. When this system is out of sync, circadian sleep disorders emerge.

The role of the circadian system's biological clock in providing prolonged alerting influences to oppose the sleep tendency during the day was described in previous chapters. Understanding clock-dependent alerting has also allowed us to understand that certain sleep disorders are directly caused by dysfunctions of the circadian timing system. Some of these sleep disorders are a consequence of an imbalance between our natural tendencies and socio-cultural expectations. This results in the problems such as seen with jet lag and shift work.

Disorders involving the biological clock can be generally grouped into two categories: those in which the problem is imposed by the external environment or schedule and those which involve some abnormality in the circadian timing system itself.

The circadian timing system includes all components that are internal to the organism. For example, blindness would represent an abnormality in the circadian system if it interfered with the ability of light to entrain the biological clock to a 24-hour schedule. The circadian timing system may be divided into (a) input signals to the clock, (b) the internal clock resetting processes, (c) the mechanisms by which the clock actually keeps time, and finally, (d) the output pathways by which the clock induces circadian rhythms in most of our physiological and psychological functions.

Our biological clock has the capacity to phase advance or phase delay. Most of us do this without much difficulty in response to traveling crossing one or two time zones. Yet even the relatively small change of one hour associated with the annual changes to and from Daylight Saving Time can be problematic. A phase delay as compared to a phase advance is usually somewhat easier to accomplish because the natural circadian period is typically slightly longer than 24 hours in humans. For example, because the circadian period is just longer than a day, it is usually easier for humans to stay up one hour later for the "fall back" conversion from daylight savings time, but more difficult to "spring forward" and go to sleep one hour earlier after changing to Daylight Saving Time. Instead of falling asleep earlier, we either get less sleep or wake "later" per the new clock time.

When jet travel or schedule changes produce a substantial mismatch of the internal phase of the clock and the desired sleep-wake schedule, disturbed sleep and impaired alertness can occur. Sleep experts believe that the final common path to disturbed sleep in circadian rhythm disorders is the combination of desynchronized clock-dependent alerting, and the associated chronic sleep deprivation.

There are nine separate circadian sleep disorders listed in the International Classification of Sleep Disorders (*ICSD-2)*. The most commonly encountered and relevant to this book's audience is **delayed sleep phase syndrome.** It is likely that after reading these materials, readers will recognize the syndrome in themselves or among their classmates. The other circadian sleep disorders tend to be more important in specific populations. For example, free running type circadian disorder is more common among the blind and advanced sleep phase syndrome in the elderly. For completeness, the nine circadian disorders are the following:

1. Delayed Sleep Phase Syndrome
2. Advanced Sleep Phase Syndrome
3. Irregular Sleep-Wake Rhythm
4. Circadian Rhythm Sleep Disorder, Free Running Type
5. Jet Lag Disorder
6. Circadian Rhythm Sleep Disorder, Shift Work Type
7. Circadian Rhythm Sleep Disorder Due to a Medical Condition
8. Circadian Rhythm Sleep Disorder Due to Substance Abuse
9. Circadian Rhythm Sleep Disorder not otherwise specified

DELAYED SLEEP PHASE SYNDROME

DSPS is a circadian rhythm sleep disorder characterized by chronic sleep-onset insomnia and an inability to rise at a time in the morning that is appropriate, given the individual's commitments. Once the DSPS patient is able to fall asleep, he or she sleeps soundly and for a normal duration of approximately 8 hours. When not required to maintain a strict sleep schedule (for example, on weekends, or during vacations and holiday periods) patients will awaken spontaneously, albeit late in the morning or early in the afternoon. When they wake up consistently on their own schedule, they usually wake up feeling refreshed. DSPS can be conceptualized as resulting from a clock for which the time can only be readily adjusted in one direction, as it is intrinsically easier for patients to fall asleep later than usual, rather than earlier. People susceptible to DSPS may have more difficulty adjusting to an earlier sleep time (phase advance) than others. DSPS was recognized as a medical disorder in 1981. It was originally described in a group of young adults that had an "atypical depression". The patients in the original report were atypical because they had trouble falling asleep in the beginning of the night, so called sleep-onset insomnia and did not respond to the usual antidepressant medications. This is the opposite pattern classically seen in typical depression. Typically, people with depression have early morning insomnia. They wake up too early and cannot get back to sleep. Once DSPS patients had their sleep schedules corrected their mood improved without needing medication.

Shortly after identifying DSPS in young adults, it was reported to be present in adolescents. DSPS seems to be particularly common among adolescents. Sleep problems in teenagers can present a challenging situation, both for the family and healthcare workers. Many healthcare professionals may have limited training in pediatric sleep disorders. Adolescent sleep problems are especially concerning since it is reported that sleep problems in teenagers are associated with higher rates of suicide.

Clinical Features

The main feature of DSPS is sleep-onset insomnia, a common disorder among teenagers, and DSPS should be included in the differential diagnosis whenever an adolescent is evaluated for these symptoms. After the initial description of DSPS, the clinical features of the syndrome were described in a case series of 22 adolescents by Thorpy et al. in 1988. The ability to experience refreshing sleep is a key factor in a pure case of DSPS; if a teenager does not report waking refreshed after sleeping for several days on a stable schedule of his or her own choosing, a co-morbid condition should be suspected. Sleeping late at a weekend may not provide sufficient time to establish whether or not sleep is truly refreshing. Furthermore, it is crucial for everybody, but teenagers in particular, to realize that chronic sleep restriction over five weekday nights cannot, be "made up" for over a 2-day weekend.

The sleep patterns of teenagers differ from those of children and adults. Younger children are unlikely to stay awake after their parents have gone to sleep, whereas teenagers may be allowed to remain awake after their parents have retired for the night and frequently spend hours awake in their bedroom. It is very unusual for a younger child to sleep in on weekends. Young children wake up at about the same time on the weekdays compared with weekends. As these children get older and start to have different schedules on weekends compared to school nights, those with tendencies toward DSPS will emerge. This may be due to a combination of external societal influences and physiological changes. Extensive research on adolescent sleep has been undertaken in the US, and the external influences on a teenager's sleep are numerous and pervasive. Parents may not be teaching children to value sleep sufficiently. The parents themselves may not be good sleep role models.

When children enter their teenage years, the levels of schoolwork and extracurricular activities expected of them increase significantly. The teenager is generally given more autonomy and the bedtime schedule is often further relaxed. Sleep simply becomes less of a priority. Since the teenager is becoming more adult, parents may incorrectly assume that he or she needs less sleep. It is increasingly apparent that adolescence may be a time of heightened sleep need coupled with a natural tendency to stay awake later. As they get older, teenagers may stay up after their parents are in bed, therefore being allowed to set their own bedtime. Teenagers often want greater freedom, and what better opportunity when they are in their homes than when their parents are asleep? Therefore, teenagers have a number of incentives to stay awake later.

DSPS is associated with depression and should be considered when evaluating teenagers with this disorder. In one study, patients with DSPS manifested increased nervousness, depression, and difficulty regulating emotional displays when compared with a control group without insomnia or psychiatric symptoms, and were defensive, compulsive, introspective, and overly abstract in their thinking.

Broad generalizations made from the results of a single study should be avoided; however, it is important to consider any factors that may influence the motivation of DSPS patients to modify their behavior. Changing sleeping habits can be challenging. The motivation to improve sleep patterns may not be as profound in patients with DSPS compared with those without the condition. As with most chronic conditions, a vicious circle is established.

Treatment

Diagnosing DSPS is not difficult; however, treating DSPS in theory should be easy but in practice it is difficult. The biggest difficulty is that the patient may lack the motivation to make lifestyle changes necessary to reverse the sleep pattern. Both behavioral and pharmacological treatments have been employed, either separately or in combination, to treat DSPS, but attempts to correct the sleep schedule will not be effective unless the patient chooses to alter lifestyle factors influencing the delayed bedtimes, particularly on weekends. The importance of motivation cannot be overemphasized.

One of the benefits of a behavioral treatment approach is the ability of an individual to gain control over his or her sleep, with greater autonomy acquired as the condition improves. A helpful mnemonic in explaining a behavioral approach is "SELF correction". SELF is an acronym for the recognized zeitgebers: Social interactions, Exercise, Light, and Food. By controlling these 4 variables on a regular schedule, the circadian pattern will stabilize over time. Chronotherapy was originally suggested for the treatment of DSPS. Chronotherapy resets the patient's sleep cycle using a series of consecutive, 3-hour delay adjustments of bedtime and wake time over several days. To maintain the readjusted sleep pattern, the patient is encouraged to strictly adhere to the new sleep onset and wake times every day of the week, including weekends. However, this treatment can be impractical, as the progressive forward bedtime shifts will involve the patient temporarily sleeping in the daytime, and constant supervision must be ensured to prevent him or her sleeping at the wrong time. This treatment, although physiologically sound, therefore presents difficulties in its execution

Another technique, phototherapy, resets the sleep– wake rhythm using bright morning light combined with evening light restriction to phase-advance the patient's sleep time. Phototherapy is based on the principle that bright light in the morning, at the end of the habitual sleep period, can phase-advance the circadian clock and hence wake onset, while bright light in the evening can phase-delay the circadian clock and sleep onset. Different phototherapy protocols have been used with varying success, and a pragmatic approach for the individual patient must be developed in each case. However, phototherapy is not advised in patients who are also bipolar, as it may aggravate mania.

Evening light suppresses melatonin activity. This reinforces the importance of minimizing evening light in patients with DSPS. This is in theory a good practice but in reality difficult to accomplish. In 2011 the National Sleep Foundation reported that over 90% of Americans are in front of a TV or computer screen within one hour of bedtime.

Instead of severely limiting light at night, a more practical approach for DSPS is to maximize light exposure upon awakening. A common practice is to use a bright light box or lamp with an intensity of 10,000 lux for 30–45 minutes immediately after awakening, over a period of several weeks. In addition, bright lights should be minimized in the last 2 hours before the expected sleep time, including restricting the use of computer monitors. These bright light devices have shown a phase shift in research subjects.

A practical compromise in teenagers may be to minimize or restrict the use of computers before bedtime, but as a tradeoff to allow their use promptly after awakening. This serves as an incentive for the adolescent to get up on time or early. Adolescents are often delighted and parents perplexed when we recommend as a therapeutic option that they play video games as soon as they wake up. Although the amount of light they receive from the morning screen time is less than the usually recommended light boxes, it does provide more motivation to change behavior.

Treatment of patients with DSPS using melatonin has been described in a number of studies. Melatonin is a hormone secreted by the pineal gland and helps regulate the circadian system. The timing of the melatonin dose must be individually determined. Melatonin is an over-the-counter product and therefore does not have the same degree of regulation as prescription pharmaceuticals. This lack of regulation may cause the quality of over-the-counter melatonin to be highly variable.

Jet Lag

Since the 1950s and the advent of widespread commercial airline travel, increasingly larger numbers of people travel distances which entail rapidly crossing a number of time zones. It is estimated that several hundred million people per year fly on trips that cross one or more time zones. Furthermore, the majority of airline routes are in the east-west direction rather than north and south because so many of the main centers of international commerce are in Europe, the United States, and Asia at similar latitudes. Jet lag can develop in all ages.

What is the jet lag syndrome? It is the collection of symptoms induced in humans by a major and rapid shift in environmental time by "jetting" to a new time zone. The internal biological clock cannot shift equally rapidly and thus will be out of sync with the desired schedule of sleep and wakefulness in the new environment. As stated in, the evening period of clock-dependent alerting in most younger adults is so powerful that it has been called "the forbidden zone" with respect to the likelihood of sleep occurring. Assume this strong period of clock-dependent alerting for a young person is from 6 P.M. to midnight. Assume this young person travels from Chicago to London. Due to the 6-hour time change, the period of clock-dependent alerting would now be from midnight to 6 A.M., and sleep during these conventional bedtime hours would be very difficult if not impossible.

Although we are emphasizing sleep disturbance, there are other consequences of jet lag which, in order of reported frequency, include gastrointestinal disturbances, a general feeling of malaise with headache and lethargy, and finally, urinary difficulties, mainly frequency and diuresis. The sleep disturbance seems clear enough, but other problems have been attributed to a variety of mechanisms, particularly hormone release at the wrong time or not being released at the right time. Almost every frequent jet traveler has a system by which they try to beat jet lag, yet they are rarely able to do it successfully.

The major factor that seems to determine whether or not travelers develop serious insomnia in the new time zone is the amount of sleep deprivation they have experienced before the trip begins. Acute phase shifts in a sleep laboratory setting have shown that sleep satiated subjects have more sleep disturbance after a phase shift than severely sleep deprived subjects. In either case, the large sleep debt will cause the travelers to be tired and sleepy in the daytime. It is also well known that traveling in an eastward direction is usually more difficult than traveling in a westward direction. Again, this is due to the circadian period being slightly longer than 24 hours which makes a phase delay easier to adjust to than a phase advance.

Circadian Rhythm Sleep Disorder, Shift Work Type

Shift work schedules vary almost infinitely. It is almost impossible to have a simple categorization. However, if one schedule is the most common, it is probably the rotating 8-hour shift in which working 8 A.M. to 4 P.M. is the day shift, 4 P.M. to midnight is the swing shift, and midnight to 8 A.M. is the night or graveyard shift. Although the number of days that workers maintain a particular schedule and the specific schedules and directions of rotation vary widely from company to company, the most common rotation is day to night to swing. Even if a worker does not rotate, and stays on a non-daytime shift indefinitely, a good adjustment is extremely difficult because environmental conditions including exposure to sunlight and noise all conspire to maintain a misalignment between the circadian rhythm and the time set aside for sleeping. In addition to this, workers who are on either a night or a swing shift will typically return to a daytime schedule on their off days. They will try, with only partial success, to sleep at night so they can spend more time with their families and participate in conventional social activities.

Shift workers are likely to be profoundly sleep deprived because they never get enough sleep. The impact of sleep loss is compounded by the fact that they do not work at a time of the day when the circadian system is fostering alertness. In one study, over seven thousand rotating shift workers in petroleum, chemical, and manufacturing industries were surveyed. They were asked, "Have you ever fallen asleep on the job?" The data were divided into night shift (12 A.M.-8 A.M.), evening shift (4 P.M.-12 A.M.), and day shift (8 A.M.-4 P.M.).

Overall, slightly more than 40 percent of these shift workers reported they had actually fallen asleep one or more times while working. At worst, such unintended sleep episodes can cause catastrophic accidents. At best, optimally productive work is certainly not being accomplished by workers who are falling asleep on the job. While on the night shift the overall figure was 62 percent, on the day shift, the overall figure was 35 percent, and on the swing shift 25 percent reported falling asleep on the job. It is of interest that these data reflect the circadian pattern of alertness described earlier in this book. Even with the schedule disruption and fatigue that typically accompanies shift work, the effect of strong clock-dependent alerting in the evening was apparent in producing the lowest percentage of unintended sleep episodes on the swing shift. Excessive sleepiness is an occupational hazard of shift work. Shift worker disorder is recognized as such a large problem in industrialized society that the Food and Drug Administration, FDA, has approved the use of medications containing modafinil to combat excessive sleepiness during the shift work in patients diagnosed with a circadian disorder. If a patient is diagnosed as having excessive sleepiness due to shift work, prescribing modafinil is an approved indication and can also be covered by health insurance. You would think that if a person had a medical condition caused by shift work the recommended cure would be to stop doing shift work. However, the FDA recognizes that this impractical solution and our society depends on shift workers maintaining alertness and therefore needs solutions to combat excessive sleepiness.

Advanced Sleep Phase Syndrome

The **advanced sleep phase syndrome** (ASPS) is essentially the mirror image of DSPS. The usual complaint is waking too early and being unable to return to sleep. Further questioning should reveal a tendency to become unusually tired or sleepy in the evening. Usually, however, individuals with ASPS are prevented from going to bed when they are tired because of the need to partake in the evening meal, engage in social activities, and so on. If an individual with ASPS is awakening at 4 A.M., adequate sleep would dictate at least an 8 P.M. bedtime, but this is almost impossibly in modern society.

One important aspect of ASPS is its higher prevalence in older individuals. Some investigators have reported a shortening of the period of the circadian clock with advanced age in animals. Studies of older humans, however, have not shown clear-cut results in this regard. At this time, the tendency of the circadian system to be advanced in older humans has no widely accepted explanation. ASPS is relatively rare in college students and young adults.

Close Up

In 1970-71, I was the resident fellow in a Stanford University dormitory that housed 90 undergraduates. We actually set up a sleep laboratory in the basement. I trained a small group of very bright students to attach electrodes and to prepare volunteers for sleep recordings. We had just opened the Stanford Sleep Disorders Clinic and I was interested in sleep onset insomnia. I was interested in how long it would take students to fall asleep. Thirty-six students volunteered to go to bed in the sleep laboratory at 10:00 P.M. Thirty-one took longer than 30 minutes to fall asleep, which conformed to the then conventional definition of insomnia. Four fell asleep within 5 to 10 minutes, and one required about 20. We then had the 31 "insomniacs" go to bed at midnight. And still 18 required 30 or more minutes to fall asleep. Thirteen fell asleep within 10 minutes. I don't think any of the students considered themselves insomniacs, but they had as much trouble getting to sleep as patients whose complaint was difficulty falling asleep. Of course, we now know that the inability of the students to fall asleep was entirely due to late-night, clock-dependent alerting. - William C. Dement, The Promise of Sleep

Conclusions

Circadian sleep disorders are a diverse group of conditions that have in common a mismatch between inner biological rhythms and environmental expectations. These disorders can occur in all ages. Of particular interest is delayed sleep phase in adolescents. Both behavioral approaches and medications can be effective in the treatment of these conditions. The prevalence of circadian sleep disorders is likely to become more prevalent as we continue to become increasingly active throughout our entire 24-hour society.

"I dream there were means of borrowing sleep by sleeping longer than normal, ahead of time." -John Neves

CHAPTER 22 Sleep Education & Public Policy

"There is a tremendous gap between public opinion and public policy," -Noam Chomsky

Close Up #1:

For more than 40 years I have fought to change the way society deals with sleep. During this time, I have made countless appearances on radio and TV. I have authored articles for magazines and newspapers and have been interviewed for publications countless times. I have written a book called <u>The Promise of Sleep</u> *whose purpose, still pretty much unfulfilled to this day, was to educate the general public. It is now available in paperback. I have provided expert consultation a myriad of times to federal and state agencies and have done my best to mobilize colleagues and citizens to become activists and to campaign for elected officials that are likely to help in this crusade.*

Notably, I have made several hundred trips to Washington D.C. chairing a congressional commission mandated to conduct a major study of American society addressing problems related to sleep and the potential solutions. Unfortunately, while the U.S. Congress created the National Commission on Sleep Disorders Research it failed to appropriate funding for its mandated studies of sleep deprivation and sleep disorders in American society (this Commission was one of the first of the so-called and despised "unfunded Congressional mandates"). Because of this, it was necessary for personnel from the Stanford University Sleep Disorders Center to carry the ball.

In this chapter, I will describe some of the current societal initiatives counteracting widespread societal fatigue and exigent lack of awareness. Because much of the data and many of the examples that I have described come from my direct experience of what I learned in my many forays into the nation's capital and on the National Commission, I have decided not to labor and strain to ensure that the text is always in the passive voice or in the third person. In the other chapters, there are "close ups," direct experiences or excerpts from other material. This chapter may be regarded as more of a first-person narrative, a personalized account of the state of sleep awareness in this country. - William C. Dement

CHAPTER 22 Sleep Education & Public Policy

Sleep Is Not Yet On Society's Front Burner

We live in a highly programmed and competitive 24-hour society. If we need more time to satisfactorily organize our lives and achieve our goals, most of us readily sacrifice our sleep. Survey after survey has shown that Americans place a high priority on work, study, diet, exercise, and recreation, but not on sleep. As a result, we continue to be a sleep deprived society, the negative consequences of which can be seen everywhere.

In today's world, stress and pressure are ubiquitous. At the same time, both stress and pressure affect sleep. In urban centers and inner cities, there are crowds and violence. One can imagine that getting a good night's sleep may someday become nearly impossible as the population swells and our cities and towns become even more crowded and violent. Sooner or later, individuals and society as a whole must receive adequate education about sleep-related matters. Only then will each person be able to manage his or her life in light of the known facts about sleep and the impact of inadequate or unhealthy sleep on waking function. Only then will each individual be able to resist excessive demands from the workplace, the classroom, and his or her social life in favor of obtaining adequate, healthy sleep.

The quality of life, health, and safety of every human being depends upon being reasonably alert when awake. This alertness should be the number one concern of all people for themselves and those around them. As we have noted, many people, both because of cultural/societal factors and because of the unreliability of subjective awareness, do not realize how sleep deprived they are. It is imperative that everyone have a clear understanding about why they are sometimes persistently tired although they seem to be getting enough sleep. They should also clearly understand various conditions and pathologies that affect their ability to be alert and attentive. Once again, I must emphasize that sleep occupies a third of our lives, and that if our sleep is inadequate, unhealthy, or both, the quality and health of the other two-thirds of our lives are seriously undermined.

Over the years, a great deal has been learned about sleep, yet even today (circa 2014) the facts about sleep deprivation and sleep disorders are not at all widely known. As a consequence, what could be done to change the way society views sleep has not been achieved. Although impaired alertness causes human beings to be inattentive, to make mistakes, to have accidents, to be less productive, to be less able to learn, to be less motivated, and to be less happy, the pervasive problem of sleep-related fatigue remains largely unaddressed.

At the present time, our society and its citizens remain almost entirely uneducated about sleep. If there is a prevailing societal attitude regarding sleep loss, it is that giving in to the need for sleep is a weakness. We must change this attitude. We now have the requisite knowledge to develop more rational guidelines, policies and rules about sleep management and fatigue for all components of society. The highest priority should be accorded to educating individuals who are at high risk (military) for sleep deprivation and suffering the consequences of disturbed sleep. High-risk individuals include shift workers, health and emergency professionals, law enforcement personnel, and transportation workers.

Regarding the more general issues related to public policy, the most urgent item to be addressed is the negative impact of sleep disorders and sleep deprivation on public safety. There are countless situations in which a lapse of consciousness can spell disaster. These range from a driver simply drifting off the road to sleepy workers at a power plant allowing a nuclear meltdown. Although epileptic seizures and syncopal attacks are occasionally responsible for lapses of consciousness, the number one cause today is inadequate sleep. A large sleep debt is associated with inattention, impaired performance, and unintended sleep episodes. As society's tolerance of smoking and drunk driving has substantially diminished, society's tolerance of sleep deprivation and its negative consequences must also come to an end!

The National Transportation Safety Board
(NTSB) is the single federal agency that has unambiguously tackled this challenge. The Board has developed guidelines for identifying sleep deprivation as the direct cause or a contributing cause of accidents. These guidelines include assessment of the uninterrupted duration of wakefulness immediately prior to the accident and the sleep/wake schedule during the 72-hour period of time before the accident, as well as circadian factors and possible sleep disorders. It is with great pride that I write that a leader at the NTSB is the Honorable Mark R. Rosekind who spearheaded efforts to increase the efforts of the importance of fatigue and transportation safety. Mark R. Rosekind, Ph.D., was sworn in as the 15th Administrator of the National Highway Traffic Safety Administration (NHTSA) on December 22, 2014.

CHAPTER 22 Sleep Education & Public Policy

He was nominated by President Obama and confirmed by the U.S. Senate. Before becoming NHTSA Administrator, Dr. Rosekind served as the 40th Member of the National Transportation Safety Board (NTSB) from 2010 to 2014. Honorable Mark Rosekind is not just a Stanford Alumni but also a past teaching assistant for Sleep and Dreams. Other federal and state agencies should follow NTSB's lead.

Approximately 100,000 vehicular accidents a year in the United States are officially attributed to fatigue and sleepiness (because automobile accidents are rarely investigated in a manner that will document the causal role of fatigue, the numbers of accidents which are attributed to sleep deprivation are always underestimated). In a substantial portion of these accidents, the result is severe injury or death to one or more individuals and extensive damage to property. Occasionally, accidents caused by sleepiness or inappropriate sleep episodes can have truly disastrous consequences. One of the most well-known catastrophes is the grounding of the giant oil tanker, the Exxon Valdez. Because these accidents were the result of human error, and because they occurred at night and involved shift workers, we assume that other catastrophes such as the Chernobyl nuclear meltdown, the Three-Mile Island near meltdown, and the Bhopal chemical disaster in India, though not properly investigated, were largely due to sleep deprivation. Finally, the erroneous decision to launch the space shuttle, Challenger, leading to its tragic and costly explosion, was made by very severely sleep deprived NASA managers.

Technological Advances Undermine Sleep

The increasing complexity of society has sharply increased the pressure to accomplish an increasingly larger number of tasks in ever-shorter periods of time. The advent of electricity and the light bulb has allowed the functional, productive day to be greatly extended beyond our biological capacity. The rotating shift work schedule is clearly the product of this industrial evolution. The transition of society to 24-hour operations occurred without taking into account scientific knowledge of sleep and biological rhythms. Yet today the applications of fundamental sleep and circadian principles have been thoroughly investigated and the resulting benefits have been demonstrated.

As technological advances, have nearly eliminated equipment breakdown and mechanical sources of accidents, sleep deprivation and fatigue have become a major cause of failure, error, and accidents throughout society. Automation has further magnified the negative effects of reduced alertness by increasing the monotony involved in many jobs. Maintaining vigilance in a monotonous and unchallengeable situation is especially difficult when we are sleep deprived. At present, the National Transportation Safety Board estimates that human error is involved in 90 percent of all industrial and transportation accidents.

The National Sleep Foundation 2011 poll exploring connections between technology and sleep shows that artificial light exposure between dusk and the time we go to bed at night suppresses release of the sleep-promoting hormone melatonin, enhances alertness and shifts circadian rhythms to a later hour—making it more difficult to fall asleep. Americans report very active technology use in the hour before trying to sleep. Almost everyone surveyed, 95%, uses some type of electronics like a television, computer, video game or cell phone at least a few nights a week within the hour before bed.

Unfortunately, cell phones and computers, which make our lives more productive and enjoyable, may also be abused to the point that they contribute to getting less sleep at night leaving millions of Americans functioning poorly the next day. Cell phones were sometimes a sleep disturbance. About in one in ten say that they are awakened after they go to bed every night or almost every night by a phone call, text message or email. Over the last 50 years, we've seen how television viewing has grown to be a near constant before bed, and now we are seeing new information technologies such as laptops, cell phones, video games and music devices rapidly gaining the same status.

Direct Evidence for Pervasive Sleep Deprivation

As a matter of public safety and health we cannot treat sleep an inconvenience. Sleep deprivation cannot be excused as a necessary evil of modern society. Investigators, utilizing the Multiple Sleep Latency Test (MSLT) have shown that severe sleepiness is pervasive in a number of communities and workplace environments. In the typical study, random samples of individuals are asked to volunteer to undergo MSLT testing. Almost without exception, very high percentages of volunteers are found to be in the "twilight zone," with sleep latency scores between 0 and 5 minutes. One group of investigators at the Henry Ford Hospital in Detroit, Michigan, has tested unusually large numbers of "normal" volunteers (well over 1,000 young adults) chosen because they claimed to be fully alert throughout the entire daytime. In the Henry Ford studies, fewer than 20 percent of those tested actually had optimal levels of alertness. We may assume that if subjects had been tested who actually admitted they experienced daytime fatigue and sleepiness, the results would have been much worse. This is one reason why we conclude that America is a sleep-deprived society, and that nearly all of us need more sleep than we get.

There is an unfortunate tendency for uninformed individuals to suggest that it is normal to be sleepy in the daytime. To test this, the above investigators selected a subset of their twilight zone subjects and persuaded them to spend 10 hours in bed for seven to fourteen consecutive nights. As described in Chapter 4, increasing the time in bed led to a progressive improvement in daytime alertness with the degree of improvement closely related to the overall amount of extra sleep. Furthermore, cognitive performance, which had been exhaustively measured in the baseline period, also improved. From this we may draw the following far-reaching conclusion: Many individuals who think they are functioning at peak alertness and at their full intellectual capacity are actually impaired due to periods of reduced sleep time. In order to realize our full physical and mental potential, we need to reduce our sleep indebtedness to the minimum level and then stay at this level by sleeping the amount we need each night.

The Challenge: Sleep and Public Policy

The challenge for our society and its future is to develop the comprehensive approaches necessary for resolving the thorny economic, social, and health issues associated with people obtaining adequate sleep. Such approaches are essential if the scientific principles of sleep research are to exert a positive influence on public policy. In a more or less stepwise fashion, the requirements for change are research, education, strategic planning and leadership. A glance at other examples of public policy change may be instructive. After the Surgeon General of the United States reported that cigarette smoking was a major cause of lung cancer, strategies were developed to warn consumers and to foster smoke-free environments. Similarly, a possible reduction in automobile crash fatalities and injuries required the development of effective seat belts and legislation demanding their use. There are also numerous examples from society's efforts to control communicable diseases. These range from required social programs, such as widespread vaccination against small pox and polio, to monitoring the safety of foods and drugs by a number of federal and state agencies.

As we review areas of our modern lives that are especially vulnerable to severe sleepiness and inappropriate sleep episodes, we can see the negative effects of the absence of adequate conceptual guidelines. For example, a driver who causes a serious accident will almost always have a blood alcohol test. However, the quality and quantity of the driver's sleep in the last 72, or even 24 to 48 hours, will almost never be assessed. Effectively dealing with the dangers of driving when alertness is impaired by sleep deprivation and/or sleep disorders will require more research on countermeasures such as roadside warning signs, alerting technology, safe sleeping areas, and tests to assess level of impairment.

These ideas are relatively new and quite radical relative to the lax attitude society has toward sleep. Not only will society need to develop these approaches, but it will also need to demand the acquisition of accurate information on the prevalence of sleep-related mishaps and detailed information on the consequences of inadequate sleep. This information is vital to design specific solutions and to target populations to whom these solutions will be the most beneficial.

Presently, much of what we know or suspect about the role of sleep and sleepiness in public and private catastrophes is gleaned inferentially from retrospective studies and time-of-day data. In spite of these caveats, we know enough to take action and start saving lives now! This is an important message, and one emphasized throughout the book. The most daunting challenge faced by the advocates of wise public policy involving sleep is the great difficulty of persuading legislators and policy makers to act when they are completely unacquainted with the issues and underlying scientific information.

There is a marked tendency for people to be inappropriately skeptical of such concepts as sleep debt and the dangers of snoring. Imagine that an audience completely unaware of the atomic structure of matter was told that all matter is made of six quarks named "up, down, charm, flavor, top, and bottom." The audience would surely be skeptical. However, physics in some form or other is an essential part of every middle and high school curriculum along with a large number of students in higher education. Accordingly, whatever physicists tell us about the ultimate structure of matter is readily accepted.

On the other hand, the concept of sleep debt is commonly rejected because the lack of education about sleep leads to a tendency to retain mythological and skeptical thinking. Most people continue to believe that the noon meal causes them to be sleepy, and find it hard to believe they are wrong. Along with efforts to change the way society deals with sleep, enacted by the U.S. Congress and federal agencies such as the Department of Transportation and Centers for Disease

Control, a simultaneous effort to penetrate the mainstream educational system at every level should always be part of the crusade. Until we are working with educated policy makers, we will have enormous difficulty in competing with all the other public health concerns and obtaining our fair share of attention and resources.

Sleep, Safety, Productivity: Areas of Impact

Students

As future productive citizens, workers, and leaders of our society, students deserve special mention. Almost all high school and college students are very sleep deprived. Numerous studies have documented the performance and learning impairments associated with sleep deprivation. Moreover, students are at risk for a number of serious consequences, including poor academic performance, increased incidence of automobile accidents, increased moodiness, and increased use of stimulants and alcohol. Adolescents typically go to bed at later bedtimes than adults, but school start times require that they continue to rise early. Some teenagers and young adults develop unambiguous Delayed Sleep Phase Syndrome, a problem that is clearly incompatible with school schedules.

Because so many students beyond the age of ten do not get enough sleep, the sleep-deprived student is the norm. Knowledge of the effects of sleep deprivation in this group has been available for decades, but no comprehensive action has been taken to reverse the situation. We must prime parents and educators for constructive changes in the home and the classroom. Among the potential remedies are year-round schools with daily schedules that permit suitable bedtime hours. There is no precise way to determine how much students' lives and careers are adversely affected by current schedules, but the incidence of poor grades and the frequency of dropouts may be good indicators. National Sleep Foundation 2006 found that more than a fourth of high school students fall asleep in class and rarely parents are not notified. In the same survey, 22% admit to falling asleep doing homework. Obviously if you are falling asleep during homework you are not learning anything. More than one half (51%) of adolescent drivers have driven drowsy during the past year. In fact, 15% of drivers in 10th to 12th grades drive drowsy at least once a week.

Although it has been reported in the scientific literature that young children and prepubertal adolescents generally have regular hours in bed on both weekdays and weekends and as a consequence are wide awake and alert all day long, an alarming new trend has recently appeared. Elementary school children in many communities are also chronically sleep deprived and falling asleep in class. This appears to be due mainly to 24-hour access to cable television, the Internet, and the failure of parents to understand the cumulative consequences of chronic sleep loss. The principal of an inner city elementary school reported major improvement in disposition and alertness in the classroom when she insisted that parents enforce an 8 P.M. curfew on television watching by their children, and require a 9 P.M. bedtime.

Transportation: Highway Safety

A number of studies throughout the world have reported that motor vehicle accidents do not occur randomly throughout the 24-hour day. Rather, they tend to peak during the early morning and mid-afternoon in accordance with peaks in human sleep tendency. A study in Great Britain estimated that 27 percent of drivers who lost consciousness behind the wheel fell asleep. The remainder of the lapses were caused by fainting, having a seizure, or experiencing a heart attack. However, falling asleep had disproportionately negative consequences. The 27 percent experiencing a sleep episode while driving accounted for 83 percent of the fatalities! The increased fatality rate is probably due to the fact that once a driver has fallen asleep, there is a complete loss of consciousness and thus no attempt to brake or otherwise avoid a collision.

Trucking

In 1990, on the heels of their report identifying fatigue as the direct cause of the Exxon Valdez grounding, the National Transportation Safety Board reported its studies showing that fatigue is the most frequent direct cause of truck accidents in which the driver is killed. It doesn't take a rocket scientist to realize that long-haul truck drivers are almost by definition severely sleep deprived. However, as the National Commission on Sleep Disorders Research delved into these issues, we also began to wonder how many long-haul truck drivers also suffered from the prevalent sleep disorder, obstructive sleep apnea.

In 1991, the second year of the National Commission's existence, we persuaded the safety manager of a company whose trucks traveled all over the United States to allow a sleep study of the drivers. In retrospect, it is clear he did not know what he was getting into. After making appropriate arrangements, several Commissioners established themselves in the company's transportation hub along with portable sleep monitoring equipment and began the study. We would enter the dormitory or cafeteria and ask drivers if they would sleep one night with our equipment. We also had them answer a series of questions about sleep, alertness, and driving. We were able to question 602 drivers (90 percent males), as well as carrying out overnight sleep recordings on a subset of 200 drivers.

The first big surprise was their response to the question "When do you feel you should stop driving?" Eighty-two percent of the drivers replied that they would stop driving when they had a startle resulting from a head drop, or when they saw something in the road that wasn't there (a hypnagogic hallucination). Studies have shown that these experiences are associated with microsleeps. In other words, for these truck drivers the signal to stop driving was falling asleep at the wheel.

The second big surprise was the huge proportion of drivers who had obstructive sleep apnea. Well over 70 percent were diagnosed with this disorder. This figure is about three times higher than the prevalence found in the general male population. In about 13 percent of drivers, the obstructive sleep apnea was very severe. Given all the other information we have uncovered about the high rates of accidents among sleep apnea patients, this was alarming news.

When we told the company about our findings, an "iron curtain" fell between us. Everyone in the company abruptly became unavailable to us. The reasons are not clear, but likely included skepticism about such dramatic findings and a perceived threat to important economic interests. Of course, it did not help that not one driver we encountered, nor one manager, nor one safety manager, nor even the three onsite company physicians had ever heard of obstructive sleep apnea or were aware of its significance. Although we did not make any headway with the trucking company, we were far enough along in our research to put a paragraph about this study in the final report of the National Commission which we submitted to the U.S. Congress in October 1992. Fortunately, Senator Mark Hatfield of Oregon, whom I am convinced actually read our final report, reserved $750,000 for sleep apnea research in truck drivers.

The National Transportation Safety Board subsequently reported "fatigue" as the leading cause of all heavy truck accidents in addition to the leading cause of fatal truck accidents. While the term "fatigue" has been used throughout federal agencies to describe human performance failure attributable to a variety of factors, it is clear from the National Transportation Safety Board's text that their intended meaning of the term "fatigue" is most congruent with what sleep specialists mean by the consequences of sleep deprivation.

There are approximately 5,000 fatal accidents each year involving trucks. When a truck driver dies in such an accident, approximately eight other people on average are also killed. Depending upon the involvement of other vehicles and the nature of the cargo, associated damages may be enormous. The mean cost of a crash involving a single large vehicle in today's society is over $100,000, but when fatalities result, the direct and indirect costs per crash average almost over $4 million.

The trucking industry employs over two million drivers and countless other maintenance and management individuals. It operates at least a million motor carriers and travels at least ten billion miles a year over the nation's highways. Trucking accounts for approximately ten percent of the gross national product of the United States.

The economic incentives for driving long hours are compelling but the time-on-task guidelines and regulations promulgated in the 1930s are inadequate in the face of current knowledge of sleep science, circadian rhythms, sleep disorders, and the economic forces affecting trucking. Perhaps the understandable fear of harming such a vital industry has, until recently, prevented the initiation of studies to help reduce accidents due to sleep deprivation and driving drowsy by truck drivers.

Automobiles

Long distance driving is routine for truckers, but for ordinary drivers, long trips usually represent a serious disruption of their normal schedule. A group of researchers in France, with the help of the French highway patrol, randomly stopped 2,197 automobile drivers one holiday weekend. They found that 80 percent of the drivers were on vacation, and half had decreased their total sleep time before departure that morning. A quarter of these had lost more than three hours of sleep before their departure. Therefore, long distant driving may be hazardous.

Poor scheduling with a reduced amount of sleep is not the sole danger. Many drivers are fatigued because they have a sleep disorder that makes it impossible to get an adequate night's sleep. A study of 6,000 patients with sleep apnea, for example, found that 15.6 percent had experienced at least one car accident, as opposed to only 6.7 percent for drivers in the non-apnea control group who reported ever having an accident. Stated differently, people with sleep apnea are more than twice as likely to be involved in at least one car accident over the course of their lifetime as people without apnea.

Even worse, the combination of alcohol use (two or more drinks per day) with severe apnea was associated with a fivefold increase in sleep-related accidents compared to healthy drivers with the same amount of alcohol use. When patients' apnea is treated, accident rates fall to the level of the population at large. This is yet another reason to diagnose and treat sleep apnea.

Railroads

Sleep deprivation has been implicated in a number of major accidents on the railroads. The National Transportation Safety Board has concluded that sleep deprivation and related impairment was a primary cause in at least ten catastrophic railroad accidents between 1987 and 1998. Scheduling work is the major problem in the railroad industry. More than in other modes of transportation, railroad personnel are on call and are never able to predict when their sleep will be interrupted. It is very difficult for them to make plans far enough in advance to schedule adequate time for sleep. Most recently in December 2013, a train in New York City derailed. Four passengers died and 75 others were injured. Preliminary reports are that the locomotive engineer suffered from sleep apnea and may have been a factor in the accident. The engineer said he "basically nodded off" just before the train derailed at 82 mph.

Maritime

Maritime transportation typically involves 24-hour operations and sleep deprivation has been identified as the cause of several disasters. As mentioned earlier, one of the most widely publicized of such incidents involved the grounding of the supertanker Exxon Valdez on the Bligh Reef in Alaska's Prince William Sound in March 1989. The media flooded us with allegations of the captain's alcohol consumption and subsequent culpability for the accident. After more than a year of investigation, the National Transportation Safety Board determined that the "probable cause of the grounding of the Exxon Valdez was the failure of the third mate to properly maneuver the vessel because of fatigue and excessive workload..." This man was so sleep-deprived that he was unable to respond appropriately to a variety of signals warning that the tanker was headed for the reef.

Exxon Valdez Oil Spill

The night of March 24, 1989 was cold and calm and the air crystalline as the giant Exxon Valdez oil tanker pulled out of Valdez, Alaska into the tranquil waters of the Prince William Sound. In these clearest of possible conditions, the ship made a planned turn out of the shipping channel, but did not turn back in time. The huge tanker ran aground, spilling over ten million of gallons of crude oil into the sound. The cost of the cleanup effort was over $2 billion. The ultimate cost of continuing environmental damage is incalculable. Furthermore, when the civil trial was finally over in the summer of 1995, the Exxon Corporation was assessed an additional $5 billion in punitive damages. Nearly everyone I query in my travels vividly recalls the accident, and most have the impression that it had something to do with the Captain's alcohol consumption. I never encounter anyone who is aware of the true cause of the tragedy. In its final report, the National Transportation Safety Board (NTSB) found that sleep deprivation and sleep debt were direct causes of the accident. This stunning result unfortunately received only the briefest of mentions in the back pages of a few newspapers.

The Exxon Valdez disaster offers a good example of how sleep debt can create a tragedy and how the true villain - sleep indebtedness – can remain concealed. I am sure that I was just as shocked as anyone when I learned about up to that time America's worst oil spill. The TV coverage of the dead birds and oil-coated seals filled me with outrage over the extent to which the environment had been devastated. One of my friends went to Alaska and participated in the cleanup. He brought back photos and a big jar of crude oil. If you haven't been previously exposed to crude oil, keep away from it. It isn't the purified stuff that goes into your car. It's awful. It stinks so badly that you feel sick to your stomach.

When I watched Congressional debates over what to do about the spill, I had no idea that it would have a special meaning for me a year later. The National Commission on Sleep Disorders Research, mandated by Congress in 1988, finally convened for the first time in March 1990 in Washington, D.C. After the first meeting, I decided to visit a friend, Dr. John Lauber, who had been confirmed by the Senate as one of five members on the National Transportation Safety Board. I had worked with John a few years earlier on a study exploring layover sleep in cockpit crews on Pan American, Lufthansa, British Airways, and Japan Airlines.

John was then head of human factors research at NASA Ames Research Center in Moffett Field, CA and at the beginning of the layover study knew little about "sleep debt." By the end of the study, he was one of the few real sleep experts in the world. When I visited him after the Commission meeting, John told me that the board would very likely identify sleep deprivation as the "direct cause" of the grounding of the Exxon Valdez. Two months later he sent me the NTSB's final report.

The report noted that there were ice floes across part of the shipping lane on the March night when the Exxon Valdez steamed out of Valdez, forcing the ship to change course to avoid them. Once the ship began this maneuver,

the captain turned over command to the third mate and instructed him to turn the tanker back into the regular channel when it was a beam of a well-known landmark, Busby Island. Although news reports linked much of what happened next to the captain's alcohol consumption, the captain was in fact off the bridge well before the accident. The direct cause of America's worst oil spill was the behavior of the man at the helm, the third mate, who had slept only 6 hours in the previous 48 hours and was severely sleep deprived.

As the Exxon Valdez passed Busby Island, the third mate ordered the helm to starboard, but he did not notice that the ship was still on autopilot and was not responding. Instead it plowed farther out of the channel. Twice lookouts told the third mate about the position of lights marking the reef, but he did not change or check his previous orders. His brain was not interpreting the extreme danger of what they were telling him. Finally, he noticed that he was still far outside the regular channel. He turned off the autopilot, and tried hard to get the ship pointed back to safety, but it was too late.

For several years, I would ask every audience that I addressed if there was anyone in the audience who had not heard the words "Exxon Valdez." A hand was never raised. Then I would say, "Who knows what caused the grounding?" Many hands would be raised, and the answer would always be "alcohol." Clearly, the potential impact of this catastrophe in getting knowledge about sleep into the mainstream was not exploited because of the media emphasis on the captain's drinking. When the report of NTSB's investigation was finally released, most of the public's interest had faded. Even at the civil trial in the summer of 1995, the true cause of the accident received little attention. Everyone should be discussing how to deal with and avoid sleep deprivation, both in the transportation industry and in society at large, citing such examples as the Exxon Valdez crash. Instead, people cite alcohol as the paramount factor behind the tragedy.

World Prodigy Grounding

The June 1989 grounding of the oil tanker, World Prodigy, off the coast of Rhode Island is another example of a major maritime accident involving sleep deprivation. In that case, the National Transportation Safety Board determined that the "probable cause of the grounding of the World Prodigy was the captain's judgment impaired from acute fatigue..." The captain had been continuously awake for 36 hours at the time of the accident.

Star Princess and Sleep Apnea

In a first ever attribution of an accident to a sleep disorder, the National Transportation Safety Board found that the grounding of the huge passenger ship, Star Princess, in Juneau Sound, Alaska on June 23, 1995 was the direct result of the pilot's chronic fatigue caused by obstructive

sleep apnea. On the evening of June 22, the Star Princess left the dock at Skagway bound for Juneau and was under the temporary control of a professional marine pilot. At 1:42 A.M., the pilot steered the vessel straight into the gigantic and well known "Poundstone Rock" that rises hundreds of feet above the water surface at a location far removed from the designated shipping lanes. Fortunately, the vessel did not sink; otherwise, it might have been the worst marine disaster in history with 1,500 passengers drowning in the freezing waters of Juneau Sound. Of course, the vacation plans of the passengers were drastically altered and they all required emergency housing. The damages were assessed at more than $27 million.

While the NTSB investigation was ongoing, several lawsuits were initiated. In an extremely lucky coincidence, the attorney defending the marine pilot had previously been diagnosed and successfully treated for obstructive sleep apnea. He recognized the very obvious symptoms in his client and sent him to a sleep disorders clinic for testing. After the positive test results became available, the NTSB was informed. While the defense attorney successfully defended his client, the pilot nonetheless lost his marine piloting license.

He felt that the whole episode was a blessing in disguise, however, because after treatment for his severe sleep disorder, he "could not remember ever having felt so good."

Aeronautics and Space

Flying planes used to be a relatively straightforward task. Airplanes had a few basic controls, and flights were slow and short. However, multi-engine commercial aircraft have grown progressively more complicated. Automated aircraft controls, including autopilots and satellite navigation, make the pilot's job both easier and more complex, creating new possibilities for error. In fact, commercial passenger carrying airplanes themselves are generally so thoroughly designed and tested, and weather tracking so improved, that pilot error and not equipment failure is now the most common cause of airplane accidents. This fact has increased the importance of learning much more about how humans perform when they are operating complex machinery, along with how and why humans make mistakes, both of which have been synthesized into a discipline known as human factors.

Now that airplanes fly faster and farther, sleep deprivation and circadian rhythms play a major role in how pilots perform in the cockpit. Any international flight crew can talk about the scheduling problems they have experienced and the burden of feeling sleepy in the cockpit on overseas flights.

An American Airlines flight to Cali, Columbia in December 1995 illustrates the tragic consequences of human error caused by sleep deprivation. As the plane approached the airport in Cali, the crew tried to program the plane's autopilot to lock onto a beacon called "Rozo." The copilot brought up a list of the navigation beacons starting with "R" by typing that letter into the computer. He selected the first beacon on the list, which is usually the closest. However, the sleep-deprived navigator overlooked the fact that the first was not "Rozo" but "Romeo" - a beacon over 100 miles away in Bogotá. The sleep-deprived human pilot didn't catch his navigator's mistake, and the plane's autopilot did what it was designed to do. It obeyed the command and, unbeknownst to the crew, made a slow turn to the left, toward Bogotá. Minutes later they realized something was wrong, but it was too late. The plane smashed into a mountain, killing 149 people.

The "black box" recording of that Cali flight is a very sad and poignant record of pilot fatigue.
Captain: I called tracking one day and said, "Hey, this [expletive] international is doing me...I want you to spell out the legal rest." And that's where I got this from, and I wrote it down very explicitly. Ten hours' crew rest.
First Officer: That's on international?
Captain: Yeah, if you fly less than 5 1/2 hours.
First Officer: Which in this case...
Captain: That's our scenario. Ten hours crew rest, 30-minute debrief, and 1-hour sign-in. And you can't move that up at all, because it's an FAA thing. You roll those wheels before 11 1/2 hours, you're [expletive].
[The captain mentions a pilot friend who had trouble flying international routes, then continues.]
Captain: He said he didn't mind it, he didn't mind driving back home at five o'clock in the morning. But to me, I'm like... it's torture.
First Officer: Yeah.

Captain: Torture in the [expletive] car, trying to keep awake and stay alive, uh-huh. I discussed this with my wife. I said, "Honey, I just don't want to do this, I hope you don't feel like I'm [unintelligible.] She said, "No way, forget it.
You don't need to do that [expletive]. [Sound of yawning.]

It was a cruel irony that the pilot and first officer didn't hear the warnings in their own words. Their mistake is typical of those made by people who are too sleepy to pay attention to details that can spell the difference between life and death. For example, in the first major airplane crash to be officially attributed by the National Transportation Safety Board to crew fatigue, the sleep-deprived crew inexplicably pursued a challenging approach instead of an easier one, which caused their plane to crash on approach to Guantanamo Naval Base in Cuba.

Charles Lindbergh, in his book about the first solo flight across the Atlantic Ocean in 1928, describes this state eloquently. Lindbergh started on the 33-1/2 hour flight after being continuously awake for more than a day and a half prior to departure, which meant that by the time he completed the flight, he had been awake for almost 70 hours. Describing the flight, he wrote, "My mind clicks on and off... My whole body argues dully that nothing, nothing life can attain, is quite as desirable as sleep." Lindbergh actually did fall asleep on his flight and nearly crashed into the ocean, but woke up in time.

A much more dramatic and highly visible tragedy was the explosion of the space shuttle Challenger. The Rogers Commission appointed by President Reagan, which officially investigated the tragedy, declared that the decision to launch the rocket in the absence of adequate data was an error. Those of us who saw this terrible tragedy on television over and over know the ghastly consequences of that error. But not well known at all is the fact that the Human Factors Subcommittee attributed the error to the severe sleep deprivation of the NASA managers. This conclusion was included in the committee's final report, which noted that top managers in such situations are generally the ones who sacrifice the most sleep.

Transportation Safety: Concluding Comment

Although relatively little research has been conducted to identify effective measures countering sleep deprivation and sleep disorders in any mode of transportation, we continue to campaign for effective awareness and more research. Impetus to expand countermeasure research will be enhanced as more attention focuses on accidents attributed to human error and as top management integrates the real costs of sleep-based fatigue accidents into their overall plans for risk management. It is long past time to make an effective effort. To re-emphasize, we know enough to take action now and finally the NTSB is making this a priory.

Physicians and Other Health Care Workers

It is alarming but readily understandable that pilots crossing multiple time zones have fatigue problems. What is more disturbing to me is that the health care professionals in whose hands we literally place our lives - surgeons, anesthesiologists, and emergency room doctors and nurses - are likely to be chronically and severely sleep deprived. In the past, it was not unusual for medical residents and interns, the foot soldiers of hospital and emergency room care, to work 100 hours or more during a week that includes two nights on call.

Unfavorable attention was first focused on these issues nationally in 1984 when Libby Zion, an 18-year-old college freshman, died in a hospital in New York City. A grand jury found that her death was due to an undiagnosed infection and they blamed resident physician fatigue for the failure to institute proper treatment. As a result of this incident and a few other highly publicized malpractice cases due to sleepiness, New York has enacted laws to regulate the work schedules of residence. In brief, the ACGME regulations prescribe no more than an 80-hour work week, a maximum shift duration of 24 hours, at least 10 hours off between shifts, 1 day off per week, and overnight on-call assignment no more than every third night.

The scene inside operating rooms is frightening to contemplate. Surgeons, anesthesiologists and nurses are all on call multiple nights a week. Doctors may be called into emergency surgery at 2:00 A.M., operate for five hours, and go straight into a previously scheduled 7:00 A.M. surgery. Patients have no idea that the surgeons preparing to cut into them are working on two or three hours' sleep, or are fatigued by a long operation the night before.

Here is one story out of the scores that I have heard that is particularly striking for the light it sheds on our serious neglect of sleep deprivation. In 1993, a team of Colorado doctors brought an eight-year-old boy into the operating room for what should have been a minor surgical operation. The anesthesiologist assigned to the case was notorious for working while extremely fatigued, and he admitted that he sometimes fell asleep during operations. In the past, one doctor had suggested to the anesthesiologist that he drink a cup of coffee and walk around the operating room to stay awake. Another doctor conjectured that the anesthesiologist had a drug problem. But despite a seemingly universal awareness among other doctors and nurses of the anesthesiologist's frequent sleepiness at work, he was allowed to administer general anesthesia to the boy.

Surgeons avoid using general anesthesia whenever possible because there is only a small margin of error between a dose that produces deep unconsciousness and one that causes death. Adding to this, everyone has a slightly different reaction to anesthetic medications, and a given dose has different effects on different people. These factors make anesthesiology the medical equivalent of flying an airplane. Both are highly complex, technological fields that follow a predictable routine most of time. Usually anesthesiologists have little to do except monitor the instruments. When things go wrong, however, they can go wrong in a hurry. More than any other medical profession, anesthesiology consists of days or weeks of routine and somewhat tedious activity and then fleeting moments of sheer terror.

For the eight-year-old boy, things did go wrong. And in the critical moments that followed, the anesthesiologist failed to perform his duties properly. The boy died. In the resulting civil court trial, the doctor admitted he had made fatal errors during the operation but denied that he fell asleep. Nonetheless, the jury determined that the physician was guilty of "grossly negligent medical care."

This and similar tragedies highlight the enormity of the medical establishment's almost willful ignorance about the dangers of fatigue. If physicians were ostensibly drunk when arriving at the operating room, they would never be allowed to work. Being drunk impairs performance significantly, even if intoxication isn't so extreme that the drinker actually passes out. Similarly, to say that someone can only be held responsible for fatigue-induced errors if they fall asleep is like saying that a drunk driver cannot be cited unless he actually passed out at the wheel. Yet trials and investigations of accidents caused by fatigued drivers, doctors, and pilots usually focus on whether or not the person actually fell asleep, rather than on their state of mind in the moments before the accident.

Of course, we must accept that doctors will occasionally make mistakes. No one is perfect. On the other hand, it is unacceptable for a doctor to make a mistake solely because of fatigue. Some years ago, in an anonymous survey of a San Francisco hospital house staff, 42 percent of resident physicians admitted killing at least one patient by making a fatigue-related mistake!

Over the years, we have run the Multiple Sleep Latency Test many times on nurses and resident physicians; I can only remember one that was not in the twilight zone of extreme sleepiness. Many health care workers are on duty for 10 or 12-hour shifts, with day and night shifts alternating as frequently as every week. In any work situation, a tired and sleepy worker is less effective and more at risk for an accident than an alert one. The nature of the risk depends upon the nature of the work. When 24-hour operations are involved, the likelihood of severe sleep deprivation and other biological rhythm effects is very great and the associated risk for accidents and diminished productivity is consequently much higher.

While the New York example of limiting resident physician duty to an average of 80 hours per week is one possible solution to the fatigue of resident physicians, it may not be as good a solution as other approaches such as encouraging napping during prolonged work periods. The merits of legislated restrictions on work hours for physicians continue to be debated in medical journals and in state legislatures, but relatively little progress has been made. This is particularly ironic considering that in many other areas physicians have led the way in making fundamental changes necessary to accommodate newly understood principles of physiology and medicine. The widespread use of sterile technique in surgery and the modern emphasis on preventive medicine are two good examples. Yet today, physicians themselves remain divided on the importance of spreading awareness about sleep, sleep deprivation, circadian rhythms, and sleep disorders.

Hazardous Work

Particularly vulnerable workplaces include nuclear power plants and 24-hour operations involving the production of hazardous or toxic chemicals. The early morning human errors that led to the near-disaster at Three Mile Island and the worst accidental nuclear catastrophe in world history, Chernobyl, bear all the earmarks of fatigue related accidents, including sluggish response to warning signs followed by delayed inappropriate action.

It would be impossible here to provide an exhaustive list of jobs in which excessive sleepiness constitutes a public health hazard. Although a number of high-risk jobs are outlined below, some additional examples include: military personnel, firefighters, police, emergency utility workers, rescue workers, and medevac personnel.

Law Enforcement and First Response

Perhaps our most urgent concern should be for those who endure very demanding schedules and high occupational risk because they must watch over and protect the rest of us. For first responders and the military, the demand may be episodic; for firefighters and police officers, it is a 24-hour-a-day, seven-days-a-week proposition.

Recent studies on fatigue in law enforcement personnel have documented terrible work schedules combined with high levels of stress. Many police officers suffer overwhelming fatigue and often have difficulty obtaining adequate amounts of sleep due to Post Traumatic Stress Disorder (PTSD). Each year, police accidents, injuries, and misconduct exact a heavy human and economic toll.

Not only are police officers expected to respond to emergencies that are often life-and-death and that entail significant work, but they are also quintessentially shift workers, since protecting their community is a 24 hour a day commitment. It is manifestly unfair for society to expect a uniformly high conduct and performance level from law enforcement officers and, while at the same time, allowing police officers to endure perhaps the worst schedules and greatest stress of any profession in our society. A recent book published in 2000 by the Police Executive Research Forum, *Tired Cops*, dramatically documents the aforementioned conditions and negative consequences of the law enforcement community.

Shift Workers

The shift work industries are probably the largest areas affected by chronic sleep loss. In carrying out 24hour operations, workers rotate schedules and never obtain enough sleep. Other workplace situations frequently require exceedingly long periods of work with very little rest. A variety of important issues involving shift workers were discussed at length in Chapter Nine and need not be repeated here.

Sleepiness Dulls the Competitive Edge

Elite athletes are also vulnerable to sleep deprivation and disrupted circadian rhythms. At the Summer Olympics in Atlanta, Georgia in 1996, investigators studied the effects of jet lag on 12 Korean athletes. After the plane trip from Korea to Atlanta, which spanned 10 time zones, nearly all the Korean athletes experienced fatigue and one-third complained of decreased strength. On average, the athletes required four days of adjustment before they could achieve athletic results similar to their baseline performance in Korea. Notably, half of the athletes were adversely affected by ambient noise in their sleeping quarters, and half also snored regularly or had pauses in their breathing while sleeping.

It has also been demonstrated that circadian rhythms are an important hidden factor in professional football games as well. The "home-field advantage" is often talked about, but people always have assumed that this advantage lies in the familiarity of the field and the support of the home fans. In actuality the advantage goes to the team that is playing at the time of peak circadian alerting. In other words, if the game is played during the mid-afternoon physiological dip in alertness for the local west coast team, the visitors from the east coast are already in their period of evening alerting and the visitors will have the advantage. For example, in December 1997 the Green Bay Packers (Wisconsin) kept the San Francisco 49ers out of the Super Bowl by beating them in a playoff game in San Francisco. The 1 P.M. game was played just as the Packers' circadian rhythm was rising to its evening peak (4 to 7 P.M. EST) and the 49ers' clock was falling into the afternoon dip in alertness (1 to 4 P.M. PST).

Concluding Close Up: Key Areas of Impact

I can think of no better way to conclude and reinforce the material in this "key areas" section than by relating my own near death as a result of obstructive sleep apnea. In November 1996, I flew to Portland, Oregon on the final leg of an advocacy tour to speak at a conference on traffic safety sponsored by the Washington State Department of Transportation. The actual site of the conference was Skamania Lodge, on the north side of the Columbia River about forty miles east of Portland. The plane landed around noon, and I decided to hail a taxi rather than wait for a limousine. I did not really get a good look at the driver as I climbed into the cab. We crossed the bridge and headed east on a two-lane highway above the Columbia River gorge. I had planned to take a short nap, but some guardian angel kept me awake and saved my life. About two thirds of the way to the lodge, I suddenly heard the gravel noise of the taxi going off the road. Responding instantly, I yelled and reached over the seat to grab the steering wheel. We veered into the westbound lane, but fortunately there were no oncoming vehicles. I won't quote my angry words to the driver, but when we got to the lodge, I immediately understood. My driver was obese with a relatively small jaw and a thick neck. Using the lobby of the lodge as my clinic I took a brief medical history:

"Do you snore?" Answer: "Yes. My wife won't sleep in the same room."

"Are you tired in the daytime?" Answer: "I don't have any pep at all. I'm driving a cab because the stop and go keeps me awake."

"Do you have high blood pressure?" Answer: "Yes, I'm taking pills."

"Have you complained about being tired to your doctor?" Answer: "Yes, I've even seen other doctors besides my own primary care physician."

"What did they do?" Answer: "Nothing, really." I had a camera with me, so I took his picture. I made the picture into a slide, and now show it almost every time I lecture. My point is that sleep disorders are all around us. Make no mistake! Sleep apnea is part of your life in one way or another. No one is safe. If my taxi had crashed into the Columbia River gorge, the accident would have been blamed on brake failure or some other nonsense. That my death was sleep-related, let alone that it was caused by obstructive sleep apnea, would never have been known. How supremely ironic such a purposeless death would have been.

After our brush with death, I warned the taxicab driver that he was abnormally sleepy because of a sleep disorder called obstructive sleep apnea, and made him promise he would get off the road any time he felt the least bit drowsy. The next day I arranged for him to be tested in a Portland, Oregon sleep center. The results were as expected - very severe obstructive sleep apnea.

We each can make a difference and save lives. All over America, the taxicab driver I encountered is your school bus driver, your policeman, your pilot, your doctor, or even yourself.

Possibilities for Progress

Napping

We should particularly encourage napping to avoid undue hazard or to improve performance in critical situations. When irresistible drowsiness occurs in a hazardous situation, taking a nap should be viewed as a laudable emergency response. The person who pulls over and takes a nap to avoid falling asleep at the wheel is showing an admirable sense of responsibility. A person who takes a nap in such circumstances should be praised while a person who refuses to take a nap and attempts to keep going at great risk to self and others should be denounced.

A campaign to make napping more respectable should be launched. A feeling of drowsiness should be recognized as an urgent warning by everyone engaged in hazardous tasks. In most cases, the only effective response to this warning will be to take a nap immediately. Napping must be made acceptable and encouraged. Unfortunately, overt napping in the office environment is not yet condoned by society, and relatively few professors welcome napping in the classroom. On the other hand, napping appears to be highly acceptable in certain situations. On airplanes for example, one often sees more than half of the passengers sleeping in their seats. One sleep specialist has suggested that all offices be equipped with a nap box, a comfortable coffin-like chest in which someone could climb in and take a nap and at the same time be out of sight. One caveat must be mentioned with regard to drowsy driving. In today's society, it is not always safe to pull off the road. It is therefore urgent to create secure rest stops other places where drivers can pause and nap along our highways.

Rational and Irrational Responses: Aviation

Although accidents associated with fatigue impair all modes of transportation, aviation has had a long history of research on the nature and development of countermeasures to fatigue. The Federal Aviation Administration is cooperating in research ventures under the scientific direction of the National Aeronautics and Space Administration. One completed study in particular illustrates what might be done to reduce sleepiness on the job when there is a genuine interest and commitment by all parties (government, industry, labor, and the public).

For transcontinental and transoceanic flight crews, inadequate sleep is a major source of fatigue. A multidisciplinary research effort was undertaken to evaluate the benefits of preplanned naps while in the air for long-haul flight crews. The results indicated that not only could crew members safely rotate taking a relatively brief (40-minute) nap in flight, but also that the naps enhanced alertness and performance of crews during flight. Interestingly, while the brief naps did not eliminate the cumulative sleep debt of the crews, they did appear to provide substantial relief from inflight fatigue, especially on night flights.

The irrational aspect of what began as a very rational approach to pilot fatigue is that the Federal Aviation Administration has not approved napping in the cockpit. On the other hand, several European airlines have transferred the benefits of the napping research to the cockpit and have reaped the benefits of improved safety and relief of fatigue. Even though American pilots are always napping unintentionally in the cockpit, the U.S. regulatory agencies are too timid about the image of encouraging napping in the cockpit to emulate the more rational Europeans.

CHAPTER 22 Sleep Education & Public Policy

Dealing Publicly with the Problems of Sleep

Sleep-related fatigue is one of the most important causes of problems in the aforementioned industries, but public action on sleep-based fatigue is lagging. While the public understands cause-and-effect relationships such as contact with bacteria leading to infection or smoking leading to lung disease, the public does not adequately extend this type of cause-and-effect understanding to sleep loss leading to accidents. As discussed earlier, society needs conceptual tools for dealing with the causal relationship between sleep loss and human error. Without such conceptual tools, it is difficult to influence public opinion. In short, our society faces huge problems related to sleep deprivation but lacks the education and knowledge to understand the problems and to formulate solutions. Imagine asking your professor to take the final exam later because you must catch up on your sleep, or calling in "sleep deprived" to one's supervisor at work! You would be ridiculed.

In today's world with the Internet, cable television, large movie theaters, and hundreds of periodicals all clamoring for our attention, it seems improbable to command the attention of the public sufficiently enough to make a lasting impact. Nonetheless we must never give up, however futile it may seem. Had a highly visible person, such as the President of the United States, Andrew Luck or Stephen Colbert, called attention to the real cause of the Exxon Valdez grounding or the major contributing cause of the Challenger explosion, it might have made a difference. We hope that sooner or later, while we simultaneously attempt an effective penetration of the educational system, celebrities will also join in the campaign to bring sleep awareness to the American public.

Late to School

It has been pointed out that social pressure and biological changes combine to foster a substantially later bedtime in teenagers. In contrast, schools are starting progressively earlier as students advance from elementary to middle to high school. Although students often cannot fall asleep until well after midnight, many high schoolers must get up at six or earlier to be at school on time, and even earlier if they are involved in an extracurricular activity that takes place before the first class. Traditionally, swimming and water polo practice in many high schools take place in the early morning.

As concern over sleep-deprived teenagers has grown, a movement to change high school start times to a later hour has taken form. The first major change took place in Edina, Minnesota where high school start times were pushed back to 9 A.M. In the several years since this change took place, accumulated data reveal that the students' moods improved, alertness in the first two periods is no longer a problem, and previous concerns regarding various relocations of janitors, cafeteria workers, coaches, and bus drivers did not come to pass. By now, several other school districts have initiated this change for the better. Legislation has finally been introduced in the Congress of the United States by an enlightened Representative, Congresswoman Zoe Lofgren, that authorizes the Secretary of Education to award a grant of $25,000 to any high school not starting earlier than 9:00 A.M

One of the most impressive studies thus far of the negative effect of early start times was carried out in San Diego by an alert teenage researcher. Due to a shortage of buses, over 200,000 5th grade students are transported to school at three different times each separated by an hour. Looking at the results of the standardized SAT the researcher found the scores of students being bused the latest to be very significantly better than the scores of the students who were bused two hours earlier. This study clearly implicates sleep with the ability to learn and reveals sleep deprivation's negative effect on academic performance.

Parallels with Driving Under the Influence of Alcohol

The use of alcohol is governed by established policies and laws. These laws exist because it is scientifically established and widely known that alcohol impairs judgment and reflexes on the highway or in other hazardous situations. This creates a risk for the intoxicated person, but more importantly, an unacceptable risk for innocent victims. Enforcement is possible because there is objective, rapid, cheap testing for the presence of alcohol and its concentration in the body: breathalyzer and blood tests.

Sleep loss, depending on its degree, also leads to impairment and creates risk on the highway and in other hazardous situations. In a court of law, if authorities determine that impairment from sleep deprivation has caused an accident, the driver is considered negligent and is liable for civil and criminal penalties. However, the similarities between accidents caused by intoxication and those caused by sleep loss ends here. Accident investigators have no easy, inexpensive, and reliable measurement of the degree of sleep deprivation. Errors due to falling asleep or impaired performance due to sleep deprivation must always be inferred from the nature of the accident. As knowledge about sleep-related human error has increased, the involvement of sleep specialists in litigation has also increased. Because accidents related to sleep are a large public health problem, society will inevitably be forced to confront issues surrounding the adjudication of sleep-related accidents.

In the past several years, sleep specialists committed to changing public policy have exploited the societal position that driving or performing any other hazardous duty under the influence of alcohol is completely unacceptable. In particular, specialists drew an analogy between sleep deprivation and the specific levels of blood alcohol that have been deemed illegal for driving in every state in the U.S.

For example, Dr. Nelson Powell at Stanford University compared a group of 80 legally drunk normal adults with 200 patients diagnosed with mild-to-moderate obstructive sleep apnea using a variety of performance measures. The patients were generally as impaired as the intoxicated comparison group and, on some measures, the patients were even more impaired. In a second study, Dr. Powell compared the driving skills of normal adults who went without sleep for one night or were allowed only four hours of sleep for six nights with the driving skills of normal adults who were legally drunk. The subjects drove real cars over a very large course in Mesa, Arizona owned and operated by the General Motors Corporation. Miles of roadway were lined with orange cones in addition to a variety of obstacles designed to suddenly appear in front of the drivers. The sleep-deprived groups hit more cones and were less able to avoid obstacles than the drivers who were legally drunk.

Researchers in Australia also found a similarity between sleep deprivation and alcohol intoxication. These investigators split 40 volunteers into two groups. One group was kept awake for 28 hours, from 8 A.M. until 2 P.M. the next day. The other group was given 10 to 15 grams of alcohol every 30 minutes starting at 8 A.M., until each volunteer's blood-alcohol level reached 0.1 percent, which is greater than legally drunk in most locales. During these periods, both groups underwent hand-eye coordination tests. The Australian researchers found that after 17 hours awake (at 1 A.M., when biological alerting is subsiding), the sleep-deprived group had the same test scores as drinking volunteers who had blood alcohol levels of 0.05 percent. After 24 hours awake, the sleep-deprived group had the same coordination deficits as those with the maximum blood alcohol level, 0.1 percent.

Since the performance impairment of relatively mild sleep loss and mild sleep disorders can now be readily equated with unacceptable levels of alcohol intoxication, it seems likely that when the general public understands these results, sleep loss and untreated sleep disorders will also become unacceptable for drivers and in other hazardous situations. Rather than punishment and incarceration, however, the approaches should be diagnosis and treatment, and education about sleep management and the avoidance of sleep deprivation.

Sleep Deprivation Potentates Alcohol Related Impairment

It is now known that sleep deprivation independently increases the sedative effects of alcohol. Since almost every driver carries a significant sleep debt, it is logical to conclude that sleep deprivation plays a role in accidents traditionally attributed solely to the use of alcohol. How great a contribution sleepiness may have in the overall number of transportation accidents involving drug and alcohol use has not been assessed. In a study of fatal-to-thedriver truck accidents, the National Transportation Safety Board found that fatigue plus alcohol or drugs accounted for the vast majority of lethal accidents. Clearly, sleep deprivation is likely a factor in a large fraction of accidents attributed to alcohol. In addition, if drivers are unaware of their sleep debt, they are also unaware of the extent to which they are susceptible to additional performance impairment from alcohol or other depressant drugs. This combination of ignorance is often deadly.

Litigation as an Impetus for Change in Public Policy

Almost everyone, including sleep specialists and the lay public, has a natural reluctance to become involved in litigation of any type - especially tort litigation. However, as should be obvious from the foregoing discussion, there is a huge gulf between the accumulated scientific knowledge about sleep and the application of this knowledge to public policy aimed at preventing personal and public catastrophes. Furthermore, the current pervasive ignorance concerning sleep and human error and the paucity of research on countermeasures for sleep-based fatigue delay the formulation of policies to reduce or prevent losses caused by sleep deprivation.

Historically, when public policy has been unclear concerning issues surrounding a potentially injurious phenomenon, individuals turn to the courts for review, judgment, and compensation.

The precedents stemming from such judicial proceedings often prompt governments to establish policies. Since sleep specialists are the only logical source of detailed information concerning sleep physiology, sleep deprivation, and sleep disorders, they are increasingly being approached by lawyers to consult and/or act as expert witnesses. Claims related to sleep deprivation and fatigue arise in many aspects of criminal, worker's compensation, and tort litigation. Such participation in the legal process is a wise and useful public policy exercise for sleep experts.

The Role of Aware Americans in Public Policy

Americans are alarmingly apathetic and uninvolved in the political process. This apathy and general unwillingness to get involved are factors that contribute to the widely publicized ineffectiveness and corruption in our public institutions. We elect representatives to manage society and to recognize and implement necessary changes as society evolves. It is important for everyone to realize that public policy and the behavior of elected leadership can be changed. This is particularly true when an issue is relatively non-controversial.

Once areas of concern and major problems are identified and solutions created, a grassroots campaign can be very effective in implementable change. There is a saying among members of Congress that 50 letters supporting an issue is a blizzard. A thoughtful letter or a well-planned visit from a constituent always has an impact. While the number of issues often seems infinite and one cannot respond to everything, one can address those issues which most significantly affect one's community and workplace. In this regard, all of the issues of sleep affect everyone, so everyone should have an interest in wise public policies that involve these sleep-related issues.

Initiating Change

There is ample precedent for aware individuals to become involved, both individually and collectively, to improve the health of the public *visà-vis* sleep-related issues. Every American can begin to participate in the political and civil processes by which society is managed and protected. Moreover, everyone should support and encourage sleep research and sleep disorders specialists to develop public policy guidelines that can foster progress in areas of sleep medicine. These areas include further development of FDA guidelines for testing of sleeping pills and stimulants, identifying sleep apnea and other sleep disorders in truck drivers and other transportation workers, and gathering data related to the sleep needs of citizens from birth to death.

More so than health professionals in general, the special expertise of sleep medicine professionals confers a unique and heightened set of obligations to address and to shape rational public policy. Sleep specialists certainly have a heavy responsibility to provide clear and authoritative materials for education at all levels. This duty extends to educating leaders of industries whose workers are entrusted with the health and safety of our citizens, along with helping to shape public policies protecting both workers and the public from sleep-related errors, accidents, and catastrophes. Every American should set the standard for public policies that foster healthy sleep. Becoming knowledgeable is the first step in transforming sleep from our adversary to our ally. When we truly start to understand the mechanisms of sleep and alertness, and apply that knowledge to our daily routines, we can begin to make our lives safer, more productive, and more fulfilling.

The second step involves taking responsibility for ensuring healthy sleep. Most sleep disorders can be effectively treated or cured, but we need to take the initiative in seeking out treatment. We need to push aside our massive and sometimes willful ignorance about sleep. We need to respect sleep enough to learn its dynamics and listen to its demands. Very simply, we need to learn to manage sleep the way we manage any other vital aspect of our lives. In addition, everyone should begin to apply their newfound awareness of sleep and wakefulness to long-range planning of their own future endeavors, travels, goals, and work.

Finally, the best thing everyone can do right now is share their newly acquired sleep awareness with their families, friends, students, and as many other individuals as possible. At the present time (circa 2013), it is almost certain that most people approached will have little or no knowledge about the fundamental nature of sleep, sleep debt, clock-dependent alerting, and the common sleep disorders.

Sleep Disorders in Primary Care Populations

There are several reasons to be concerned about sleep disorders in clinical populations. Probably the most important is that sleep disorders are frequently misdiagnosed, or they are not recognized as the cause of another condition, most importantly high blood pressure and/or congestive heart failure. It is also true that when a patient finally receives a medical diagnosis, such as refractory depression, it is almost never challenged or seriously questioned. If, in fact, the cause of the depressed mood is a sleep disorder, then the condition will simply get progressively worse as the physician continues to try to treat the misdiagnosed patient with yet another anti-depressant.

Figure 23-1 shows the results of a study in which a trained assistant administered a validated sleep disorders questionnaire to every patient walking through the door of a typical family practice clinic for one year. This clinic is located in Moscow, Idaho. The sample included 1,254 adults eighteen years of age and older. Slightly more than 60 percent (760) of the patients had unambiguous sleep disorder symptomatology. Three categories accounted for the vast majority of the diagnoses: obstructive sleep apnea, insomnia, and restless legs syndrome. Two of these sleep disorders were diagnosed in 189 of the 760 patients and another 60 of the 760 patients suffered from all three conditions.

The most lamentable and revealing fact in this study is that absolutely none of these patients were previously suspected of having a sleep disorder. Of course, the situation is now completely different due to our intervention of educating the primary care physicians in this clinic. Unfortunately, it may be assumed that the thousands of family practices and primary care practices throughout the United States and all over the world are at the same low level of sleep awareness as the Moscow clinic prior to the study.

Conclusion

I have attempted to outline the ways in which the environment, culture, and structure of society interact with our sleep. The immediate purpose is to encourage the understanding of human behavior in a more sophisticated and effective way. Attention to the requirement for adequate amounts of sleep as well as obtaining it at the appropriate chronobiological time should be a high priority for everyone. In communities everywhere, this will require the cooperation of each citizen and business, as well as society at large. To truly accomplish this goal – and to avoid the kind of tragic consequences exemplified by falling asleep at the wheel or in other hazardous situations – means not only to establish a collaborative spirit, but also to have the willingness of academia, industry, and government to commit the resources necessary to implement research and effective programs.

Sleep and sleepiness are among the most basic aspects of human existence. Sleep is made unhealthy, by both pathology, and by mismanagement and abuse. When sleep disruption occurs, the consequences for the individual and, in some circumstances for the larger community, can be very serious. Educated readers now know that there is only limited recognition of this fundamental fact and that a vast gulf still exists between gains in the scientific and medical knowledge about sleep and the application of these gains to the prevention of personal and public catastrophes. The act of falling asleep at an inopportune time has wreaked havoc on industrialized societies and will continue to do so if the problem is not properly addressed.

In the past eons of history, natural selection may have been gentle on prehistoric people who did not sleep enough. The major killers – war, natural enemies, and pestilence – were much more important in deciding who lived to conceive and care for children than were sleep problems that impaired alertness. However, the main causes of death today are radically different than they were in centuries past.

Sleep, like many other aspects of human behavior, has been irrevocably altered by the industrial revolution and the attendant development of artificial light and inexpensive energy. Around-the-clock operations are now commonplace, and the time traditionally allocated to sleep has given way to time for more waking activity. Even today, as societies continue to make the transition from agrarian cultures to industrial powers and on to the information age, some of the more dramatic societal changes that are occurring involve sleep. For example, the siesta is increasingly disappearing in many cultures. The advent of highly technological societies has made poor sleep and substandard levels of wakefulness a very real, and often life-or-death, concern to humans. Industrialized societies are increasingly dependent upon the alertness of key individuals. We now live in a society where catastrophic consequences can follow a single sleep-related error in one who operates a public transportation vehicle, supervises the control of a nuclear power station, or leads 24-hour military functions. It is long past the time to wake up to the situation and address sleep issues appropriately.

"In the unrest of the masses I augur great good." -Leland Stanford

Coda:

What Resources Are Available?

HELPFUL SLEEP-RELATED WEBSITES:

- www.sleepfoundation.org
- http://www.nhlbi.nih.gov/about/ncsdr/index.htm
- http://www.ntsb.gov/
- www.**aasmnet.org**
- http://sleep.stanford.edu/

The Bottom-line of this Book

Dear reader, you now know more about sleep and have more updated information than your parents, teachers and friends. Please share his information with them. The take away messages from this book include the description of the more serious and commonplace sleep disorders, including insomnia and obstructive sleep apnea. A few chapters addressed the other less prevalent sleep disorders, which may often be serious, and/or life threatening. The overarching purpose of this book is to communicate a permanent understanding of sleep disorders and their symptoms with the expectation that people who absorb this material will recognize these symptoms, should they occur not only in themselves, but also in their relatives, friends, and colleagues. And always remember……….

Drowsiness is RED Alert!!

"Your life is a reflection of how you sleep; and how you sleep is a reflection of your life." - *Rafael Pelayo, M.D.*

Glossary

A

Abdominal Movement - In a sleep study one measure of respiratory effort is movement of the abdomen which reflects movement of the diaphragm.

Advanced Sleep Phase Syndrome - A circadian rhythm disorder in which phases of the daily sleep/wake cycle are advanced with respect to clock time. Accordingly, the sleep phase occurs well ahead of the conventional bedtime. Also associated with a tendency to wake up too early.

Airflow Limitation (Flow Limitation) - Partial closure of the upper airway with reduced airflow.

Alpha Rhythm - An EEG rhythm with a frequency of 8-12 Hz in human adults, which is easily seen over the back of the skull when the eyes are closed. It indicates the awake state in most normal individuals.

Alpha Sleep - Sleep in which alpha activity occurs with sleep EEG patterns.

Ambulatory Monitor - Portable system used for the continuous recording of multiple physiological variables during sleep. This allows patients and research subjects to sleep in their own beds.

Apnea - Cessation of breathing during sleep. The episode must be at least 10 seconds in duration.

Apnea/Hypopnea Index - The frequency of abnormal respiratory events per hour of sleep. These events are classified as apneas or hypopneas. Apnea is when breathing (airflow) stops for 10 seconds or more. Hypopnea is a decrease in airflow of greater than 50% for 10 seconds or more. An AHI of 45 would mean that the subject is experiencing complete or partial airflow blockage an average of 45 times per hour.

Apnea Index - The frequency of apneas per hour of sleep.

Arousal - An abrupt brief change from sleep to wakefulness, or from a "deeper" stage of non-REM sleep to a "lighter" stage.

Arousal Disorder - A parasomnia disorder presumed to be due to an abnormal arousal function. The classical arousal disorders are sleepwalking, sleep terrors and confusional arousals.

Glossary

Arousal Threshold - The ease with which a sleeping person is awakened at a particular time.

Arrhythmia - An absence or irregularity of the heart rhythm caused by disturbances of the transmission of electrical impulses through the cardiac tissue.

Auto Adjusting Continuous Positive Airway Pressure Device (SmartPAP) - A type of CPAP machine that monitors changes in the individual's breathing and compensates automatically by making the appropriate adjustment in pressure on a breath by breath basis.

Awakening - The return to Consciousness On the polysomnographically defined awake state from any of the sleep stages: characterized by alpha and beta waves, rise in tonic EMG, voluntary eye movements and eye blinks.

B

Basic Sleep Cycle - The progression through an orderly succession of sleep states and stages. In a healthy adult, the first cycle is always initiated by going from wakefulness to non-REM sleep. The first REM period always follows the first period of non-REM sleep, and the two sleep states continue to alternate throughout the night with an average period of about 90 minutes. A full night of normal human sleep will usually consist of 4-6 non-REM/REM sleep cycles.

Benzodiazepines - Class of compounds developed in the 1950's which tranquilize and sedate. They have muscles relaxing, anxiolytic, anticonvulsant properties

Beta Activity - Brain waves having a frequency greater than 13 Hz (Hertz).

Bi-Level Positive Airway Pressure (Bi-level or Bi-PAP®) - A CPAP device providing two measured pressure levels of continuous airflow: one level for breathing in and a lower level for breathing out.

Biological Clock - A term applied to the brain process which causes us to have 24-hour fluctuations in body temperature, hormone secretion, and a host of other bodily activities. Its most important function is to foster the daily alternation of sleep and wakefulness. The biological clock is housed in a pair of tiny bilateral brain areas called the suprachiasmatic nuclei.

Bradycardia - A heart rhythm with a rate below 60 beats per minute in a human adult.

Brain Waves - Spontaneous electrical activity of the brain studied by method of electroencephalography (EEG).

Bruxism - Grinding one's teeth while asleep.

C

Cardiac Arrest - Sudden cessation of heart beat.

Cardiovascular - Pertaining to the heart and blood vessels.

Central Nervous System (CNS) - The brain and spinal cord.

Central Sleep Apnea - A period of at least 10 seconds without airflow, during which no respiratory effort is evident.

Chronotherapy - Treatment of a circadian rhythm sleep disorder by systemically changing sleeping and waking times to reset the patient's biological clock.

Circadian - Any periodicity which is near but not necessarily exactly 24 hours.

Circadian Rhythms - An innate daily fluctuation of physiological or behavioral functions, including sleep-wake states generally tied to the 24-hour daily dark-light cycle. Sometimes occurs at a measurably different periodicity (e.g. 23 or 25 hours) when light-dark and other time cues are removed.

Compliance - Conforming with or adhering to a regimen of treatment such as CPAP therapy.

Continuous Positive Airway Pressure (CPAP) Machine - Medical device used to treat sleep apnea. An apparatus that provides a highly effective, non-invasive therapy that eliminates blockages and prevents collapse of the upper airway by generating a prescribed level of air pressure that maintains airway patency during sleep. Air pressure is delivered through a hose to a mask that fits over the nose or the nose and mouth. The mask is secured on the face by headgear that is worn over the head. The appropriate air pressure level is determined during a "CPAP titration" sleep study. The complete system consists of a programmable pressure generator, tubing, mask and headgear. Sometimes referred to as nCPAP (nasal Continuous Positive Airway Pressure).

CPAP Pressure - Amount of pressure needed to maintain an open airway in a sleep apnea patient being treated with CPAP, expressed in centimeters of water (cm H_2O). The positive pressure may range

Glossary

from 5 to 20 cm H20. Different patients will require different pressures. This value is determined in a CPAP titration study.

D

Deep Sleep - In sleep studies, refers to combined non-REM sleep stage 3.

Delayed Sleep Phase Syndrome - A circadian rhythm disorder, which in the daily sleep/wake cycle, is delayed with respect to clock time. Accordingly, the sleep phase occurs well after the conventional bed time. Usually associated with difficulty getting up in the morning.

Delta Sleep - Stage of sleep in which EEG delta waves are prevalent or predominant (older terminology sleep stages 3 and 4, respectively). See Slow Wave Sleep.

Delta Waves - Delta waves have a frequency of 2 cycles per second (cps) or slower and amplitudes greater than 75 microvolts peak to peak (the difference between the most negative and positive points of the wave). Also known as Delta Activity.

Diagnostic Sleep Study - Continuous monitoring of several physiological activities in a sleeping individual. Usually carried out to determine the absence or presence of a specific sleep disorder. A diagnostic sleep study can be performed in a sleep disorders center or in a patient's home with portable recording equipment.

Diurnal - Active and wakeful in the daytime as opposed to active in the nighttime.

DME - Durable Medical Equipment. A DME dealer provides medical devices such as CPAP and Bi-Level machines.

Drowsiness, Drowsy - A state of quiet wakefulness that typically occurs prior to sleep onset. If the eyes are closed, diffuse and slowed alpha activity usually is present, which then gives way to early features of stage 1 sleep.

Dyssomnias - A class of sleep disorders that produce either insomnia or excessive sleepiness.

E

Electrocardiography (EKG) - Method of measuring the electrical activity of the heart. EKG is continuously recorded in both diagnostic sleep studies and titration sleep studies.

Electrodes - Small devices that transmit biological electrical activity from subject to polygraph amplification and display)

Electroencephalography (EEG) - A recording of the electrical activity generated by the brain. Frequencies expressed in hertz (Hz) or cycles per second (cps), and amplitude in microvolts. Characteristic frequency and amplitude patterns of the activity define which stage of sleep the patient is experiencing. Brain waves are recorded in both diagnostic and titration sleep studies.

Electromyography (EMG) - A recording of the electrical activity of the muscles. The absence or very low level of EMG activity indicates the presence of REM sleep. Very high levels can help identify periods of wakefulness. EMG is recorded in both diagnostic and titration sleep studies.

Electro-oculogram (EOG) - A recording of the movements of the eyeballs. If rapid eye movements are detected during sleep, the subject is in REM sleep, the state in which dreaming takes place. EOG is recorded in both diagnostic and titration sleep studies.

EPAP - Expiratory Positive Airway Pressure. The pressure prescribed for the expiratory (breathing out) phase of an individual on Bi-level CPAP therapy for OSA (obstructive sleep apnea).

Epidemiology - The scientific discipline that studies the incidence, distribution, and control of disease in a population. Includes the study of factors which effect the progress of an illness, and, in the case of many chronic diseases, their natural history.

Epoch - A standard 30 second duration of the sleep recording that is assigned a sleep stage designation; occasionally, for special purposes, longer or shorter epochs are scored.

Epworth Sleepiness Scale - An index of sleep propensity during the day as perceived by patients, and derived from the answers to 8 questions.

Esophageal Pressure - A measurement used to determine respiratory effort and by inference, airway resistance.

Excessive Daytime Sleepiness (EDS, Somnolence, Hypersomnia) - A subjective report of difficulty in maintaining the awake state, accompanied by a ready entrance into sleep when the individual is sedentary; may be quantitatively measured by use of subjectively defined rating scales of sleepiness.

Expiratory Phase - The phase of the breathing cycle in which air is expelled.

Glossary

F

Fatigue - A feeling of tiredness or weariness usually associated with performance decrements.

Flattening Index - A number that indicates the amount of airflow limitation caused by partial closure of the upper airway. 0.3 indicates an open airway, 0.15 is mildly obstructed, 0.1 is severely limited airflow, and 0.0 reflects a totally closed airway. Flattening Index is used to identify a condition known as Upper Airway Resistance Syndrome (UARS), and is continuously recorded in both diagnostic sleep studies and CPAP titrations.

Flow Limitation - Partial closure of the upper airway which impedes the flow of air into the lungs.

Forbidden Zone - Period of strongest clock-dependent alerting, usually in the evening. Prevents falling asleep.

Fragmentation (pertaining to Sleep Architecture) - The interruption of a sleep stage due to the appearance of a lighter stage, or to the occurrence of wakefulness, leading to disrupted non-REM-REM sleep cycles.

G

GABA (Gamma-Amniobutyric Acid) - A major inhibiting compound (neurotransmitter) in the brain, which is considered to be involved in muscle relaxation, sleep, diminished emotional reaction and sedation.

Gastroesphageal Reflux Disease (GERD) - The flow of stomach acid upwards into the esophagus which can cause arousals and disrupt sleep.

Genioglossus Tongue Advancement - Surgical treatment sometimes used for sleep apnea and/or snoring. This is designed to improve the airway behind the base of the tongue. The genioglossus which is the main tongue muscle, relaxes during sleep, often allowing the tongue to fall into the airway. This muscle attaches to the middle of the lower jaw. If a segment of bone containing this muscle is pulled forward and stabilized, it can open the airway space behind the tongue. This procedure does not move the teeth or jaw. It is performed under local intravenous sedation or general anesthesia and requires a one or two-day hospital stay.

H

Heart Rate or beats per minute (bpm) - Pace or speed of the heart measured in beats per minute. 60-80 is generally considered normal. Heart rate is continuously monitored in both diagnostic sleep studies and titration studies.

Hertz (Hz) - A unit of frequency; preferred over its synonym, cycles per second (cps).

Histogram - A graph showing frequency distributions.

Homeostasis- The tendency of the body to seek and maintain a condition of balance or equilibrium within its internal environment, even when faced with external changes. A simple example of homeostasis is the body's ability to maintain an internal temperature around 98.6 degrees Fahrenheit, whatever the temperature outside.

Humidification - Adding moisture to the airflow as an adjunct to CPAP (Continuous Positive Airway Pressure) therapy in the treatment of obstructive sleep apnea (OSA). Humidification is usually added to the CPAP therapy by diverting the airflow over or through a cool or heated water reservoir (humidifier) to prevent the upper airway from drying out.

Hypercapnia - An excess of carbon dioxide in the blood.

Hypersomnia - Sleeping for uncharacteristically long periods of time.

Hypersomnolence - Excessive daytime sleepiness.

Hypnophobia - Morbid fear of falling asleep.

Hypnotics - Sleep-inducing drugs.

Hypopharynx - The lowermost portion of the pharynx leading to the larynx and esophagus.

Hypopnea - An episode of shallow breathing during sleep lasting 10 seconds or longer, defined by an airflow reduction of at least 50%.

Hypoventilation - Reduced rate and depth of breathing.

Hypoxemia - Lack of an adequate amount of oxygen in the blood.

Hypoxia - A deficiency of oxygen reaching the tissues of the body.

Glossary

Inappropriate Sleep Episodes - Periods of sleep that are not planned and often occur in an unsafe situation (i.e., while driving). These episodes are always due to sleep deprivation.

Insomnia - Difficulty falling asleep and/or staying asleep.

Inspiratory Phase - Part of the breathing cycle in which air is inhaled.

IPAP - Inspiratory Positive Airway Pressure. The pressure prescribed by a physician for the inspiratory phase on a Bi-level CPAP device, used in the treatment of OSA.

J

Jet Lag - A disturbance induced by a major rapid shift in environmental time during travel to a new time zone. Typically include fatigue, sleep and impaired alertness.

K

K-Alpha - A type of microarousal where a K complex is followed by several seconds of alpha rhythm.

K Complex - An EEG wave form that has a well-delineated negative sharp wave and is immediately followed by a slower positive component.

L

Laser Assisted Uvuloplasty (LAUP) - Surgical treatment usually used for simple snoring. This procedure is designed to open the airway behind the palate. It requires multiple procedures where the laser cuts the palate and the area heals by scarring. It is an out-patient procedure done under local anesthesia. It is not typically indicated for O.S.A.

Leg Movement - Leg movement is recorded in both diagnostic sleep studies and titration studies. Left and right leg movements are recorded as independent events.

Light Sleep - A common term used to describe non-REM sleep stage 1, and sometimes, stage 2.

Light Therapy - Used to treat SAD (Seasonal Affective Disorder) and other circadian conditions. Exposing the eyes to light of appropriate intensity and duration and at the appropriate time of day to effect the timing, duration and quality of sleep.

Limit-Setting Sleep Disorder - When a child has difficulty in falling asleep due to delaying and refusing to go to bed due to inconsistent rules or limits being implemented

M

Macroglossia - Large tongue; usually a congenital disorder (present at birth). May result from other medical conditions such as acromegaly.

Maxillofacial - Pertains to the jaws and face.

Maxillomandibular Advancement - Surgical treatment usually done if previous procedures have not completely improved the obstructive breathing episodes and/or the patient has persistent symptoms of daytime sleepiness and fatigue. This procedure is designed to open the airway behind the palate as well as behind the base of the tongue. The operation cuts the bone of the upper and lower jaw and pulls these structures forward. This

is performed under general anesthesia and requires a two day hospital stay.

Melatonin - A hormone secreted by the pineal gland in the brain.

Microsleep - Brief episode of sleep, usually consisting of non-REM sleep. Microsleeps are associated with drowsiness and automatic behavior.

Mixed Sleep Apnea - Combination of central and obstructive apnea.

Monocyclic - Having a single major sleep period and a single major wake period in one 24-hour day.

Motor Activity in Sleep - Any muscular movement during sleep.

Movement Arousal - A body movement associated with an EEG pattern of arousal or a full awakening; a sleep scoring variable.

Motor Atonia - Absence of muscle activity during sleep.

Multiple Sleep Latency Test (MSLT) - The standard test used to quantify the overall daytime sleep tendency by measuring the speed of falling asleep (sleep latency) usually in 5 tests carried out at two-hour intervals.

Muscle Tone - The amount of tension in a muscle.

Glossary

Myoclonus - Muscle contractions in the form of abrupt "jerks" or twitches generally lasting less than 100 milliseconds. The term should not be confused with the periodic leg movements of sleep that characteristically have a duration of 0.5-5 seconds.

N

Nap - A short period of planned sleep generally obtained at a time separate from the daily major sleep period.

Narcolepsy - A sleep disorder characterized by excessive sleepiness, cataplexy, sleep paralysis, hypnogogic hallucinations, and an abnormal tendency to pass directly into REM sleep from wakefulness.

Nasal Airflow/Nasal Ventilation - A recording of the complete respiratory cycle by measuring inspiratory and expiratory airflow. In diagnostic sleep studies it is referred to as nasal ventilation and is measured in arbitrary units from 0-20. In titration studies it is measured in liters per minute. This value is usually recorded in both diagnostic sleep studies and titration sleep studies.

National Commission on Sleep Disorders Research - Created by the U.S. Congress in 1990, the commission conducted a comprehensive study of the social and economic impact of sleep disorders in America and made recommendations based on its findings to the Congress in January of 1993.

Neurotransmitters - Endogenuous chemical components that are released from axon terminals of one neuron and transmit the signal to the next neuron by combining with its receptor molecules. Examples of neurotransmitters that appear to be important in the control of sleep and wakefulness include: norepinephrine, serotonin, acetylcholine, dopamine, adrenaline and histamine. The process of neurotransmission may be inhibited, modulated, or enhanced by other chemical mediators within the brain, or by exogenous pharmaceuticals.

Nightmare - An unpleasant and/or frightening dream occurring in REM sleep. Occasionally called a dream anxiety attack, it is not synonymous with a night (sleep) terror.

Night Terrors - Also known as sleep terrors, or pavor nocturnus. Night terrors are characterized by an incomplete arousal from slow wave sleep. If awakened during a night terror, the individual is usually confused and does not remember details of the event. Night terrors are different from nightmares in that if an individual is awakened during a nightmare, he or she functions well and may have some recall of the nightmare.

Nyctophobia - Morbid fear of the night and darkness.

Nocturia - Excessive, and often frequent, urination during the night.

Nocturnal - "Of the night;" pertains to events that happen during sleep or the hours of darkness.

Nocturnal Confusion - Episodes of delirium and/or disorientation close to or during nighttime sleep; often seen in victims of Alzheimer's Disease and more common in the elderly.

Nocturnal Enuresis (Bedwetting) - The release of urine while asleep.

Non-Invasive - Medical procedure that does not penetrate the skin or a body cavity.

Non-Rapid Eye Movement (NREM, Non-REM) - See Sleep Stages.

NREM-REM Sleep Cycle (synonymous with Sleep Cycle) - A period during sleep composed of a NREM sleep episode followed by a REM sleep episode. Each NREM-REM sleep couplet is equal to one cycle. The average duration in adults is 90 minutes. An adult sleep period of 6.5-8.5 hours generally consists of four to six cycles.

NREM Sleep Intrusion - A brief period of NREM sleep patterns appearing in REM sleep. It is a portion of NREM sleep not appearing in its usual sleep cycle position.

NREM Sleep Period - The NREM sleep portion of NREM-REM sleep cycle. See Sleep Cycle, Sleep Stages.

NREM Sleep - Consists of sleep stages 1 through 4, and characterized by a decrease of mental activity. A state that lacks the visible motility of rapid eye movements and twitches, and exhibits a different EEG pattern than REM sleep. See Sleep Stages.

Nychthemeron – Refers to an exactly 24-hour day – no more, no less.

O

Obesity-Hypoventilation Syndrome - A term applied to obese individuals who hypoventilate during wakefulness.

Glossary

Obstructive Hypopnea - Periodic and partial closure of the throat usually during sleep resulting in reduced air exchange at the level of the mouth and/or nostril.

Obstructive Sleep Apnea - Repetitive cessation of breathing during sleep for 10 seconds or more due to complete closure (collapse) of the throat. Usually characterized by snoring, excessive daytime sleepiness, and other symptoms of fatigue.

Opponent Process Model- An experiment that describes clock-dependent altering and homeostasis.

Oscillograph - A device that displays ongoing patterns (EEGs) or other physiological variables usually tracing on moving paper or a CRT screen.

Oxygen Saturation - A measure of oxygen carried by hemoglobin in the blood. Normal values range from 90% to 100%. An important indicator of sleep disordered breathing which is directly affected by the degree of throat closure (partial vs. total) and its duration. Oxygen saturation is recorded in both diagnostic sleep studies and titration studies. Oxygen Desaturation - Less than normal amount of oxygen carried by hemoglobin in the blood. A value below 90% is considered abnormal.

Oximetry (Pulse) - The continuous monitoring of oxygen saturation of arterial blood from a pulse oximeter. sensor attached to the finger.

P

Parasomnia - Disorders of arousal, partial arousal, and sleep stage transition. These disorders intrude into the sleep process and are not primarily disorders of sleep and wake states per se. These disorders are manifestations of central nervous system activation usually transmitted through skeletal muscle or autonomic nervous system channels. They are divided into four groups: Arousal Disorders, Sleep-Wake Transition Disorders, Parasomnias Usually Associated with REM Sleep, and Other Parasomnias.

Pathological Sleep - Abnormal sleep patterns.

Pavor Nocturnus (Night Terrors) - See Night Terrors.

Perceptual Disengagement - Refers to change in consciousness at the onset of sleep when environmental stimuli are no longer perceived, and there is no longer any conscious, meaningful interaction with the environment.

Periodic Breathing - Characterized by repetitive apneic pauses, common in premature infants.

Periodic Limb Movement Disorder - Also known as periodic leg movements and nocturnal myoclonus. Characterized by periodic episodes of repetitive and highly stereotyped limb movements that occur during sleep. The movements are often associated with a partial arousal or awakening; however, the patient is usually unaware of the limb movements or frequent sleep disruption. Between the episodes, the legs are still. There can be marked night-tonight variability in the number of movements.

Persistent Insomnia - Continuing insomnia that responds poorly to treatment.

Pharynx - The area posterior to the nares and the oral cavity. A passageway for air from the nasal cavity and/or the mouth to the lungs via the larynx and the trachea, and for food and liquids from the mouth to the esophagus.

Phasic (Event/Activity) - Brain, muscle, or autonomic related event of a brief and episodic nature occurring in sleep. Characteristically occur during REM sleep, such as eye movements and/or muscle twitches, and can last from milliseconds to 1-2 seconds.

Photoperiod - The duration of light in a light/dark cycle.

Pickwickian Syndrome - Obesity accompanied by somnolence, lethargy, chronic hypoventilation, hypoxia, and secondary polycythemia (a condition marked by an abnormal increase in the number of circulating red blood cells). Usually has severe obstructive sleep apnea.

Pineal Gland - A gland in the brain that secretes the hormone melatonin.

PLMD-Arousal Index - The number of sleep-related periodic leg movements per hour of sleep that are associated with an EEG arousal. See Periodic Leg Movement.

Polycyclic - In reference to sleep, multiple sleep periods and wake
periods in a 24-hour day.

Polysomnography(PSG) - A biomedical recording for the measurement of multiple physiological variables of sleep. In sleep studies, it continuously displays the three-basic sleep stage scoring parameters, EEG, EOG, EMG, and EKG, respiratory airflow, respiratory movements, leg movements, and others depending on the situation.

Polysomnographic Technologist - Health care professional trained in performing diagnostic sleep studies.

Glossary

Post-Prandial Drowsiness - Sleepiness occurring after a meal usually lunch.

Post-Traumatic Stress Disorder - The re-experiencing of a traumatic event in the form of repetitive dreams, recurrent and intrusive daytime recollections, and/or dissociative flashback episodes.

Q

Quiet Sleep - The term frequently used instead of NREM sleep in describing the sleep of infants.

R

Radiofrequency - Electromagnetic radiation in the frequency range 3 kilohertz (kHz) to 300 gigahertz (GHz). It is generally considered to include microwaves and radio waves. Microwaves occupy the spectral region between 300 GHz and 300 MHz, while RF or radio waves include 300 MHz to 3 kHz. The primary health effects of radiofrequency energy are considered to be thermal. The absorption of radiofrequency energy varies with frequency.

Rapid Eye Movement Sleep (REM Sleep) - See REM Sleep, Sleep Stages.

REM Density - A function that expresses the frequency of eye movements per unit of time during REM sleep.

REM-Associated Disorders - Sleep disturbances that occur in REM sleep.

REM sleep paralysis/ Motor Atonia - The active suppression of activity in the antigravity and voluntary muscles during REM sleep. The muscles are completely flaccid and limp.

REM Sleep - The behavioral state identified by the occurrence of binocularly synchronous rapid eye movements, global paralysis of muscles and low voltage EEG patterns. Also associated with bursts of muscular twitching, irregular breathing, irregular heart rate, and generally increased autonomic activity. The state in which vivid dreaming occurs.

REM Sleep Behavior Disorder - A disorder in which REM motor atonia is partially or completely absent and the individual acts out the ongoing dream. The behavior in REM behavior disorder is often totally consonant with the ongoing, hallucinatory REM dream episode.

REM Sleep Episode - The REM sleep portion of a NREM-REM sleep cycle. Early in the first sleep period, episodes may be only several minutes in duration. Later REM episodes are almost always longer, 20 to 30 minutes up to an hour.

REM Sleep Intrusion - A brief interval of REM sleep appearing out of its usual positioning in the NREM-REM sleep cycle.

REM Sleep Latency - The interval from sleep onset to the first appearance of REM sleep.

REM Sleep Onset - The designation for the first epoch of a REM sleep episode.

REM Sleep Rebound - A compensatory increase in REM sleep following experimental reduction. Extension of time in, and an increase in frequency and density of REM sleep episodes. Usually an increase in REM sleep percent of total sleep time above baseline values.

Respiratory Disturbance Index (RDI) (Apnea/Hypopnea Index) - The frequency of abnormal respiratory events per hour of sleep. These events are classified as Apneas or Hypopneas. Apnea is when breathing (airflow) stops for 10 seconds or more. Hypopnea is a decrease in airflow of greater than 50% for 10 seconds or more. An RDI of 45 would indicate that the patient is experiencing complete or partial airflow blockage 45 times per hour. Sleep experts generally interpret RDI values according to the severity index of Sleep Disordered Breathing.

Restless Legs Syndrome - The sleep disorder characterized by a deep creeping, or crawling sensation in the legs that tends to occur when an individual is not moving. There is an almost irresistible urge to move the legs and the sensations are relieved by movement. Inability to remain at rest can result in severe sleep disturbance.

Restlessness (Referring to Quality of Sleep) - Persistent or recurrent body movements, arousals, and/or brief awakenings in the course of sleep.

Retrograde Amnesia – The loss of transfer of short-term memories to long term memory storage that occurs at the moment of sleep. The result of this loss is that one does not remember the events that transpired several minutes prior to sleep onset.

S

Scanning Hypothesis - A concept that the measured EOG activity in REM sleep refelcts the subjects scanning of the images in the dream world

Glossary

Seasonal Affective Disorder - A mood disorder occurring in the winter months which is characterized by diminished energy, hypersomnia, overeating and depressed mood. Exposure to bright light in the morning hours may be effective in alleviating or decreasing symptoms. Sedatives - Compounds that tend to calm, and reduce nervousness or excitement and foster sleep.

Septoplasty - Surgery on the nasal septum (the partition that divides the nasal passage).

Serotonin - A neurotransmitter in the brain that modulates mood, appetite, sexual activity, aggression, body temperature and sleep.

Shiftwork - Working hours other than the conventional daytime hours of 9:00 a.m. to 5:00 p.m.

Sleep - Naturally occuring phyisological process characterized by disengagment

Sleep Apnea - Cessation of breathing for 10 or more seconds during sleep.

Sleep Architecture - The NREM-REM sleep states and stages and cycle infrastructure of sleep. Often displayed in the form of a histogram.

Sleep Debt - The result of recurrent sleep deprivation which occurs over time when an individual does not experience a sufficient amount of the restorative daily sleep that is required to maintain a sense of feeling rested and refreshed. This accumulation of "lost sleep" may contribute to a decreased quality of life, the onset of related health problems, and the increased risk of injury and/or accident. See Sleep Deprivation.

Sleep Disorders - A broad range of illnesses arising from many causes, including, dysfunctional sleep mechanisms, abnormalities in physiological functions during sleep, abnormalities of the biological clock, and sleep disturbances that are induced by factors extrinsic to the sleep process.

Sleep Efficiency (or Sleep Efficiency Index) - The ratio of total sleep time to time in bed. The proportion of sleep in the period potentially filled by sleep.

Sleep Fragmentation - Brief arousals that occur throughout the night, that reduce the total amount of time spent

in the deeper levels of sleep. Frequent episodes of sleep fragmentation lead to sleep deprivation.

Sleep Hygiene - Behavioral activities that either contribute to or detract from restorative sleep. Good sleep hygiene would include activities such as going to bed the same time each night, restricting caffeine intake, avoiding napping during the day, etc.

Sleep Hyperhydrosis - Excessive sweating during sleep.

Sleep Inertia - Feelings of grogginess and/or sleepiness that persist longer than 10 to 20 minutes after waking up.

Sleep Latency - The length of time it takes to go from full wakefulness to the moment of sleep.

Sleep Log (Diary) - A daily, written record of a person's sleep wake pattern containing such information as time of retiring and arising, time in bed, estimated total sleep time, number and duration of sleep interruptions, quality of sleep, daytime naps, use of medications or caffeine beverages, nature of waking activities.

Sleep Onset - The transition from awake to sleep, normally to NREM stage1 sleep but in certain conditions, such as infancy and narcolepsy, into REM stage sleep.

Sleep Onset Imagery - A kaleidoscope of images and experiences during the moments following the transition from wake to sleep.

Sleep Onset REM Period (SOREMP) - REM periods within 15 minutes of sleep onset, considered to confirm the diagnosis of narcolepsy.

Sleep Restriction - A limitation of the number of hours in bed.

Sleep Spindle - A synchronized, rhythmic EEG waveform, with a frequency of 7 to 14 Hz which waxes and wanes in a highly characteristic fashion over 1-2 seconds.

Sleep Stage 1 (NREM Stage 1) - A stage of NREM sleep that ensues directly from the awake state. Its criteria consist of a lowvoltage EEG tracing with well-defined alpha activity and

theta frequencies in the 3 to 7 cps range, occasional vertex spikes, and slow rolling eye movements (SEMs), and includes the absence of sleep spindles, K complexes, and REMS. Stage 1 normally represents 4-5% of total sleep.

Sleep Stage 2 (NREM Stage 2) - A stage of NREM sleep characterized by the advent of sleep spindles and K complexes against a relatively low-voltage, mixed frequency EEG background. High voltage delta waves may comprise up to 20% of stage 2 epochs. Stage 2 usually accounts for 45-55% of total sleep time.

Sleep Stage 3 (NREM Stage 3) - A stage of NREM sleep defined by at least 20% and not more than 50% of the period consisting of EEG waves of 2 cps or slower, with amplitudes of more than 75 mV (high-amplitude delta waves). It constitutes "deep" NREM sleep, or, so-called slow wave sleep (SWS), and is often combined with stage 4 NREM sleep because of the lack of documented physiological differences between the two. It normally appears usually only in the first third of the sleep episode, and usually compromises 4-6% of total sleep time.

Sleep Stage 4 (NREM Stage 4) - All statements concerning NREM sleep stage 3 apply to stage 4 except that high-voltage, slow EEG waves cover 50% or more of the record. NREM sleep stage 4 usually represents 12-15% of total sleep time. Sleepwalking, sleep terrors, and sleep-related enuresis episodes generally start in stage 4 or during arousals from this stage. The distinction between stages 3 and 4 have been minimized and they are collectively referred to slow wave sleep (N3)

Sleep talking - Talking in sleep that usually occurs in the course of transitory arousals from NREM sleep. Can occur during REM sleep, at which time it represents a motor breakthrough of dream speech. Full consciousness is not achieved and no memory of the event remains.

Sleepwalker or Sleepwalking - One who is subject to somnambulism (one who walks while sleeping). Sleepwalking typically occurs in the first third of the night during deep NREM sleep (stages 3 and 4).

Sleep-Wake Cycle - The clock hour relationships of the major sleep and wake episodes in the 24-hour cycle. See Phase Transition, Circadian Rhythm.

Sleep-Wake Shift (Change, Reversal) - When sleep as a whole or in part is moved to a time of customary waking activity, and the latter is moved to the time of the major sleep episode. Common in jet lag and shift work.

Slow Wave Sleep (SWS) - Stage of sleep in which EEG delta waves are present. Synonymous with sleep stages 3 and 4 combined.

Snoring - Sounds made during sleep caused by vibrations in the pharynx on inspiration. In the diagnosis of obstructive sleep apnea, snoring volume and frequency of occurrence often correlate with the severity of the condition. Snoring noise is recorded in both diagnostic sleep studies and titration studies.

Soft Palate - The membranous and muscular fold suspended from the posterior margin of the hard palate and partially separating the oral cavity from the pharynx.

Somnambulism - Walking while asleep.

Somnolence - Prolonged drowsiness or sleepiness.

Stanford Sleepiness Scale (SSS) - A 7-point Likert rating scale consisting of seven numbered statements that describe subjective levels of sleepiness/alertness.

Suprachiasmatic Nucleus - Area of the brain in the hypothalamus that regulates circadian rhythms. It is located above the optic chiasm. This area of the brain is thought to be the location of the biological clock.

Synchronization - A chronobiological term used to indicate that two or more rhythms recur with the same phase relationship. In an EEG tracing, the term is used to indicate an increased amplitude with an occasional decreased frequency of the dominant activities.

T

Tachycardia - Rapid heart rate, usually defined by a pulse rate of over 100 beats per minute (bpm).

Thermoregulation - The regulation of body temperature in mammals.

Theta Activity - EEG activity with a frequency of 4-8 Hz, generally maximal over the central and temporal cortex.

Tidal Volume - The amount of air that passes in and out of the lungs in an ordinary breath. Usually expressed in liters.

Titration - Refers either to the automatic or manual determination the level of air pressure that is required to eliminate or inhibit airway instability that might otherwise cause either a partial or complete blockage of the flow of air to a patient's lungs. A method of finding the optimal or lowest pressure that eliminates most of the apneas and hypopneas.

Glossary

Tonic (Event/Activity) - Generally, brain, muscle, or autonomic events, which are continuous. Usually refers to continuous activity (e.g. muscle atonia) during REM sleep.

Tonsillectomy - Surgical removal of the tonsils.

Total Recording Time - The duration of time from sleep onset to final awakening. In addition to total sleep time, it is comprised of the time taken up by wake periods and movement time until wake-up.

Total Sleep Time - The amount of actual sleep time in a sleep episode. Equal to total sleep episode less awake time. Total sleep time is the total of all REM and NREM sleep in a sleep episode.

Tracheostomy - Refers to the opening in the trachea. As a treatment for severe obstructive sleep apnea, a tube to assist oxygenation and ventilation and/or to overcome an obstruction in the airway located superiorly.

Tricyclic Antidepressants - Medication for depression. Most tricyclic antidepressants also reduce REM sleep. They are used to control cataplectic attacks, hypnogogic hallucinations, and sleep paralysis.

Turbinate - Small, shelf-like, cartilaginous structures covered by mucous membranes, which protrude into the nasal airway to help warm, humidify, and cleanse inhaled air on its way to the lungs.

Twilight Zone - A slang popular term to describe the waking state of individuals whose MSLT scores are 5 minutes or less. Such individuals are usually sleep deprived or suffer from a sleep disorder.

U

Unattended CPAP Titration Study - A sleep study that is usually performed in the home, after determining that a patient has a sleep related breathing disorder such as OSA or Upper Airway Resistance Syndrome, and is likely to benefit from CPAP therapy. A type of automatic calibration test performed with a CPAP machine that measures the patient's airway pressure requirements during sleep. These findings are then used to prescribe the appropriate pressure the patient will use with their CPAP therapy.

Upper Airway - The part of the respiratory anatomy that includes the nose, nostrils, sinus passages, septum, turbinates, the tongue, jaws, hard and soft palate, muscles of the tongue and throat, etc.

Upper Airway Resistance Syndrome - Part of the spectrum of obstructive sleep-related breathing disorders in which repetitive increases in resistance to airflow in the upper airway lead to brief arousals and daytime fatigue. Usually associated with loud snoring. Apneas and hypopneas (see RDI) may be totally absent. Blood oxygen levels can be in the normal range.

Uvula - Small soft structure hanging from the bottom of the soft palate in the mid-line above the back of the tongue. It is composed of connective tissue and mucous membrane.

Uvulopalatopharyngoplasty (UPPP) - Surgical treatment for obstructive sleep apnea and/or snoring. This procedure is designed to open the airway behind the palate. If tonsils are present they are removed, and redundant palatal tissue

is also removed. The incision is closed with sutures. The procedure is performed under general anesthesia and usually requires a one or two-day hospital stay.

W

Wake Time - The total time scored as wakefulness in a polysomnogram occurring between sleep onset and final awakening.

White Noise - A mixture of sound waves extending over a wide frequency range that may be used to mask unwanted noise that may interfere with sleep. Also, called white sound.

Wilkinson Addition Test - A performance test in which the individual will add numbers for one hour. Often included in a battery of tests to measure the impact of acute or chronic sleep loss.

Withdrawal - The effects experienced when a patient stops taking
sleeping pills.

Z

Zeitgeber - An environmental time cue that entrains biological rhythms to a specific circadian periodicity. Known zeitgebers include light, melatonin and physical activity. To be effective, these signals must occur when the biological clock is in a responsive phase.

Index

Advanced Sleep Phase Syndrome, 109, 401, 408, 444

alcohol, *111, 129, 137, 141, 153, 166, 311, 312, 323, 337, 349, 356, 376, 388, 416, 417, 421, 422, 423, 424, 436, 437, 438*

alpha rhythm, *61, 451*

Apnea/Hypopnea Index, *333, 444, 458*

biological clock, *56, 57, 81, 84, 85, 86, 87, 90, 91, 92, 93, 95, 96, 97, 98, 100, 102, 104, 108, 109, 110, 111, 137, 143, 155, 162, 163, 164, 165, 167, 169, 302, 400, 406, 445, 446, 460, 462, 465*

Bruxism, *319, 446*

cataplexy, *347, 348, 350, 351, 352, 353, 354, 355, 356, 357, 359, 389, 390, 391, 453*

circadian rhythm, *85, 89, 92, 108, 109, 141, 162, 163, 188, 302, 401, 407, 432, 444, 446, 447*

clock-dependent alerting, *57, 84, 86, 89, 91, 92, 93, 95, 96, 97, 98, 99, 101, 102, 104, 105, 106, 108, 109, 111, 113, 129, 130, 134, 142, 144, 156, 163, 165, 166, 168, 169, 400, 401, 406, 407, 409, 440, 449*

Clock-Dependent Alerting, *57, 84, 163*

Cognitive Behavioral Therapy, *308*

Confusional Arousals, *362*

CPAP, *322, 337, 338, 340, 342, 343, 344, 388, 389, 445, 446, 447, 448, 449, 450, 451, 464*

delayed sleep phase syndrome, 138, 141, 380, 393, 401

depression, *75, 118, 145, 148, 305, 306, 321, 344, 402, 403, 441, 463*

dream, *121, 169, 170, 171, 172, 173, 174, 175, 176, 178, 179, 180, 181, 182, 183, 184, 185, 186, 187, 188, 189, 190, 191, 192, 193, 194, 195, 196, 197, 198, 199, 200, 201, 202, 203, 204, 205, 206, 207, 208, 209, 210, 211, 212, 213, 214, 215, 216, 217, 218, 219, 222, 223, 224, 225, 226, 227, 228, 229, 230, 231, 232, 233, 377, 380, 409, 454, 458, 459, 461*

electroencephalogram, *321, 399*

fatigue, *54, 57, 58, 82, 84, 87, 95, 110, 112, 115, 118, 130, 131, 145, 151, 152, 155, 158, 160, 303, 304, 314, 319, 323, 326, 339, 344, 346, 407, 410, 412, 413, 414, 418, 419, 420, 421, 424, 426, 427, 428, 429, 430, 431, 434, 435, 438, 439, 451, 452, 455, 464*

free-running, *85, 90*

GABA, *313, 449*

hypnagogic, *177, 180, 347, 348, 350, 352, 380, 383, 389, 396, 419*

hypopnea, *331, 332, 334, 340, 342, 383, 385*

insomnia, *108, 109, 138, 141, 144, 146, 162, 300, 301, 302, 303, 304, 305, 306, 307, 308, 309, 311, 312, 313, 314, 319, 391, 393, 394, 397, 401, 402, 403, 406, 408, 441, 443, 447, 456*

jet lag, *84, 96, 109, 162, 163, 164, 165, 166, 300, 400, 406, 431, 462*

Lucid Dreaming, *201*

MSLT, *59, 60, 61, 62, 63, 64, 65, 67, 68, 69, 70, 71, 76, 77, 79, 80, 94, 95, 99, 100, 101, 102, 105, 109, 122, 123, 135, 150, 151, 157, 353, 354, 414, 453, 464*

MWT, *69*

napping., *66, 156, 158, 309*

nightmare, *231, 398, 454*

Obstructive sleep apnea, *322, 323, 333, 334, 345, 394, 399*

Opponent Process Model, *57, 97, 102*

parasomnia, *399, 444*

Polysomnography, *457*

restless leg syndrome, *306, 397*

Salience Hypothesis, *208, 209*

saw-tooth waves, *199*

scanning hypothesis, *196, 198, 199, 200, 201*

sleep debt, *56, 58, 63, 67, 68, 69, 70, 71, 72, 73, 74, 75, 78, 79, 80, 81, 82, 86, 89, 92, 94, 95, 96, 97, 98, 99, 102, 104, 105, 106, 107, 108, 111, 112, 113, 114, 117, 128, 129, 130, 132, 134, 136, 137, 138, 139, 140, 143, 144, 145, 146, 147, 148, 151, 152, 153, 154, 155, 157, 158, 159, 160, 161, 163, 164, 165, 167, 169, 304, 406, 412, 416, 422, 434, 438, 440*

sleep homeostasis, *55, 57, 62, 80, 82, 93, 96, 97, 110, 169*

sleep paralysis, *347, 348, 350, 352, 389, 397, 453, 458, 463*

sleep spindles, *461*

sleep terror, *398*

sleeping pill, *164, 166*
sleepwalking, *361, 362, 376, 380, 384, 397, 398, 399, 444*
slow wave sleep, *58, 380, 384, 391, 397, 398, 399, 454, 461*
snoring, *158, 232, 322, 323, 325, 326, 327, 332, 334, 335, 340, 342, 346, 383, 384, 385, 387, 388, 389, 416, 449, 451, 455, 462, 464*
suprachiasmatic nuclei, *57, 81, 87, 89, 93, 110, 445*
twilight zone, *59, 62, 64, 68, 77, 79, 94, 103, 109, 134, 153, 414, 415, 430*
uvulopalatopharyngoplasty, *339, 340*
yawning, *119, 120, 427*
zeitgebers, *141, 404, 465*

CPSIA information can be obtained
at www.ICGtesting.com
Printed in the USA
LVHW012046120721
692484LV00011B/762